The Institute of Mathematics
and its Applications
Conference Series

The Institute of Mathematics
and its Applications
Conference Series

Previous volumes in this series were published by
Academic Press to whom all enquiries should be addressed.
Forthcoming volumes will be published by
Oxford University Press throughout the world.

NEW SERIES
1. *Supercomputers and parallel computation* Edited by D. J. Paddon
2. *The mathematical basis of finite element methods*
   Edited by David F. Griffiths
3. *Multigrid methods for integral and differential equations*
   Edited by D. J. Paddon and H. Holstein
4. *Turbulence and diffusion in stable environments* Edited by J. C. R. Hunt
5. *Wave propagation and scattering* Edited by B. J. Uscinski
6. *The mathematics of surfaces* Edited by J. A. Gregory
7. *Numerical methods for fluid dynamics II*
   Edited by K. W. Morton and M. J. Baines
8. *Analysing conflict and its resolution* Edited by P. G. Bennett
9. *The state of the art in numerical analysis*
   Edited by A. Iserles and M. J. D. Powell
10. *Algorithms for approximation* Edited by J. C. Mason and M. G. Cox
11. *The mathematics of surfaces II* Edited by R. R. Martin
12. *Mathematics in signal processing*
    Edited by T. S. Durrani, J. B. Abbiss, J. E. Hudson, R. N. Madan,
    J. G. McWhirter, and T. A. Moore
13. *Simulation and optimization of large systems*
    Edited by Andrzej J. Osiadacz
14. *Computers in mathematical research*
    Edited by N. M. Stephens and M. P. Thorne
15. *Stably stratified flow and dense gas dispersion*
    Edited by J. S. Puttock
16. *Mathematical modelling in non-destructive testing*
    Edited by Michael Blakemore and George A. Georgiou
17. *Numerical methods for fluid dynamics III*
    Edited by K. W. Morton and M. J. Baines

Numerical methods for fluid dynamics III

Based on the proceedings of a conference
organized by The Institute for Computational
Fluid Dynamics of The Universities of Oxford
and Reading in association with the Institute of
Mathematics and it Applications on Numerical
Methods for Fluid Dynamics, held in Oxford
in March 1988

Edited by

K. W. MORTON
University of Oxford

and

M. J. BAINES
University of Reading

CLARENDON PRESS · OXFORD · 1988

Oxford University Press, Walton Street, Oxford OX2 6DP

Oxford New York Toronto
Delhi Bombay Calcutta Madras Karachi
Petaling Jaya Singapore Hong Kong Tokyo
Nairobi Dar es Salaam Cape Town
Melbourne Auckland

and associated companies in
Berlin Ibadan

Oxford is a trademark of Oxford University Press

Published in the United States
by Oxford University Press, New York

British Library Cataloguing in Publication Data
Numerical methods for fluid dynamics III. –
(The Institute of Mathematics and its
Applications conference series. New series; no. 17
1. Fluids. Dynamics. Mathematics. Numerical Methods.
I. Morton, K. W. II. Baines, M. J. (Michael
John). 1935 –. III. Institute of
Mathematics and its Applications IV. Series
532'.05'015194
ISBN 0-19-853632-1

Library of Congress Cataloging in Publication Data
Numerical methods for fluid dynamics III: based on the proceedings of
a conference organized by the Institute for Computational Fluid
Dynamics of the Universities of Oxford and Reading in association
with the Institute of Mathematics and Its Applications on numerical
methods for fluid dynamics, held in Oxford in March 1988 / edited by
K. W. Morton and M. J. Baines.
p. cm. – (The Institute of Mathematics and Its Applications
conference series; new ser., 17)
1. Fluid dynamics – Mathematics – Congresses. 2. Numerical
calculations – Congresses. I. Morton, K. W. II. Baines, M. J.
(Michael John) III. Institute for Computational Fluid Dynamics.
IV. Institute of Mathematics and Its Applications. V. Series.
TA357.N8726 1988 620.1'06'015194 – dc 19 88–28223
ISBN 0-19-853632-1

Printed in Great Britain by
St. Edmundsbury Press
Bury St. Edmunds, Suffolk

Preface

The Third Conference on Numerical Methods for Fluid Dynamics was held at Oxford University in March 1988: it was the second such conference organised by the ICFD (Institute for Computational Fluid Dynamics) in association with the IMA (Institute of Mathematics and its Applications); and it was the first to be held at Oxford, the earlier conferences being at Reading.

As with the earlier conferences, the aim was to bring together mathematicians and engineers working in the various fields of computational fluid dynamics and aerodynamics, to review recent advances in mathematical and computational techniques, and to promote the cross fertilisation of ideas across the differing applications areas.

Recent rapid growth in CFD, and in the appreciation of its utility, makes it even more necessary to give coherence to such a conference by focusing attention on some specific themes. At this meeting three main themes were selected:

Numerical algorithms specific to CFD together with studies of their behaviour and performance.
Grid generation techniques, adaptive grids and domain decomposition.
Unsteady flows, such as those which occur in aircraft flutter, tidal forcing, blade row interaction in turbines, vortex flows, separation etc.

We would like to thank our colleagues on the organising committee, Dr. D. A. H. Jacobs (CEGB), Dr. I. P. Jones (Harwell), Dr. P. Stow (Rolls Royce), for their suggestions of these themes, for names of invited speakers and many other contributions to the success of the Conference; all the speakers for taking up these themes so enthusiastically - and with only two exceptions, for submitting their contributions in time for the prompt compilation of this volume; Dr. J. Rae (Department of Energy) for his admirable after dinner speech at the Conference Banquet; and Nicolette Boult, our ICFD Secretary, who with the sterling support of administrative staff at Oxford organised the Conference so effectively.

We are grateful for financial support from the Royal Society and the United States Air Force.

The Proceedings is arranged in two sections, the first consisting of the Invited Papers, in the order of the Conference Programme. The second section contains the Contributed Papers, organised in the same way. We made so bold as to invite authors to submit their papers in machine-readable form and specifically in LaTeX, undertaking to retype the remainder ourselves. In the event, the majority of papers were retyped in the Computing Laboratory at Oxford and we are enormously grateful to our Computing Officer Didier Delaplace for organising this, to his team of DPhil students who did the retyping, of whom Joe Kolibal should be singled out for special thanks, and to Jean Reynolds and Mike Field for all their help throughout the whole organisation of the Conference and the Proceedings.

Finally our thanks must go to Martin Gilchrist and his colleagues at Oxford University Press who have supported us admirably in this new approach to publishing.

K. W. Morton.
M. J. Baines.

Table of contents

List of contributors   xiii

Invited Papers

CFD applications to the aero-thermodynamics of tur-   1
bomachinery.
  P. Stow

Computational fluid dynamics in the automobile in-   25
dustry.
  A. D. Gosman

Developments in the calculation of unsteady turboma-   45
chinery flow.
  M. B. Giles

Efficient solution of the steady Euler equations with a   65
centered implicit method.
  A. Lerat and J. Sidès

Some current trends in numerical grid generation.   87
  J. F. Thompson

A strategy for the use of hybrid structured–   101
unstructured meshes in computational fluid dynamics.
  N. P. Weatherill

Implicit methods in CFD.   117
  T. H. Pulliam

The cell vertex method for steady compressible flow.   137
  K. W. Morton, P. N. Childs, and M. A. Rudgyard

Multigrid, defect correction and upwind schemes for   153
the steady Navier–Stokes equations.
  P. W. Hemker and B. Koren

Recent developments of the Taylor–Galerkin method   171
for the numerical solution of hyperbolic problems.
  J. Donea, V. Selmin, and L. Quartapelle

Numerical grid generation in 3-D Euler-flow simula-   187
tion.
  J. W. Boerstel

An approach to geometric and flow complexity using   215
feature-associated mesh embedding (FAME): strategy
and first results.
  C. M. Albone

Lax-stability vs. eigenvalue stability of spectral meth-   237
ods.
  L. N. Trefethen

Acceleration of compressible Navier–Stokes flow calcu-   255
lations.
  M. O. Bristeau, R. Glowinski, B. Mantel,
  J. Périaux, and G. Rogé

Contributed Papers

Steady incompressible and compressible solution of   273
Navier–Stokes equations by rotational correction.
  F. El Dabaghi

Comparison of implicit methods for the compressible   282
Navier–Stokes equations.
  Y. Marx and J. Piquet

Implicit finite difference methods for computing dis-   289
continuous atmospheric flows.
  M. J. P. Cullen

Numerical simulation of unsteady flows using the   296
MUSCL approach.
  P. Guillen, M. Borrel, and J. L. Montagne

Computation of viscous separated flow using a particle   310
method.
  J. M. R. Graham

A streamwise upwind algorithm for the Euler and   318
Navier–Stokes equations applied to transonic flows.
  P. M. Goorjian

Computation of diffracting shock wave flows.   325
  R. Hillier

Multiple mesh simulation of turbulence.   332
  P. R. Voke

Some experiences with grid generation on curved sur-   341
faces using variational and optimisation techniques.
  C. R. Forsey and C. M. Billing

Adaptive orthogonal curvilinear coordinates.   353
  R. Arina

An approximate equidistribution technique for un-   360
structured grids.
  P. K. Sweby

Multiblock techniques for transonic flow about complex   367
aircraft configurations.
  S. E. Allwright

Cartesian grid methods for irregular regions.   375
  R. J. LeVeque

Numerical characteristic decomposition for compress-   383
ible gas dynamics with general (convex) equations of
state.
  P. Glaister

A hybrid scheme for the Euler equations using the Ran-   391
dom Choice and Roe's methods.
  E. F. Toro and P. L. Roe

A variational finite element formulation for three-   403
dimensional incompressible flows.
   P. Ward, R. Desai, W. Kebede, and A. Ecer

A comparison of multigrid methods for the incompress-   410
ible Navier–Stokes equations.
   S. Sivaloganathan, G. J. Shaw, T. M. Shah,
   and D. F. Mayers

Multigrid calculations of jet flows.   418
   S. A. E. G. Falle and M. J. Wilson

The accurate approximation and economic solution of   425
steady-state convection dominated flows.
   P. H. Gaskell, A. C. K. Lau, and N. G. Wright

A 3D finite element code for industrial applications.   432
   J. P. Chabard and O. Daubert

The behaviour of Flux Difference Splitting schemes   442
near slowly moving shock waves.
   T. W. Roberts

A Total Variation Diminishing scheme for computa-   449
tional aerodynamics.
   D. M. Causon

Properties of two computational methods for shallow   458
water flow problems.
   Th. L. van Stijn, P. Wilders, G. A. Fokkema,
   and G. S. Stelling

Consistent boundary conditions for cell centred upwind   464
finite volume Euler solvers.
   H. Deconinck and R. Struys

An optimistic reappraisal of computational techniques   471
in the supercomputer era.
   G. Moretti

Evaluation of a parallel conjugate gradient algorithm.   478
  R. W. Leland and J. S. Rollett

Mixed finite elements for highly viscoelastic flows.   484
  J.-M. Marchal

Multiphase flow—a self consistent approach.   492
  D. F. Fletcher and A. Thyagara

Chebyshev collocation methods for the solution of the   500
incompressible Navier–Stokes equations in complex geometries.
  T. N. Phillips and A. Karageorghis

Far field boundaries and their numerical treatment: an   507
unconventional approach.
  S. Karni

Moving element methods for time dependent problems.   513
  M. J. Baines

Non-existence, non-uniqueness and slow convergence in   519
discrete conservation laws.
  P. L. Roe and B. van Leer

List of contributors

Albone, C. M. , *AE5 Division, Aerodynamics Department, Royal Aircraft Establishment, Farnborough, Hants, GU14 6TD*

Allwright, S .E. , *BAe Multiblock Technical Leader, Research Department, British Aerospace plc, Civil Aircraft Division, Hatfield, Hertfordshire, AL10 9TL*

Arina, R. , *CNR-CSDF, Politecnico di Torino, Corso Duca degli Abruzzi, 24, 10129 Torino, Italy*

Baines, M. J. , *Department of Mathematics, Reading University, P.O. Box 220, Whiteknights, Reading RG6 2AX*

Billing, C. M. , *ARA Aircraft Research Association Limited, Manton Lane, Bedford, MK41 7PF*

Boerstel, J. W. , *National Aerospace Laboratory NLR, Anthony Fokkerweg 2, 1059 CM Amsterdam, The Netherlands*

Borrel, M. , *Aerodynamics Department, ONERA, 29, Avenue de la Division Leclerc, 92322 Chatillon, France*

Bristeau, M. O. , *INRIA, B.P. 105, Rocquencourt, 78153 Le Chesnay, France*

Causon, D. M. , *Department of Mathematics and Physics, Manchester Polytechnic, John Dalton Building, Chester Street, Manchester M1 5GD*

Chabard, J. P. , *Eléctricité de France, Direction des Etudes et Recherches, Laboratoire National d'Hydraulique, 6, Quai Watier, 78400 Chatou, France*

Childs, P. N. , *Oxford University Computing Laboratory, 8–11 Keble Road, Oxford OX1 3QD*

Cullen, M. J. P. ,*Meteorological Office, London Road, Bracknell, Berks. RG12 2SZ*

Dabaghi, F. El , *INRIA, B.P. 105, 78153 Le Chesnay Cedex, France*

Daubert, O. , *Electricite de France, Direction des Etudes et Recherches, Laboratoire National d'Hydraulique, 6, Quai Watier, 78400 Chatou, France*

Deconinck, H. , *von Karman Institute for Fluid Mechanics, Chaussée de Waterloo, 72, 1640 Rhode-Saint-Genèse, Belgium*

Desai, R. , *SDRC Engineering Services Ltd.,York House, Stevenage Road, Hitchin, Herts. SG4 9DY*

Donea, J. , *Joint Research Centre, 21020 Ispra, Italy*

Ecer, A. , *Purdue University, School of Engineering & Technology at Indianapolis, Indianapolis, Indiana, USA*

Falle, S. A. E. G. , *School of Applied Mathematical Studies, University of Leeds, Leeds LS2 9JT*

Fletcher, D. F. , *General Physics & Theory Division, Culham Laboratory, UKAEA, Abingdon, Oxon. OX14 3DB*

Fokkema, G. A. , *RWS, Data Processing Division, Mathematics Group, P.O. Box 5809, 2280 HV Rijswijk, the Netherlands*

Forsey, C. R. , *ARA Aircraft Research Association Limited, Manton Lane, Bedford, MK41 7PF*

Gaskell, P. H. , *Department of Mechanical Engineering, University of Leeds, Leeds LS2 9JT*

Giles, M. ,*Department of Aeronautics and Astronautics, M.I.T., Cambridge, MA 02139, USA*

Glaister, P. , *Department of Mathematics, University of Reading, P.O. Box 220, Whiteknights, Reading RG6 2AX*

Glowinski, R. , *University of Houston, Texas, USA*

Goorjian, P. M. , *Applied Computational Fluids Branch, NASA Ames Research Center, M.S. 258-1, Moffett Field, CA 94035, USA*

Gosman, A. D. , *Mechanical Engineering Department, Imperial College of Science and Technology, London SW7 2BY*

Graham, J. M. R. , *Department of Aeronautics, Imperial College of Science and Technology, London SW7 2BY*

Guillen, P. , *Aerodynamics Department, ONERA, 29, Avenue de la Division Leclerc, 92322 Chatillon, France*

Hemker, P. W. , *CWI, Centre for Mathematics and Computer Science, P.O. Box 4079, 1009 AB Amsterdam, The Netherlands*

Hillier, R. , *Department of Aeronautics, Imperial College of Science and Technology, London SW7 2BY*

Karageorghis, A. , *The University College of Wales, Penglais, Aberystwyth, Dyfed SY23 3BZ*

Karni, S. , *College of Aeronautics, Cranfield Institute of Technology, Cranfield, Beds. MK43 0AL*

Kedebe, W. , *SDRC Engineering Services Ltd.,York House, Stevenage Road, Hitchin, Herts. SG4 9DY*

Koren, B. , *CWI, Centre for Mathematics and Computer Science, P.O. Box 4079, 1009 AB Amsterdam, The Netherlands*

Lau, A. C. K. , *University of Leeds, Leeds LS2 9JT*

van Leer, B. , *University of Michigan, Aerospace Engineering, Ann Arbor, MI 48109-2140, USA*

Leland, R. W. , *Oxford University Computing Laboratory, 8–11 Keble Road, Oxford OX1 3QD*

Lerat, A. , *ENSAM, 151 Boulevard de l'Hôpital, 75013 Paris, France*

LeVeque, R. J. , *University of Washington, Department of Mathematics, GN-50, Seattle, Washington 98195, USA*

Mantel, B. , *AMD–BA Industries, 78 quai Marcel Dassault, 92214 St-Cloud, France*

Marchal, J.-M. , *University of Louvain-la-Neuve, Place du Levant, 2, B – 1348 Louvain-la-Neuve, Belgium*

Marx, Y. , *CFD Group, UA 1217, E.N.S.M., 1 Rue de la Noë, 44072 Nantes, France*

Mayers, D. F. , *Oxford University Computing Laboratory, 8–11 Keble Road, Oxford OX1 3QD*

Montagne, J.-L. , *Aerodynamics Department, ONERA, 29, Avenue de la Division Leclerc, 92322 Chatillon, France*

Moretti, G. , *G.M.A.F., P.O. Box 184, Freeport, NY 11520, USA*

Morton, K. W. , *Oxford University Computing Laboratory, 8–11 Keble Road, Oxford OX1 3QD*

Périaux, J. , *AMD–BA Industries, 78 quai Marcel Dassault, 92214 St-Cloud, France*

Phillips, T. N. , *The University College of Wales, Penglais, Aberystwyth, Dyfed SY23 3BZ*

Piquet, J. , *CFD Group, UA 1217, E.N.S.M., 1 Rue de la Noë, 44072 Nantes, France*

Pulliam, T. H. ,*Computational Physics Section, CFD Branch,, NASA Ames Research Center, MS 202A-1, Moffett Field, CA 94035, USA*

Quartapelle, L. , *Istituto di Fisica del Politecnico di Milano, Piazza Leonardo da Vinci, 32, 20133 Milan, Italy*

Roberts, T. W. , *The Aeronautical Research Institute of Sweden, P.O. Box 11021, S-161 11 Bromma, Sweden*

Roe, P. L. , *College of Aeronautics, Cranfield Institute of Technology, Cranfield, Beds. MK43 0AL*

Rogé, G. , *AMD–BA Industries, 78 quai Marcel Dassault, 92214 St-Cloud, France*

Rollett, J. S. , *Oxford University Computing Laboratory, 8–11 Keble Road, Oxford OX1 3QD*

Rudgyard, M. A. , *Oxford University Computing Laboratory, 8–11 Keble Road, Oxford OX1 3QD*

Selmin, V. , *Aeritalia, Corso Marche, 41, 10146 Turin, Italy*

Shah, T. M. , *Oxford University Computing Laboratory, 8–11 Keble Road, Oxford OX1 3QD*

Shaw, G. , *Oxford University Computing Laboratory, 8–11 Keble Road, Oxford OX1 3QD*

Sidès, J. , *ONERA, 29, Avenue de la Division Leclerc, 92320 Chatillon, France*

Sivaloganathan, S. , *Oxford University Computing Laboratory, 8–11 Keble Road, Oxford OX1 3QD*

Stelling, G. S. , *Nederlandse Philips Bedrijven B.V., Corporate CAD-Centre, P.O. Box 218, 5600 MD Eindhoven, the Netherlands*

van Stijn, Th. L. , *Rijkswaterstaat, Dienst Informatieverwerking, Postbus 5809, 2280 HV Rijswijk, The Netherlands*

Stow, P. , *Chief of Theoretical Science, Rolls-Royce plc, P.O.Box 31, Derby DE2 8BJ*

Struys, R. , *von Karman Institute for Fluid Mechanics, Chaussée de Waterloo, 72, 1640 Rhode-Saint-Genèse, Belgium*

Sweby, P. K. , *Department of Mathematics, Reading University, P.O. Box 220, Whiteknights, Reading RG6 2AX*

Thompson, J. F. , *Mississippi State University, Department of Aerospace Engineering, Drawer A, Mississippi State, Mississippi 39762, USA*

Thyagara, A. , *Culham Laboratory, UKAEA, Abingdon, Oxon. OX14 3DB*

Toro, E. F. , *College of Aeronautics, Cranfield Institute of Technology, Cranfield, Beds. MK43 0AL*

Trefethen, L. N. , *Department of Mathematics, M.I.T., Cambridge, MA 02140, USA*

Voke P. R. , *Turbulence Unit in the School of Engineering, Queen Mary College, University of London, Mile End Road, London E1 4NS*

Ward, P. , *SDRC Engineering Services Ltd.,York House, Stevenage Road, Hitchin, Herts. SG4 9DY*

Weatherill, N. P. , *Institute for Numerical Methods in Engineering, University College of Swansea, Singleton Park, Swansea*

Wilders, P. , *Delft University of Technology, Dept. of Mathematics and Informatics, P.O. Box 356, 2600 AJ Delft, the Netherlands*

Wilson, M. J. , *School of Applied Mathematical Studies, University of Leeds, Leeds LS2 9JT*

Wright, N. G. , *University of Leeds, Leeds LS2 9JT*

CFD applications to the aero-thermodynamics of turbomachinery

P. Stow

Rolls-Royce plc, Derby

1. Introduction

Computational Fluid Dynamics is being applied widely within Rolls-Royce to the various design and analysis aerodynamic and thermodynamic problems in turbomachinery. This paper presents some of the advances being made in a number of the application areas and also indicates areas for further development of the numerical and mathematical models currently available.

The development and application of Computational Fluid Dynamics for the aero-thermodynamics problems associated with turbomachinery is part of a major initiative within Rolls-Royce on Computer Aided Engineering and Manufacture. CAEM is aimed at reducing both design and development times and costs through more exact modelling at each stage in the design process and by ensuring faster interaction between the various stages.

CFD is the core of all computing systems for Aero-Thermodynamic design and is seen as a key element in ensuring future competitiveness in the design and development of turbomachinery components. If the mathematical models employed in the CFD codes are good representations of the true physical processes then using such models a designer should be able to produce better, more optimum designs than empirical approaches, and at the same time get the design right first time, saving on development costs. As a consequence a great deal of emphasis is placed on the mathematical modelling aspects of CFD. Only once adequate models have been developed, and thoroughly evaluated can they be used within their regions of validity and be exploited by a designer.

CFD is being applied to all the major aero-thermodynamic areas in the turbomachine, for example,

- Installation aerodynamics (intakes, nozzles, after-bodies).

- Compressor and Turbine aerodynamics (steady and unsteady flow).

- Turbine heat transfer (internal and external flow).

- Combustion.

- Disc cavities (flow and heat transfer).

The CFD modelling work covers all major fluid dynamics topics, eg.

- 2D/3D flow.

- steady/unsteady flow.

- subsonic/transonic/supersonic flow.

- inviscid flow.

- boundary layer modelling and inviscid coupling.

- Navier–Stokes analysis.

- Turbulence and transition modelling.

- Combustion modelling.

In the following sections of the paper a number of the application areas mentioned above will be discussed, in each case stating the engineering objectives with regard to the CFD modelling, illustrating some of the advances being made and giving an indication of areas for future work on both numerical and mathematical aspects of the modelling.

2. Compressor and turbine aerodynamics

The major phenomena that need to be modelled or accounted for in the analysis of two-dimensional flow past a blade are illustrated in Fig. 1. Although this is for a turbine blade, where the effects of heat transfer are important, the flow past a compressor blade is similar and identical mathematical models are used. In reality the flow is three-dimensional and the additional complexities caused by the presence of an annulus or end-wall boundary layer or over-tip leakage in the case of a rotating blade are illustrated in Fig. 2.

In reviewing the advances being made in mathematical modelling the emphasis will be placed on two-dimensional flow and on steady and unsteady flow separately.

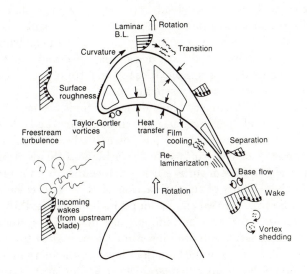

Fig. 1: 2D Phenomena

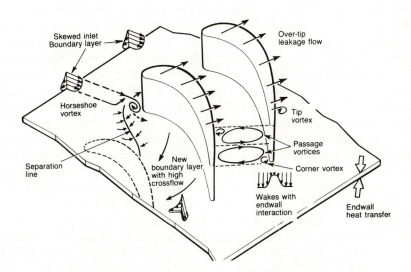

Fig. 2: 3D Phenomena

2.1   Steady flow

The objective with regard to steady flow analysis in a blade row is to calculate the performance, namely the boudary layer and mixing losses, and additionally in the case of a turbine the surface heat transfer rates. These need to be calculated at both design and off-design conditions with the latter adding modelling complexities in the form of laminar separation bubbles and turbulent boundary separation.

Traditional turbomachinery blade design systems are based heavily on the work of Wu (1952) using quasi-3D through-flow and blade-to-blade analyses. With this approach sections of a blade are designed on isolated axisymmetric stream-surfaces using a blade-to-blade program with the definition of the stream-surface being determined from a through-flow analysis. Further details of the interaction of these two calculations and benefits from linking the two can be found in Jennions and Stow (1984) .

Blade-to-blade programs in common use in design systems tend to adopt a coupled inviscid boundary layer approach, mainly from considerations of speed. The effects of the boundary layer on the inviscid mainstream can be modelled either using a displacement model or surface transpiration model. In general direct mode coupling of the boundary layer and inviscid flow is used for attached flow with inverse of semi-inverse coupling for separated flow. Further details of some of methods available in the Rolls-Royce design system and the techniques developed can be seen in (Stow and Newman 1987; Calvert 1982; Carrahar and Kingston 1986; Cedar and Stow 1985). Figs. 3.1, 3.2 and 3.3 from the former of these, show predictions for a low pressure turbine blade tested in cascade by Hodson (1984). A full potential finite element method is used with a fully coupled integral boundary layer. The integral method adopted handles laminar and turbulent flow with correlations being used to predict the start and end of transition. Laminar separation bubbles are handled with correlations for the length and reattachment conditions. The effects of the trailing edge base pressure on the development of the wake downstream of the blade are also included in the integral analysis. Fig. 3.1 shows the blade and the mesh used, with Fig. 3.2 indicating the good agreement between the calculated and measured surface Mach number. Fig. 3.3 gives a comparison of the suction surface momentum and displacement thicknesses; it can be seen that good agreement is found. In the flow the leading edge velocity over-speed creates a laminar separation bubble with almost immediate re-laminarisation after re-attachment and later natural transition towards the blade trailing edge. In this case direct mode coupling can be adopted because of the weak effect of the boundary layer on the inviscid flow. Results for cases where stronger interaction exists and semi-inverse mode coupling needs to be employed are given in Newman and Stow (1985); in general

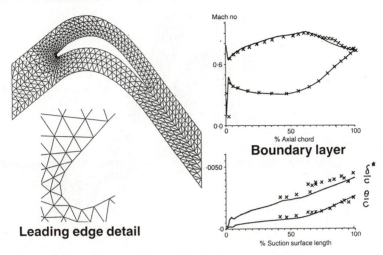

Fig. 3: LP Turbine

good predictions are found.

The main limitations with the coupled approach lie not in the numerical coupling procedures but in the mathematical models employed in the boundary layer analysis whether integral or finite difference methods are employed. The major limitations lie in the treatment of transitional flow and laminar separation bubbles where correlations need to be employed, turbulent separation, and trailing edge losses where empirical relationships for base pressure are often employed.

It is in trying to remove some of the limitations with current coupled techniques, especially for off-design loss prediction, that interest has been generated in solving the Reynolds averaged Navier–Stokes equations.

A number of techniques are being developed within Rolls-Royce based mainly around the pressure correction technique, see for example (Moore 1985), and time-marching either cell centred, see (Jameson et al. 1981; Norton et al. 1984), or cell node based schemes, see (Ni 1982; Carrahar and Kingston 1986).

Currently simple turbulence models are being employed, for example the Cebeci-Smith or Baldwin-Lomax mixing length model or a one equation kinetic model, see (Birch 1987). Figs. 4, 5 and 6 from Connell (private communication) show results for a high pressure turbine blade using a fully implicit cell centred time-marching method with a mixing length model, see (Norton et al. 1984) for more details; similar results are found with the one equation model. Fig. 4 shows the O-C-H grid system used with about

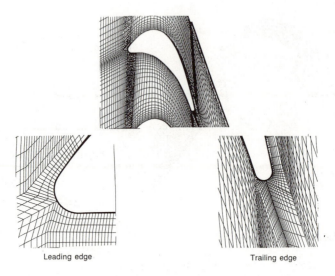

Leading edge   Trailing edge

Fig. 4: Turbine grid system

7,000 grid points including 20 through the boundary layers. Fig. 5 shows the predicted isentropic *surface* Mach number against that measured by Nicholson et al. (1982), Fig. 6 shows the good agreement with surface heat transfer. In these predictions the start and end of transition was specified. In general the position and extent of transition is not known in advance; it is, however, crucially important in determining the boundary layer development and hence performance of a blade. The start and end can be predicted either using a correlation, of the type used in integral boundary layer approaches, or using the kinetic energy model developed by Birch (1987). Fig. 7 shows results from such models for the high pressure turbine blade shown in Fig. 4 but at a higher Reynolds number than in Fig. 6. It can be seen that fairly good agreement exists for the kinetic energy model on the suction surface whereas the correlation model predicts the start of transition to be too late. It can be seen that on the pressure surface there is some effect of free stream turbulence on the laminar boundary layer which is accounted for in part by the kinetic energy model; by the very nature of the correlation model no effect can be included. At low free stream turbulence levels it is found that the correlation model performs better than in Fig. 7 and the kinetic energy less well, indicating that although encouraging results are being produced further development of the models is needed.

Another area where modelling is important in determining the quality

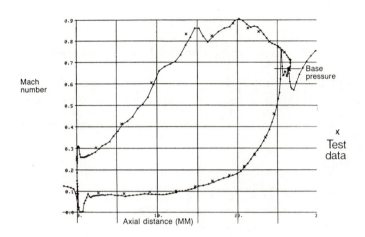

Fig. 5: Blade surface Mach number

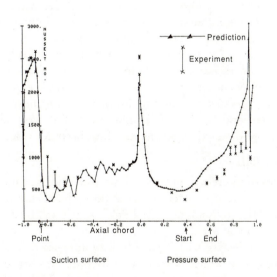

Fig. 6: Blade surface heat transfer

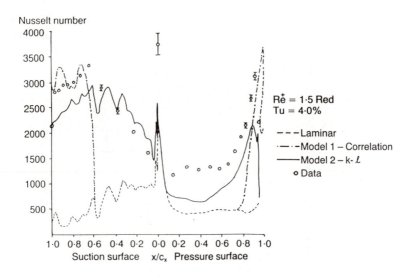

Fig. 7: Blade surface heat transfer

of loss prediction is the blade trailing edge region. For many blades the base pressure has an important effect on the overall loss. In many cases, e.g. subsonic flow, it is found that the flow in the trailing edge region is unsteady and this is felt to have an important influence on the base pressure and overall loss. As in general mean flow effects are of main interest, mechanisms for producing the overall mean effects from Navier–Stokes codes, where steady flow turbulence models are being employed, are being considered. Some results showing that in general good predictions of the profile, base pressure and mixing losses can be achieved can be found in Stow, Northall and Birch (1987); more modelling work and evaluation is needed as more detailed experimental data becomes available.

2.2   Unsteady flow

The main areas of interest in unsteady flow prediction are flutter and blade row interaction, with the latter covering both the effects of an upstream blade wake and the inviscid, potential field, which will be rotating relative to the following blade row.

The current objective with regard to flow analysis is to predict the unsteady blade loading which is then used as part of a structural stability analysis; in many cases inviscid methods are perfectly adequate. Ultimately the objectives are to calculate the unsteady blade performance and blade surface heat transfer; it can be appreciated that this is a much harder task demanding unsteady models of turbulence and transition.

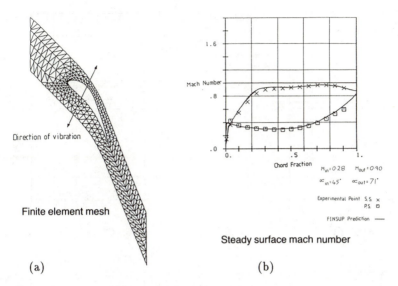

Fig. 8: EPFL turbine blade

Flutter analysis is carried out using a finite element full potential model, see (Whitehead 1982), in which the unsteady flow is taken as a small perturbation of prescribed frequency (namely that of the blade motion) about the steady or mean flow. Fig. 8a shows the grid for a cascade of vibrating turbine blades tested by Bolcs and Franson (1986) at different inter-blade phase angles; the direction of vibration is indicated. Fig. 8b shows the close agreement between the predicted mean or steady blade surface Mach number and that measured. In general the blade unsteady pressure distribution will be out of phase with the blade vibration and consequently is represented by a complex quantity. Fig. 9 gives the predicted and measured amplitude and phase angle of the unsteady pressures for both the suction and pressure surfaces for different inter-blade phase angles. It can be seen that in general good agreement is found.

The main limitations with the approach adopted lie in the neglect of the blade surface boundary layers, which are important in stall flutters and in the neglect of the effects of incoming upstream wakes or blade interaction. The latter are being investgated using an unsteady cell vertex Euler code due to Giles (1987).

Fig. 10 shows typical flow calculations for the case of an incoming rotating wake from an upstream blade row. The figure shows the velocity vectors at an instant of time with the mean flow velocity subtracted so

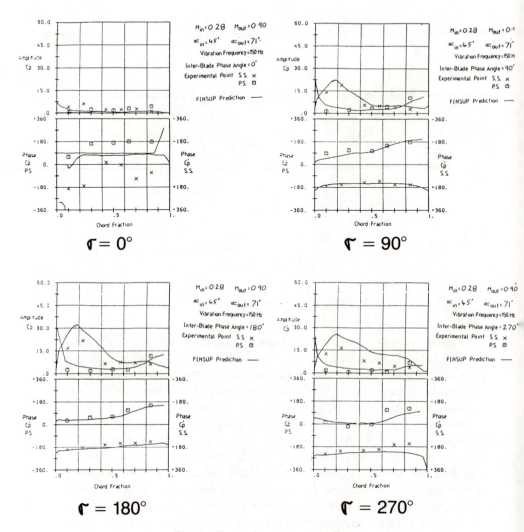

Fig. 9: Unsteady pressure coefficient

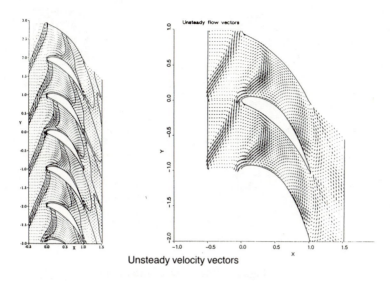

Unsteady velocity vectors

Fig. 10: Blade wake interaction

that the wakes at inlet appear as incoming negative jets. It can be seen that the number of wakes is different from the number of blades. The incoming wake is convected with the flow, being convected faster on the suction surface than the pressure surface and becoming stretched. The outline of the jet is highlighted; also noticeable is the unsteady flow towards the suction surface and away from the pressure surface with counter rotating vortices supporting this. The main interest is in the calculation of the unsteady blade loading due to the effects of the incoming wakes as part of a structural stability analysis.

The unsteady loading due to the inviscid, potential blade interaction is important and this is best studied using a full rotor-stator analysis in which one blade row rotates relative to the other. The interacting wave systems produced are crucially important in determining the unsteady blade loading. Also important in ensuring that spurious waves are not generated or reflected are the implementation of boundary conditions at the inlet and outlet to the flow domain and especially the blade row interface plane, see Fig. 11. Further discussion of this and more detailed results can be seen in (Giles 1988).

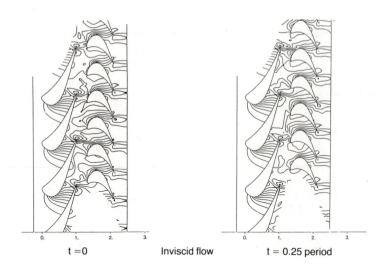

t =0   Inviscid flow   t = 0.25 period

Fig. 11: Rotor-Stator interaction

3.   Installation aerodynamics

One area of interest in installation aerodynamics is in calculating the flow
in engine intakes, nozzles, mixers and over after-bodies. The objectives are
to calculate flow details such as thrust and discharge coefficients as well as
component performance (losses) and surface heat transfer rates.

Although some of the geometric configurations are relatively simple the
flow details can be complex. Fig. 12 shows a simple grid system used in
a Navier–Stokes analysis of flow in a 30^0 convergent axisymmetric nozzle.
Fig. 12b shows Mach number contours in the nozzle and the surface pressure distribution versus experimental results, taken from Kingston (private
communication); this is the same cell centred implicit time-marching code
presented earlier and described in (Norton et al. 1984). It can be seen
from Fig. 12b that the flow chokes and expands around the throat region;
the over expansion causes compression waves from the nozzle wall downstream of the throat which form a reflected shock system on the axis of the
nozzle. Futher comparisons have been made for 40^0 and 50^0 nozzles where
the shocks become stronger. In all cases good agreement with experiment
is found.

The application of this code to the flow in nozzles and after-bodies is
seen in Fig. 13 taken from Kingston (private communication). The external flow Mach number of 0.8 with the internal nozzle flow being choked.
Comparison with experimental data from Carson and Lee (1981) given

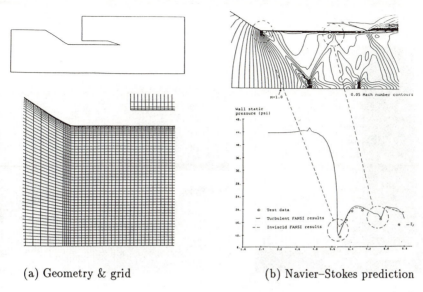

(a) Geometry & grid   (b) Navier–Stokes prediction

Fig. 12: 30⁰ convergent nozzle

in Fig. 14 shows good agreement on the nozzle and after-body surfaces. Fig. 15 shows results for cases of higher external flow Mach number where now shocks exist on the after-body surfacr and boundary layer separation occurs; it can be seen that the agreement is less satisfactory. The discrepancies are attributed to short-comings in the mixing length model in regions of separation and indicate an area where further work is needed.

The same Navier–Stokes code is also being applied very effectively to the analysis of engine intake flows and to full powerplant configurations i.e. combined intake, nozzle, after-body flow. It can be appreciated that in cases where an external flow is present treatment of the outer boundary condition is important both from the quality of the solution as well as computational efficiency. The boundary must be far enough away to avoid any spurious effects in the regions of importance; however, with a conventional grid system the consequences of this can be seen from Fig. 13 where a large number of grid points exist in regions of the flow where little is happening. This is an application where patched or embedded grids will have significant impact.

4.   Turbine blade internal heat transfer
Some parts of the turbine operate in regions of very high temperature, especially just downstream of the combustion chamber, and the blades need

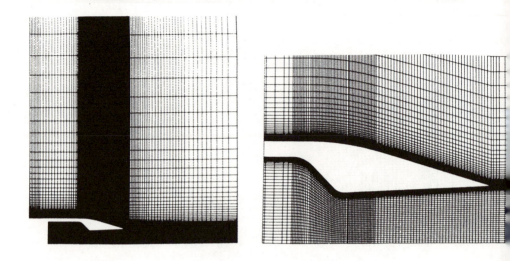

Fig. 13: Nozzle-afterbody grid system

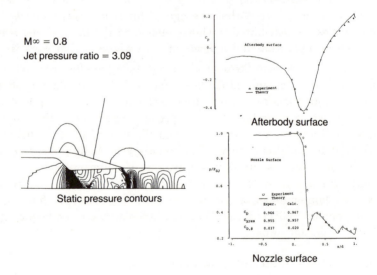

M∞ = 0.8
Jet pressure ratio = 3.09

Static pressure contours

Afterbody surface

Nozzle surface

Fig. 14: Nozzle-afterbody modelling

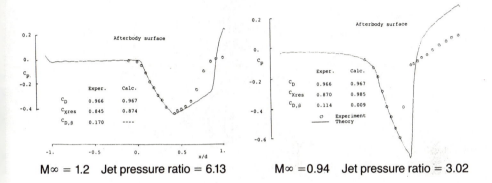

Fig. 15: Nozzle-afterbody modelling

to be cooled in order to have acceptable lives. Fig. 16 gives an indication of the internal and external cooling features that are employed. With film cooling, jets of cold air are ejected on the external surface of the blade in order to reduce the heat transfer from the hot gas. With internal cooling, cold air pumps up and down the blade to increase the heat transfer internally and reduce the blade metal temperatures; heat transfer enhancement devices are used within the internal cooling passages to improve this process.

The major objectives with the analysis of the flow within the internal cooling passages is to calculate the surface heat transfer rates, which are used as input to a blade heat conduction and thermal stress analysis, and to assess the losses within the cooling system. Areas of particular interest are the effects of blade rotation and flow or passage curvature.

The effects of blade rotation on the flow patterns and associated heat transfer are illustrated in Fig. 17 from a Navier–Stokes calculation by Birch (private communication) for the flow in a simple cooling passage. The effects of the circumferential pressure gradient set up by the coriolis force is to create a secondary flow away from the leading face of the passage towards the trailing face; the temperature contours associated with this indicate a reduced heat transfer on the leading face and enhanced heat

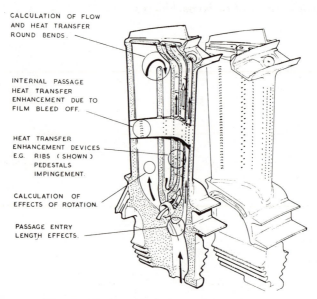

CALCULATION OF FLOW
AND HEAT TRANSFER
ROUND BENDS.

INTERNAL PASSAGE
HEAT TRANSFER
ENHANCEMENT DUE TO
FILM BLEED OFF.

HEAT TRANSFER
ENHANCEMENT DEVICES
E.G. RIBS (SHOWN)
PEDESTALS
IMPINGEMENT.

CALCULATION OF
EFFECTS OF ROTATION.

PASSAGE ENTRY
LENGTH EFFECTS.

Fig. 16: Turbine blade internal heat transfer.

transfer on the trailing face.

Further effects of blade rotation are given in Fig. 18 from Baines (private communication) for flow in a 180^0 bend of rectangular cross-section which rotates out of the plane of the figure. The results are from a 3D fully elliptic pressure correction scheme due to Moore (1985) and show contours of heat transfer coefficients on the leading and trailing faces for the rotating case against the symmetrical results for the non-rotating case. It can be seen that in this case the effects of the rotation on the heat transfer enhancement alters as the flow direction changes round the bend, changing the sign of the associated Coriolis force, and thus affecting the secondary flow patterns. Experimental results for this example indicate reasonably good agreement with the calculation. The other area of interest with this example is the effects of passage curvature on both heat transfer and flow separation. Tests with the turbulence model against measured heat transfer and flow visualisation results indicate that this is important in predicting the heat transfer and flow separation in the region of the bend and that further development is needed.

5. Disc air flow

The interest in this application area is in calculating the flow details in air systems associated with disc cavities. The effects of disc rotation and leakage flows are important in determining the levels of heat transfer to the discs and shrouds as well as the losses associated with the flows. The

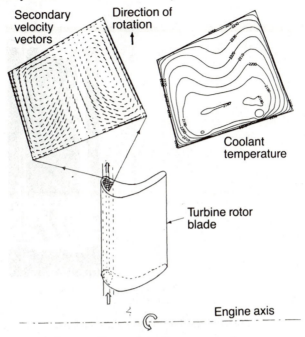

Fig. 17: Internal cooling passage flow.

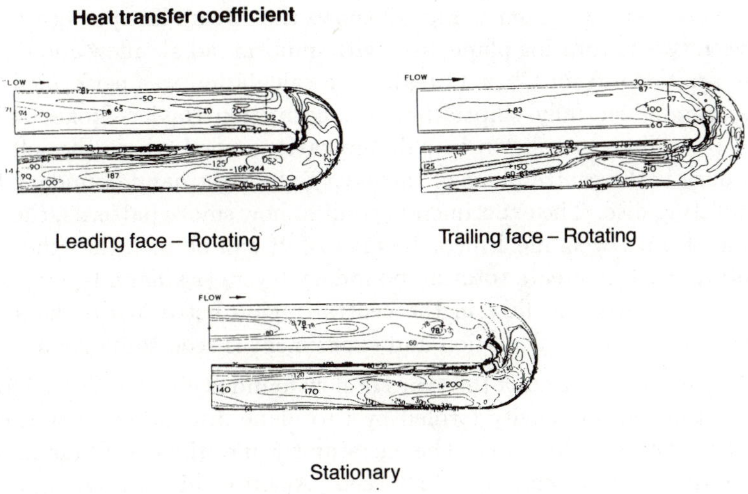

Heat transfer coefficient

Leading face – Rotating   Trailing face – Rotating

Stationary

Fig. 18: Square section rotating duct.

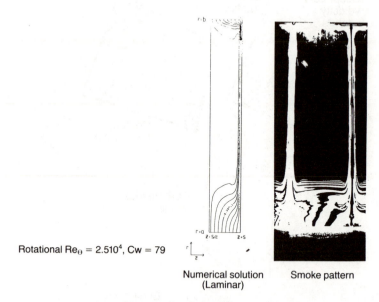

Rotational Re$_\theta$ = 2.510^4, Cw = 79

Numerical solution   Smoke pattern
(Laminar)

Fig. 19: Radial inflow-radial outflow.

heat transfer forms an input to heat conduction calculations which are
performed as part of stress lifing calculations on the discs.

The effects of disc rotation can be illustrated by considering the flows in
simple geometrical systems. Fig. 19 shows the stream-line pattern for flow
between two co-rotating plane discs with uniform radial inflow and outflow;
these are taken from Chew (1979). The calculation was performed using
an axisymmetric fully elliptic pressure-correction Navier–Stokes scheme.
The streamlines are shown in half the cavity region where the left hand
boundary is the centre plane of the cavity, the right-hand boundary being
the rotating disc. The experimental results show smoke patterns, the white
regions, from Owen and Pincombe (1979). It can be seen that the effects
of rotation are to create rotating boundary layers (Ekmann layers) on the
disc surfaces with no flow in the central region between the discs. The
agreement between calculation and experiment is seen to be good.

Another illustration of the types of flow found is given in Fig. 20 for the
flow in the rotating cavity formed by two plane discs where now the flow
enters in the axial direction. The entrainment into the two Ekmann layers
on the disc surfaces can be clearly seen. Again good qualitive agreement
is seen with the experimental flow pattern. More detailed quantitative
comparisons are given in (Chew 1979).

Calculations at higher rotational speeds where the flow is turbulent

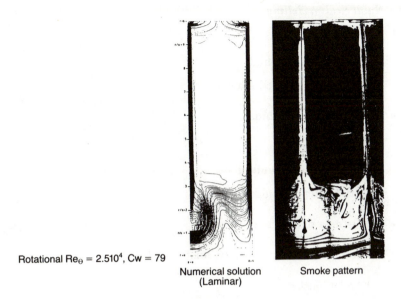

Rotational $Re_\theta = 2.510^4$, $Cw = 79$

Numerical solution (Laminar)

Smoke pattern

Fig. 20: Axial inflow-radial outflow.

have been performed using a simple mixing length model and shown to agree closely with experimental results, see for example Chew (1987). It seems very likely, however, that for more complex flow configurations higher order differential turbulence models will be needed but it is clear from earlier work that further development of these will be needed for this application.

As the disc rotation speed is increased then the convergence rate of standard pressure correction techniques, e.g. SIMPLE, decreases. As a consequence there is interest in improving calculation times using multi-grid techniques with existing schemes, see for example, the advances shown in Chew and Vaughan (1988); it is also likely that more strongly coupled correction techniques will have an impact on the problems encountered.

6. Summary

Computational Fluid Dynamics is being applied widely within Rolls-Royce to the various design and analysis aerodynamic and thermodynamic problems encountered in turbomachinery. In this paper some of the advances being made in a number of the applications areas have been presented.

It is clear from these applications that in order to extend the impact of CFD there are a number of areas for development aimed at improving both the quality of the solution and the cost of obtaining it. The main areas for numerical and mathematical modelling are summarised below.

6.1   *Numerical modelling*

- Speed of solution

  1. multi-grid method
  2. strongly coupled scheme

- Accuracy of algorithms

  1. grid restrictions
  2. smoothing formulations
  3. time-stepping algorithms

- Robustness of solvers

- Boundary conditions (steady/unsteady flow)

  1. inlet/outlet
  2. far field

- Grid systems

  1. generation
  2. embedded grids
  3. adaptive grids

6.2   *Mathematical modelling*

- Transition modelling

  1. free stream turbulence
  2. pressure gradient effects
  3. incoming wakes

- Turbulence modelling

  1. flow separation
  2. flow curvature
  3. rotation
  4. unsteady flow

References

Birch, N. T. (1987) Navier–Stokes prediction of transition, loss and heat transfer in a turbine blade. *American Society of Mechanical Engineers*, Paper 87-GT-22.

Bolcs, A. and Franson, T. H. (1986) Aeroelasticity in turbomachines: comparison of theoretical and experimental cascade results. *Communication du Laboratoire de Thermique Appliquée et de Turbomachines*, No. 13, EPFL, Lausanne.

Calvert, W. J. (1982) An inviscid-viscous interaction treatment to predict the blade-to-blade performance of axial compressors with leading edge normal shock waves. *American Society of Mechanical Engineers*, Paper 82-GT-35.

Carrahar, D. and Kingston, T. R. (1986) Some turbomachinery blade passage analysis methods-retrospect and prospect. *Transonic and Supersonic Phenomena in Turbomachines*, AGARD, Munich.

Carson, G. T. and Lee, E. E. (1981) *Experimental and analytical investigation of axisymmetric supersonic cruise nozzle geometry at Mach number 0.6 to 130*. NASA Technical Paper 1953.

Cedar, R. D. and Stow, P. (1985) A compatible mixed design and analysis finite element method for the design of turbomachinery blades. *International Journal for Numerical Methods in Fluids*, 5 : 331–345.

Chew, J. W. (1979) Computation of flow in rotating cavities. Part 1: Isothermal laminar source -sink flows. *Thermo-Fluid Mechanics Research Centre Report FRMRC/5*. School of Engineering and Applied Sciences, University of Sussex.

Chew, J. W. (1987) Computation of flow and heat transfer in rotating disc systems. *Proceedings American Society of Mechanical Engineers, Thermal Engineering Conference*, Hawaii, pp.361–367.

Chew, J. W. and Vaughan, C. M. (1988) Numerical predictions for the flow induced by an enclosed rotating disc. *33rd American Society of Mechanical Engineers Conference*, June 1988, Amsterdam.

Giles, M. B. (1987) Calculation of unsteady wake/rotor interaction. *American Institute of Aeronautics and Astronautics*. Paper 87-0006.

Giles, M. B. (1988) Calculations of unsteady turbomachinery flow. *Conference on Numerical Methods for Fluid Dynamics*, 21–24 March, 1988, Oxford.

Hodson, H. P. (1984) Boundary layer transition and separation at the leading edge of a high speed turbine blade. *American Society of Mechanical Engineers*, 84-GT-241.

Jameson, A., Schmidt, W. and Turkel, E. (1981) Numerical solution of the Euler equations by finite volume methods using Runge-Kutta time stepping schemes. *American Institute of aeronautics and astronautics.* Paper 81-1259.

Jennions, I. K. and Stow, P. (1984) A quasi three-dimensional turbomachinery blade design system. Part 1: Through-flow analysis; Part 2: Computerised system. *American Society of Mechanical Engineers*, 84-GT-26; 84-GT-27.

Moore, J. and Moore, J.G. (1985) Performance evaluation of a linear turbine cascade using a 3D viscous calculation. *Transactions of American Society of Mechanical Engineers, Journal of Engineering for Gas Turbines and Power.* Volume 107, 969-975.

Newman, S. P. and Stow, P. (1985) Semi-inverse mode boundary layer coupling. *Institute of Mathematics and its Applications. Conference on Numerical Methods for Fluid Dynamics.* University of Reading, England.

Ni, R. H. (1982) A multiple-grid scheme for solving the Euler equations. *American Institute of aeronautics and Astronautics Journal,* Vol. 20, No. 11.

Nicholson, J. H., Forest, A. E., Oldfield, M. L. G. and Schultz, D. L. (1982) Heat transfer optimised turbine rotor blades—an experimental study using transient techniques. *American Society of Mechanical Engineers,* 82-GT-304.

Norton, R. J. G., Thompkings, W. T. and Haimes, R. (1984) Implicit finite difference scheme with non-simply connected grids—a novel approach. *American Institute of Aeronautics and Astronautics Journal 22nd Aerospace Sciences Meeting,* January 1984, Reno.

Owen, J. M. and Pincombe, J. N. (1979) Velocity measurements inside a rotating cylindrical cavity with a radial outflow of fluid. *University of Sussex, School of Engineering and Applied Sciences.* Report No. 79/me/101.

Stow, P. and Newman, S. P. (1987) Coupled inviscid-boundary layer methods for turbomachinery blading design. *Joint IMA/SMAI Conference on Computational Methods in Aeronautical Fluid Dynamics.*

Stow, P., Northall, J. D. and Birch, N. T. (1987) Navier–Stokes methods for turbomachinery blade design. *Joint IMA/SMAI Conference on Computational Methods in Aeronautical Fluid Dynamics.*

Whitehead, D. S. (1982) The calculation of steady and unsteady transonic flow in cascades. *Cambridge University report CUED/A– Turbo/TR118.*

Wu, C. H. (1952) A general theory of three-dimensional flow in subsonic and supersonic turbomachines of axial, radial and mixed flow types. *Transactions, American Society of Mechanical Engineers* 1363–1380.

Computational fluid dynamics in the automobile industry

A. D. Gosman

Imperial College of Science and Technology

1. Introduction

In this paper the areas of existing and potential application of CFD techniques to automobile design are reviewed and the nature of the CFD methodology requirements are outlined. The applications are numerous and include external aerodynamics; passenger compartment ventilation; engine combustion, cooling and lubrication; and intake and exhaust system flow. Example calculations are shown for some of these areas. The CFD methodology requirements are identified to be similar in many respects to those of other industries, but there is a particular need for techniques capable of dealing with highly complex geometries, moving boundaries, non-stationary turbulent flow and combustion and sprays. Some existing methods possessing such capabilities will be identified and characterised.

During the past decade the pattern and extent of industrial penetration by CFD has significantly altered from one of exploitation in a relatively small number of high-technology areas (notably aerospace and nuclear) to much more extensive usage in a growing number of industries covering a much wider range of products. These changes have been brought about by three main factors: improvements to the efficiency and capabilities of CFD methodology; dramatic reductions in the cost:performance ratios of computers; and the emergence of integrated commercial code systems. These developments have collectively contributed to a situation where CFD is becoming to thermofluids analysis as computational solid mechanics is to structural analysis, although it has to be said that CFD has some way to go before it will be capable of comparable accuracy and reliability.

The automobile industry, which is the focus of this paper, is an example of a comparatively new field of application of CFD. It is a particularly appropriate example, because the growth in its utilisation of this methodology has verged on the explosive. At first sight this might seem surprising, since this industry has not traditionally been perceived as either 'high tech' or strongly thermofluids-oriented; however the reverse is actually the case. Automobile design is strongly fluids-influenced in many areas, including aerodynamics, engine and brake cooling, passenger compartment heating and ventilation, engine combustion and numerous others. The related research and development costs based on traditional 'cut-and'try' approaches are very high and their pace is often too slow. These factors have impelled the industry to strongly embrace CFD, even in areas where it is relatively unproven; and it is clear that this enthusiasm will lead to a degree of utilisation which will soon match, if not exceed, that in the traditional high-tech industries.

In what follows, further details will be provided of the major areas of application of CFD in the industry. For each area an outline will first be provided of the thermofluids processes involved and the related design problems; then the nature of the CFD requirements will be discussed; and finally the current status of the methodology, according to available information, will be assessed with the aid, where possible, of example applications. An indication will then be given of the directions of future development.

2. Automotive CFD applications areas

2.1 Aerodynamics

Design Considerations - The flow of air around the vehicle exterior determines, among other factors, the degree of aerodynamic drag, which has received considerable publicity in recent years as manufacturers have strived to lower their vehicles' drag coefficient below the levels of competitors. Drag is indeed an important factor in determining fuel economy, stability and wind noise, as are other aerodynamic factors such as lift and yaw (ie crosswind) effects.

There are however a host of additional external aerodynamic effects which, although less fashionable, are also important in vehicle design: they include aerodynamic loads on body panels, which determine how strong they must be to avoid deflection; the flows into and out of the passenger and engine compartments, the design and siting of the apertures for which influence ventilation/cooling efficiency and drag; and the flows in the vicinity of the wheels, which influence brake cooling.

Nature of CFD Task - The exterior of an automobile is inherently more complex in form than that of an aircraft, the aerodynamic design of which is both in principle and practice a much easier task. The reason is, simply, that the automobile cannot be made as streamlined, despite the effort which has gone into smoothing the exterior surface: for the latter must house its occupants within an envelope of acceptable height and length; and it also includes the vehicle wheels and underbody structure, with its wheel arches, suspension and exhaust components, etc, all of which are exposed to the air stream. The external flows are also inseparably coupled, to varying degrees, to the flows through the engine and passenger compartments.

The above factors collectively contribute to the need, ideally, to analyse the entire problem, including the flows in the underbody structures and engine compartment as a single entity: thus they entail a task of considerable size and geometrical complexity.

Complexities also exist in the fluid physics. Flow separation (which generally has adverse effects on drag and other factors) is virtually unavoidable around the wheels, underbody and engine compartment structures; and it is even difficult to avoid on the streamlined surfaces, especially towards the rear of the vehicle and in the presence of a crosswind. Moreover the onset of separation can be strongly sensitive to small-scale features, such as the dimensions and shape of a moulding surrounding a window. A further complexity is that, when separation does occur, it can lead to semi-ordered large-scale unsteadiness (e.g. vortex shedding) in the flow field.

Current Methodology and Status - As recorded in a recent review (Buckheim et al. 1988), three main classes of CFD method have been used in automobile aerodynamic studies, namely Panel, Euler and Navier-Stokes. However although the first two are useful in some respects, such as in the calculation of the pressure distribution on the vehicle forebody, only the last-named is capable of predicting the separation phenomena referred to above. Navier-Stokes methods are therefore seen as the ultimate design tool.

Developments are taking place in the finite volume (FV) and finite element (FE) areas, although most of the activity to date has been in the former, e.g. (Buckheim et al. 1988; Shaw 1988; Adey and Greaves 1988). The FV methods employed share many features in common, viz the use of structured (ie topologically rectangular) body-fitted meshes, first- or second-order upwind differencing, implicit iterative solution to the steady state, and representation of turbulence effects by the well-known k-e model. There have however been some departures from these practices, generally confined to pilot studies in two dimensions, such as that in (Buckheim et al. 1988) where a multigrid method was used. In all cases the problem of large-scale unsteadiness has been ignored.

As an illustration of current capabilities, extracts will now be shown from a recent study reported in (Adey and Greaves 1988) of the flow around a current production vehicle, albeit with the wheels removed and the underbody and radiator intake covered with smooth panels. These simplifications to the real configuration were apparently driven by the need to limit the size and complexity of the computing mesh and are currently common practice. The mesh itself is shown in Fig. 2, which clearly illustrates an important drawback of structured meshes which contributes to the need for the forementioned simplifications: it is the necessity to propagate mesh refinement, required in the immediate vicinity of the vehicle, out to the boundaries of the calculation domain and thus to regions where it may not be required.

Fig. 1 shows a comparison between the predicted and measured (on the same simplified body shape) centreline surface presssure distributions. Agreement is quite good in the forebody region (where, it should be said, even the simpler panel methods perform quite well (Buckheim et al. 1988)) but deteriorates towards the rear of the vehicle, due in part to the fact that the calculations do not reproduce the experimentally observed separation. The overall drag and lift coefficients are predicted to be 0.234 and -0.643 respectively, as compared with the measured values of 0.218 and -0.022. Although there is perhaps fortuitously closer agreement for drag, the accuracy is still far from that achievable, at less cost, in wind tunnel studies (Buckheim et al. 1988).

The failure of the predictions is undoubtedly due to a combination of insufficient resolution and inadequacies in the modelling of the physics, in the latter case possibly due both to the turbulence model employed and the associated neglect of large-scale non-random motions. These deficiencies will undoubtedly be resolved in the future, but in the meantime there is no need to abandon external aerodynamics calculations: rather they should be exploited in areas where competing approaches are either non-existent or relatively crude, as in for example the design of engine compartment and brake cooling arrangements.

2.2   Passenger compartment ventilation

Design Considerations - This is an area where it is known that CFD is being utilised, but as yet there appear not to be any published accounts. The features of interest are the velocities and temperatures of the air in the passenger compartment, which collectively determine the 'comfort level' of the occupants and the efficiency of window defrosting and demisting. For a given compartment configuration the controlling factors are siting of the air supply and exhaust ducts; and the supply flow rates and temperatures. Increasingly the configuration itself is being designed with these requirements in mind.

Nature of CFD Task - Geometrical complexity also features here, especially when the supply ducting is taken into consideration (although in this instance there is probably no need to compute the flow in the ducting and compartment simultaneously, for the former drives the latter). Turbulence also features, as does heat transfer to the exterior. In general however the required level of prediction accuracy is probably not high and may well be within the compass of existing methodology.

2.3 Engine cooling

Design Considerations - The engine must be cooled to remove the excess heat generated by the combustion process: this is necessary to ensure that engine component temperatures do not exceed acceptable limits, among other reasons. Cooling is effected both externally, by the air flowing through the engine compartment and internally, via a network of passages in the cylinder block and head through which a liquid coolant is pumped, the accumulated heat being removed by the radiator located in the engine compartment. Some additional cooling is provided by the lubricant flow network.

In addition to its cooling effects, the engine compartment flow also determines the temperature levels to which critical external engine components like carburettors and microprocessors are exposed. For all these reasons, CFD is beginning to be utilised to calculate the engine compartment velocity and temperature distributions, although there seems to be no published work.

Of all the forementioned heat extraction mechanisms, the internal liquid coolant network is the predominant one. An impression of the extent and complexity of this network can be gained from Fig. 3(a) which pertains to a contemporary engine design. Coolant is admitted at the front of the block, with the pump being located within it, immediately behind the inlet plane. The pump impels the fluid through annular passages surrounding each of the four cylinders and thence into a complex of linked passages in the cylinder head, designed to cool the surfaces exposed to the flame. The coolant then exits via one or more outlets in the head. The design of this network is receiving increasing attention (Finlay et al. 1988; Adyagi et al. 1988; Macdonald 1988) driven in part by the continual rise in engine power, resulting in increased heat loading. The goals are, in addition to controlling peak temperatures, reducing pumping power losses and minimising thermal stresses, which have a strong bearing on engine durability. Interestingly, in this instance CFD is being used in conjunction with computational stress analysis - an example of an integration of CAD methods of which there will undoubtedly be much more in the future.

Nature of CFD Task - Here too geometrical complexity and turbulence feature, in both the exterior and interior flows, along with heat transfer. An additional complication on the liquid cooling side is the possibility of boiling which, of course, gives rise to a two-phase flow.

It is important to note that the complex geometry of Fig. 3(a) is fundamentally different in character from that of Fig. 2, in that, whereas the latter can be relatively easily accommodated within a regular body-fitted grid structure, the former is decidedly far from topologically rectangular, as a consequence of its branched structure. However, regular grids can be still used at some expense in efficiency, as recorded below.

Current Methodolgy and Status - CFD has been applied to the liquid cooling problem on at least two occasions (Adyagi et al. 1988; Macdonald 1988): in both instances it was assumed that boiling did not occur, and turbulence was represented by the standard k-e model.

In the study of (Adyagi et al. 1988) a FE method was used to analyse a small portion of the network, (consideration of the whole system being prevented by methodology and computer limitations) using estimates of the conditions prevailing at the boundaries of this subregion. An obvious drawback of this approach is that in practical applications these conditions are seldom if ever known; thus a strong additional element of uncertainty is introduced into the calculations. Notwithstanding this, the authors report reasonable agreement with some largely qualitative experimental observations which they made. The FE calculations could of course be performed on an unstructured mesh.

Recently an analysis has been carried out of an entire cooling network (Macdonald 1988), using an implicit FV method, which, although utilising a structured mesh, contained the facility for declaring portions of the mesh inactive. This device, which is currently extensively using in FV methodology, is a compromise solution for topologically non-rectangular regions which, when properly employed, can minimise the calculation overheads in inactive regions but generally entails accepting the storage overheads. The computing mesh used in this study is in fact that portrayed in Fig. 3(a): it contains around 90000 active cells, which is about 30% of the total. Fig. 3(b) is an extract of the results of the analysis, in the form of a plot of the velocity vectors in a horizontal section of the cylinder head. In this instance no experimental data were available for comparison, although undoubtedly inaccuracies do exist due to the limited mesh resolution and, probably, turbulence modelling errors.

Despite the caveats mentioned above, the few calculations performed thus far have strongly excited the interest of engine designers, probably for the simple reason that the prior sources of information available to them were extremely limited in accuracy and coverage. This is an example of a not uncommon situation in industry, where the current limitations of CFD are acceptable, but where expectations will undoubtedly rise with time.

2.4   Engine induction and exhaust systems

Design Considerations - The manifolding and ducting employed to supply air or air/fuel mixture to the engine combustion chambers and to collect and remove the exhaust gases, are also candidates for CFD application. The design of the induction system is crucial to engine performance for a variety of reasons: the overall pressure losses determine breathing efficiency and the detailed flow characteristics at the inlet valves strongly influence the combustion behaviour, as also does the degree of fuel/air mixing upstream. On the exhaust side, the usual goal is to minimise pressure losses, although on turbocharged engines the performance of the turbocharger (which is also a CFD candidate) is sensitive to the details of the supply flow.

Nature of CFD Task - For these applications, the usual features of complex geometry (here also including branched networks) and turbulence apply. Additionally, the flow is pulsating and compressible, and in spark-ignition engines the induction stream is two-phase, since it contains liquid fuel.

Current Methodology and Status - CFD has been employed in this area for many years, although until comparatively recently only in the form of one-dimensional gas dynamics methods, which are outside the scope of this survey. See e.g. (Gosman 1986), for a full account of this methodology. Multidimensional CFD calculations date from around 1979 (Chapman 1979) but although steady progress has been made since then, it appears that a full 3D unsteady flow simulation of an entire inlet or exhaust system has yet to be performed, presumably due to the size and complexity of the task.

Calculations have however been made of the flows in component parts of the systems, especially the inlet ports, e.g. (Wakisaka et al. 1988; Gosman and Ahmed 1987) using implicit FV methods and a variety of turbulence models. In the study reported in (Wakisaka et al. 1988) the geometry was approximated by a castellated cylindrical-polar mesh, but body-fitted meshes as used in (Gosman and Ahmed 1987) are clearly more appropriate and they should preferably be unstructured.

Fig. 4 illustrates the recent use of CFD at Imperial College to compute the flow past an engine exhaust valve (Naser 1988), in this case mounted in a special experimental apparatus with a steady air supply. The implicit FV method employed a body-fitted mesh and k-e turbulence modelling. Plot (a) shows the mesh (in this instance the flow is axisymmetric), plot (b) the predicted flow field and plot (c) a comparison with measured velocity profiles. Agreement is reasonable, as is the disparity between the measured (Oldfield and Watson 1983) and predicted discharge coefficients of less than 10%. It has to be said however that substantially larger errors, attributable to turbulence modelling deficiencies, have been observed in other applications e.g. (Gosman and Ahmed 1987). The overall picture is therefore one of promise, but with a clear need for improvements to speed and accuracy, before the full 3D unsteady problem can be tackled effectively.

2.5 Engine combustion processes

Design Considerations - The initial impetus for the utilisation of multi-dimensional CFD methods undoubtedly came from the designers of engine combustion systems, at a time when they were besieged by high fuel costs and increasingly stringent emissions legislation. Engine combustion, it turns out, is not a purely chemical matter: the combustion rates are strongly linked to the fluid dynamics, because turbulent mixing is usually the rate-controlling process, or nearly so. It is for this reason that the combustion performance is strongly influenced by all factors that condition the fluid dynamic behaviour: these include the shapes of the combustion chambers and induction passages; the inlet and exhaust valve configurations and motions; the engine speed and compression ratio; and in the case of the Diesel engine, the dynamic effects of the fuel sprays. To these must be added the fuel-related chemical processes, which may be influential on such characteristics as ignition, knock and emissions behaviour, especially in 'lean burn' and Diesel engines. Design of the latter also requires consideration of the fuel injection arrangement, ie number, size and orientation of injection holes, fuel delivery rate schedule, etc. With so many parameters to play with it is scarcely surprising that CFD has appeal.

Nature of CFD Task - In addition to the usual geometrical complexities (combustion chambers come in diverse shapes and sizes) the reciprocating engine has the additional feature of moving boundaries, associated with the motions of the pistons and valves. Turbulence is as usual present, but here too there is an additional complication, for the flow is non-stationary; and this gives rise to additional controversies about how to define and model the turbulence effects (Gosman 1985b). The combustion process itself must also be modelled and calculated (modelling is necessary because the flames are sub-millimeter in thickness and distorted by the turbulence) as must the fuel sprays, when present. The representation of the sprays is a major task in itself, requiring consideration of the processes of atomisation, droplet dynamics and interaction with the gas field, evaporation and deposition on the combustion chamber walls.

Current Methodology and Status - The initial emphasis in CFD for this application was in air motion prediction in the absence of combustion and a number of methods have been developed, as recorded in recent reviews (Gosman 1985b; Gosman 1985a). All are of the FV variety and employ, with one exception, some form of structured Eulerian/Lagrangian mesh to accomodate the moving boundaries.. The two main centers of development have been the Los Alamos Scientific Laboratory, from which has emanated the well-known KIVA code (Butler et al. 1985) embodying a semi-implicit method with a structured non-orthogonal mesh; and Imperial College, which has developed a series of fully-implicit FV methods

(Gosman and Johns 1980; Gosman et al. 1985; Gosman et al. 1987) cul-
minating recently in one which uses an unstructured non-orthogonal mesh
(El Tahry 1984) as illustrated in Fig. 5. Turbulence is generally treated
by computing the ensemble-averaged motion using the k-e model, although
other approaches have been tried (Butler et al. 1985; Marooney 1988). A
considerable amount of validation has been performed involving comparison wtih experiment, as in the example of Fig. 6 from (Arcoumanis et al.
1986): the degree of agreement displayed is typical, the errors being due
to a combination of insufficient resolution, deficiencies in the turbulence
modelling and uncertainties in the boundary conditions. (It is an unfortu-
nate fact of life that fully-resolved 3D unsteady flow predictions are still
difficult to achieve, despite advances in methodology and hardware).

More recently spray and combustion models have been incorporated
into these methods. On the spray modelling side, the almost-universal
approach has been to use stochastic Lagrangian representations, whereby
the equations of motion and energy for a statistical sample of droplets are
solved simultaneously with the gas-phase equations (Butler et al. 1985;
Gosman et al. 1987; Arcoumanis et al. 1986) with allowance for droplet
breakup and collision/coalescence. Since the continuum formulation is retained for the gas phase, and strong coupling exists between the phases due
to drag and other interactions, some care is necessary to implicitly couple
the two sets of equations in the calculations. Such efforts have been successful, as evidenced by Fig. 7, showing favourable comparisons between
predicted and measured droplet size, taken from (Chatwani and Bracco
1985). It should however be noted that these models have yet to be fully
tested, especially for their ability to predict evaporation rates and vapour
concentrations (important inputs for combustion calculations); and they
are known to rest on insecure foundations, notably in their representation
of atomisation.

The status regarding combustion, the process of greatest interest, is
currently uncertain, due to the absence of suitable models of turbulent
combustion (Gosman 1985c). This holds true for both spark-ignition engines, where the fuel and air are usually premixed; and for Diesel engines,
where they are introduced separately. Calculations have been performed
for both classes of engine, e.g. (Ahmadi-Befrui et al. 1981; Abraham et al.
1985; Gosman and Harvey 1982), but it is questionable whether the accuracy and width of applicability of the methods are as yet sufficient for
quantitative design purposes - this will have to await the emergence of better combustion models (a situation which is not unique to the reciprocating
engine field).

3. Future developments

The shopping list of desirable improvements to automotive CFD contains many items which no doubt are needed in other areas, notably enhanced economy through more accurate discretisation, faster algorithms and the use of parallel processing; improved representation of the physics via better mathematical models of turbulence, sprays and combustion; and greater flexibility and ease of application through the use of general unstructured and self-adaptive grids. The common items would also include improvements to the peripheral operations of surface modelling, mesh generation, and graphical display of results, as well as the integration of the whole into a CFD Computer-Aided Design system. The perhaps uncommon feature of this particular industry is that some of its applications call for all of the above in a single package!

4. Conclusions

The foregoing has clearly demonstrated that CFD has a strong and growing presence in the automobile industry. In some areas, such as external aerodynamics it is facing strong competition from traditional experimental approaches and will have to improve in accuracy and cost-effectiveness before it displaces them. In others such as engine cooling, it is currently the only viable source of information and has therefore already gained acceptance, despite its imperfections.

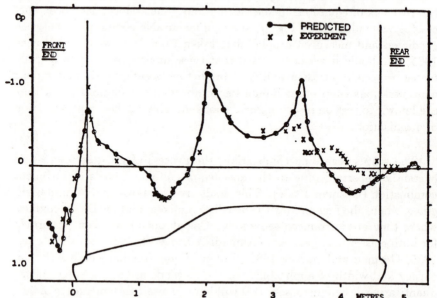

Fig. 1: Predicted and measured center body surface pressure variation, from (Adey and Greaves 1988)

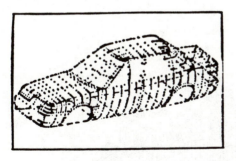

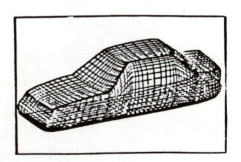

Initial body surface data

Vehicle surface mesh

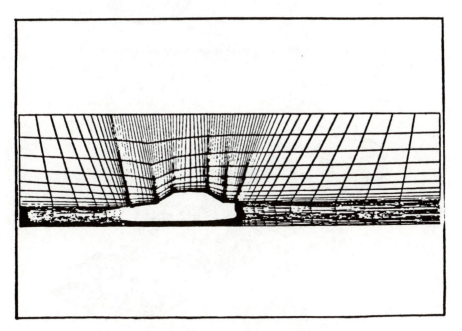

Centreline plane of complete volume mesh

Fig. 2: Structured CFD mesh for a vehicle aerodynamics calculation, from (Adey and Greaves 1988)

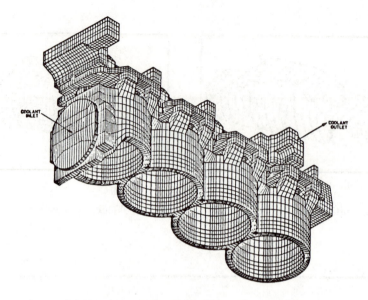

(a) Network configuration and computing mesh

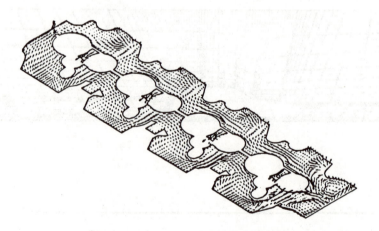

(b) Predicted flow field in section through head

Fig. 3: Calculation of engine coolant network flow, from (Macdonald 1988)

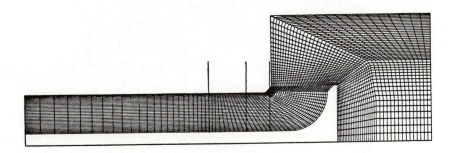

(a) Computing mesh

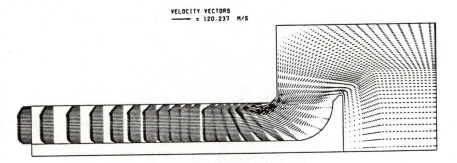

(b) Predicted velocity field

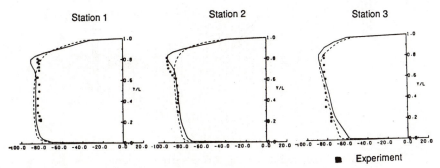

(c) Comparison with measure axial velocity

Fig. 4: Calculation of steady flow past engine exhaust valve, from (Naser 1988)

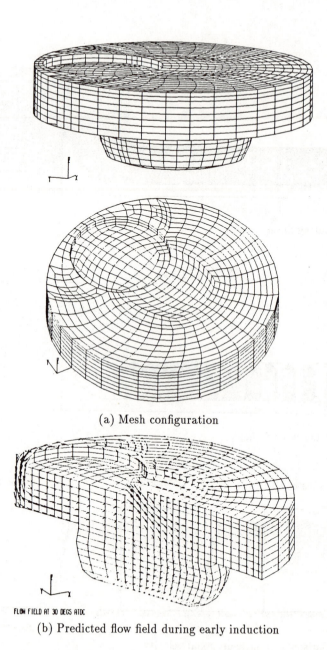

(a) Mesh configuration

FLOW FIELD AT 30 DEGS ATDC

(b) Predicted flow field during early induction

Fig. 5: Illustration of use of unstructured Eulerian/Lagrangian mesh for in-cylinder flow calculations, from (Marooney 1988)

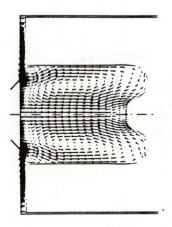

(a) Predicted velocity field near TDC compression

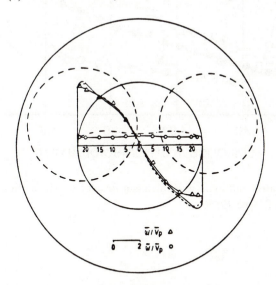

(b) Comparisons with measured swirl velocities in bowl

Fig. 6: Comparison between measured and predicted velocity distributions in a Diesel combustion chamber, from (Arcoumanis et al. 1986)

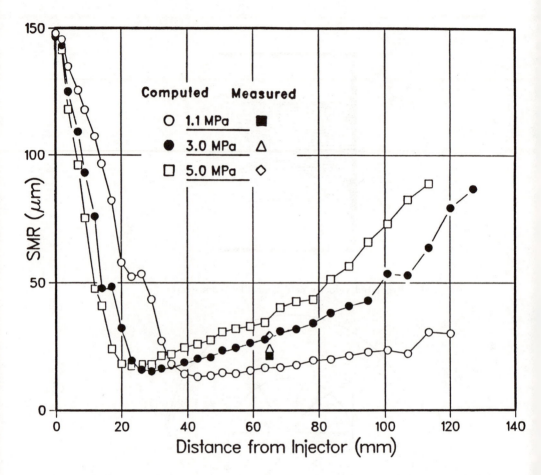

Fig. 7: Calculated and experimental mean droplet size in a Diesel spray, from (Chatwani and Bracco 1985)

References

Abraham, J., Bracco, F. V., and Reitz, R. D. (1985). Comparisons of computed and measured premixed charge engine combustion. *Combustion and Flame*, 60:309–322.

Adey, P. C. and Greaves, J. R. A. (1988). The application of a 3-D aerodynamics model to the Sterling 825 body shape and comparison with experimental data. SAE Paper 880456.

Adyagi, Y., Takenaka, Y., Niino, S., and Joko, I. (1988). Numerical simulation and experimental observation of coolant flow around cylinder liners in V8 engine. SAE Paper 880109.

Ahmadi-Befrui, B., Gosman, A. D., Lockwood, F. C., and Watkins, A. P. (1981). Multidimensional calculation of combustion in an idealised homogeneous charge engine: a progress report. *SAE Trans.*, 89(9).

Arcoumanis, C., Begleris, P., Gosman, A. D., and Whitelaw, J. G. (1986). Measurements and calculations of the flow in a research diesel engine. SAE Paper 861563.

Buckheim, R., Rohe, H., and Wustenberg, H. (1988). Experiences with computational fluid mechanics in automotive aerodynamics. In *Proc. ATA Symposium on Use of Supercomputers in Automotive Industry*, *Turin*.

Butler, T. D., Amsden, A. A., O'Rourke, P. J., and Ramshaw, J. D. (1985). KIVA: A comprehensive model for 2D and 3D engine simulations. SAE Paper 850554.

Chapman, M. (1979). Two dimensional numerical simulation of inlet manifold flow in a four cylinder internal combustion engine. SAE Paper 790244.

Chatwani, A. U. and Bracco, F. V. (1985). Computation of dense spray jets. In *Proc. ICLASS-85*, Eisenklam, P. and Yule, A., editors, Inst. of Energy, London.

El Tahry, S. (1984). Application of a Reynolds Stress model in engine flow calculations. In *Flows in Internal Combustion Engines-II*, pages 39–46. ASME, New York.

Finlay, I. C., Galleher, G. R., Biddulph, T. W., and Marshall, M. A. (1988). The application of precision cooling to the cylinder head of a small automotive, petrol engine. SAE Paper 880263.

Gosman, A. D. (1985a). Computer modelling of flow and heat transfer in engines: progress and prospects. In *Proc. JSME Symp. on Diagnostics and Modelling of Combustion in Reciprocating Engines*. Tokoyo.

Gosman, A. D. (1985b). Multidimensional modelling of cold flows and turbulence in reciprocating engines. SAE Fuels and Lubricants Trans. Vol. 1, Paper 850344.

Gosman, A. D. (1985c). The simulation of combustion in reciprocating engines. In *Proc. Symp. on Numerical Simulation of Combustion Phenomena*, Nice, INRIA.

Gosman, A. D. (1986). Flow processes in cylinders. In *Thermo-dynamics and Gas Dynamics of Internal Combustion Engines, Vol 2*, Horlock, J. H. and Winterbone, D., editors. OUP.

Gosman, A. D. and Ahmed, A. M. Y. (1987). Measurement and multidimensional prediction of flow in a axisymmetric port/valve assembly. SAE Int. Congress Trans. Paper 870952.

Gosman, A. D. and Harvey, P. S. (1982). Computer analysis of fuel-air mixing and combustion in an axisymmetric diesel. *SAE Trans.*, 91(820036).

Gosman, A. D. and Johns, R. J. R. (1980). Computer analysis of fuel-air mixing in direct-injection engines. *SAE Trans.*, 88(6).

Gosman, A. D., Tabrizi, B. S., and Watkins, A. P. (1987). Calculation of three-dimensional spray motion in engines. SAE Int. Congress Trans. Paper 860468.

Gosman, A. D., Tsui, Y. Y., and Vafidis, C. (1985). Flow in a model engine with a shrouded valve – a combined experimental and computational study. SAE Paper 850498.

Macdonald, S. (1988). Private communication.

Marooney, C. J. (1988). Private communication.

Naser, J. (1988). Private communication.

Oldfield, S. G. and Watson, N. (1983). Exhaust valve geometry and its effect on gas velocity and turbulence in an exhaust port. SAE Paper 830151.

Reitz, R. D. and Diwakar, R. (1986). Effect of droplet breakup on fuel sprays. SAE Paper 860496.

Shaw, C. T. (1988). Predicting vehicle aerodynamics using computational fluid dynamics – a user's perspective. SAE Paper 880455.

Wakisaka, T., Shimamoto, Y., and Isshiki, Y. (1988). Induction swirl in a multiple intake valve engine – three-dimensional numerical analysis. IMechE Technical Report C40/88.

Developments in the calculation of unsteady turbomachinery flow

Michael B. Giles[1]

Department of Aeronautics and Astronautics

Massachusetts Institute of Technology, USA

1. Introduction

The purpose of this paper is to present three algorithm and modelling ideas, developed in the last two years as part of a project to calculate unsteady flows in turbomachinery. A paper in this volume by P. Stow (1988) describes in much greater detail the power of computational fluid dynamic methods in calculating and understanding flows in turbomachinery, and illustrates their application to a wide range of steady and unsteady problems. In this paper I restrict my attention to unsteady flow in two dimensions, while intending eventually to proceed to calculations of unsteady three-dimensional flow when computer resources permit. As shown in Fig. 1 there are four principal sources of unsteadiness in a single stage of a turbomachine in which there is one row of stationary blades (stators) and one row of moving blades (rotors).

Wake/rotor interaction causes unsteadiness because the stator wakes, which one can assume to be approximately steady in the stator frame of reference, are unsteady in the rotor frame of reference since the rotor is moving through the wakes and chopping them into pieces. This causes unsteady forces on the rotor blades and generates unsteady pressure waves. Although the stator wakes are generated by viscosity, the subsequent interaction with the rotor blades is primarily an inviscid process and can so can be modelled by the inviscid equations of motion. This allows two different approaches in numerical modelling. The first is to perform a full unsteady Navier–Stokes calculation of the stator and rotor blades. The second is to perform an unsteady inviscid calculation for just the rotor blade row, with the wakes being somehow specified as unsteady inflow boundary conditions. This latter approach is computationally much more efficient, but assumes that one is not concerned about the unsteady heat transfer and other viscous effects on the rotor blades.

---

[1]Harold E. Edgerton Assistant Professor

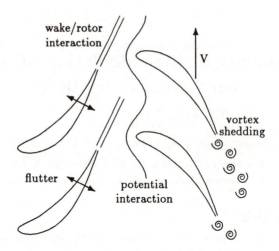

Fig. 1: Sources of unsteadiness in turbomachinery flow

Potential stator/rotor interaction causes unsteadiness due to the fact that the pressure in the region between the stator and rotor blade rows can be decomposed approximately into a part that is steady and uniform, a part that is non-uniform but steady in the rotor frame (due to the lift on the rotor blades) and a part that is non-uniform but steady in the stator frame (due to the lift on the stator blades). As the rotor blades move, the stator trailing edges experience an unsteady pressure due to the non-uniform part that is locked to the rotors, and the rotor leading edges experience an unsteady pressure due to the non-uniform part that is locked to the stators. This is a purely inviscid interaction which is why it is labelled a "potential" interaction. There are again two approaches to modelling this interaction. The first is an unsteady, inviscid calculation of the stator and rotor blade rows. The second is an unsteady, inviscid calculation of just one of the blade rows, either the stator or the rotor, with the unsteady pressure being specified as a boundary condition. The latter approach is more efficient, but unfortunately the situation in which the potential stator/rotor interaction becomes important is when the spacing between the stator and rotor rows is extremely small, and/or there are shock waves moving in the region between them, and in this case one does not usually know what to specify as unsteady boundary conditions and so one must adopt the first approach.

The first two sources of unsteadiness were both due to the relative motion of the stator and rotor rows. The remaining two sources are not. The viscous flow past a blunt turbine trailing edge results in vortex shedding, very similar to the Karman vortex street shed behind a cylinder. In fact real wakes lie somewhere between the two idealized limits of a Karman vortex street and a turbulent wake with steady mean velocity profile. It is believed that provided the integrated loss is identical the choice of model does not affect the subsequent interaction with the downstream rotor blade row. However, this is an assumption which needs to be investigated sometime in the future. The importance of vortex shedding lies in the calculation of the average pressure around the blunt trailing edge, which determines the base pressure loss, a significant component of the overall loss. There is also experimental evidence to suggest that the vortex shedding can be greatly amplified under some conditions by the potential stator/rotor interaction.

Finally, there can be unsteadiness due to the motion of the stator or rotor blades. The primary concern here is the avoidance of flutter. This is a condition in which a small oscillation of the blade produces an unsteady force and moment on the blade which due to its phase relationship to the motion does work on the blade and so increases the amplitude of the blade's unsteady motion. This can rapidly lead to very large amplitude blade vibrations, and ultimately blade failure.

The computer program UNSFLO which has been developed over the last two years is at present able to analyse both the wake/rotor interaction and the potential stator/rotor interaction. In both cases it achieves this by solving the unsteady, inviscid equations of motion. The extension to viscous flows is near completion, but has not yet been validated. The extension to oscillating blades is planned for the future.

The remainder of the paper will now describe the basic approach adopted, and the three algorithm and modelling developments which were required to achieve the above goals. The first is non-reflecting boundary conditions, which are needed to minimize spurious, non-physical reflections from numerical far-field boundaries which, despite their name, may be relatively near the blade rows. The second is the treatment of the stator/rotor calculations. Here separate calculations are performed on grids fixed in the stator and rotor frames, and these have to be coupled through an interface region. The last development is a novel technique called "time-inclined" computational planes, which overcomes a tricky problem in the specification of periodic boundary conditions when the stator and rotor blade rows have different pitches, which is the usual situation.

2. Basic approach

The Euler equations which describe the unsteady motion of an inviscid, compressible gas can be written in the following form in two dimensions

$$\frac{\partial U}{\partial t} + \frac{\partial F}{\partial x} + \frac{\partial G}{\partial y} = 0, \qquad (2.1)$$

$$U = \begin{pmatrix} \rho \\ \rho u \\ \rho v \\ \rho E \end{pmatrix}, \quad F = \begin{pmatrix} \rho u \\ \rho u^2 + p \\ \rho u v \\ \rho u H \end{pmatrix}, \quad G = \begin{pmatrix} \rho v \\ \rho u v \\ \rho v^2 + p \\ \rho v H \end{pmatrix}. \tag{2.2}$$

ρ is the density, u and v are the x and y components of velocity, p is the pressure, E is the total internal energy per unit mass, and H is the total enthaply defined by $H = E + p/\rho$. These equations are in a strong conservation form, which guarantees the correct Rankine-Hugoniot jump relations at any shocks in the flow. To complete the set of equations, an equation of state for an ideal gas is used

$$p = (\gamma - 1) \left(\rho E - \frac{1}{2} \rho (u^2 + v^2) \right). \tag{2.3}$$

The numerical method which is used to solve the Euler equations, is based upon the Lax-Wendroff method as implemented by Ni (1981). This algorithm is applied in a finite-element manner on a grid which is composed of quadrilateral cells which do not need to be arranged in any particular structured manner. This requires minor modification to Ni's method such that all operations can be applied on a cell-by-cell basis, computing the integrated flux into each cell, and then distributing it appropriately (Giles 1988a). The algorithm can be further modified to treat triangular cells (Lindquist 1988), and is very similar to the Taylor–Galerkin finite element method (Donea 1984; Lohner 1985; Morton 1987). The advantage of the unstructured algorithm over Ni's original structured algorithm is that unstructured grids offer much more flexibility in grid generation, and allow adaptive grid refinement to be used to resolve sharp features (Lohner 1985; Dannenhoffer 1986).

Boundary conditions at blade surfaces are treated very easily. Assuming that the calculation is being performed in a frame of reference in which the blade surface is stationary, there is no mass flux normal to the blade surface so the only contribution from the surface to the flux integral of a boundary cell comes from the pressure term in the momentum equations. In addition, at the end of each time-step the velocity at each surface node is corrected to be tangent to the blade surface. See (Giles 1988a) for additional details.

The next section will discuss the treatment of the inflow and outflow boundary conditions, and a later section will deal with the periodic boundary condition.

3.   Non-reflecting boundary conditions

The objective in the formulation of non-reflecting boundary conditions is to prevent spurious, non-physical reflections at inflow and outflow boundaries, so that the calculated flow field is independent of the position of the far-field boundary condition. This leads to greater accuracy and greater computational efficiency since the computational domain can be made much smaller.

The theoretical basis of non-reflecting boundary conditions stems from a paper by Engquist and Majda (1977), which discusses both ideal non-reflecting boundary conditions and a method for constructing approximate forms, and a paper by Kreiss (1970), which analyses the wellposedness of initial boundary value problems for hyperbolic systems. Many workers have been active in this area in the last ten years, but their work has been mainly concerned with scalar p.d.e.'s, with only a couple of recent applications to the Euler equations in specific circumstances. I have recently completed a lengthy report on the formulation of non-reflecting boundary conditions as applied to the Euler equations (Giles 1988b), and in this section I will briefly summarize the work, and the results section presents two examples of its effectiveness.

The analysis considers the following general unsteady, two-dimensional, hyperbolic partial differential equation

$$\frac{\partial U}{\partial t} + A\frac{\partial U}{\partial x} + B\frac{\partial U}{\partial y} = 0. \tag{3.1}$$

This has wave solutions,

$$U(x, y, t) = u^R e^{i(kx+ly-\omega t)}, \tag{3.2}$$

where k, l and ω satisfy the dispersion relation

$$\det(-\omega I + kA + lB) = 0 \tag{3.3}$$

and u^R is the corresponding right eigenvector.

$$(-\omega I + kA + lB)\,u^R = 0. \tag{3.4}$$

At a boundary at $x = 0$, U can be decomposed into a sum of Fourier modes with different values of ω and l. For one particular choice of ω and l the most general form for U is

$$U(x, y, t) = \left[\sum_{n=1}^{N} a_n u_n^R e^{ik_n x}\right] e^{i(ly-\omega t)}, \tag{3.5}$$

where k_n is the nth root of the dispersion relation for the given values of ω and l, and u_n^R is the corresponding right eigenvector.

The ideal non-reflecting boundary condition is to specify $a_n = 0$ for each n that corresponds to an incoming wave. Because of orthogonality of right and left eigenvectors, one can instead specify that

$$v_n^L U = 0 \qquad (3.6)$$

where v_n^L is a left eigenvector defined by

$$v_n^L A^{-1} (-\omega I + k_n A + l B) = 0. \qquad (3.7)$$

In principle these exact boundary conditions can be implemented in a numerical method. The problem is that v_n^L depends on ω and l and so the implementation would involve a Fourier transform in y and a Laplace transform in t. Computationally this is both difficult and expensive to implement and so instead we consider three simpler variations which use different assumptions and approximations.

3.1   Single-frequency boundary conditions

In some problems, such as flutter calculations, there is only one frequency in the unsteadiness and so ω is known. In this case the ideal non-reflecting boundary conditions can be used. The implementation for inflow and out-flow boundaries in turbomachinery applications requires a Fourier transformation along the boundary in the circumferential direction. This separates the solution into modes with a known set of (ω, l) values, to which the ideal b.c.'s are applied, and then the solution in the physical domain is obtained by an inverse Fourier transform.

3.2   Steady-state boundary conditions

The exact steady-state boundary conditions may be considered to be the limit of the single-frequency boundary conditions as $\omega \to 0$. Again the Fourier decomposition gives modes with known values of l. The modes with $l \neq 0$ are treated with the non-reflecting theory, while the average mode $l = 0$ is modified to achieve user-specified average values, such as stagnation enthaply, stagnation pressure and flow angle at the inflow and static pressure at the outflow. The flow in the Fourier domain is then transformed back into the physical domain.

3.3   Approximate, unsteady boundary conditions

When there is more than one frequency in an unsteady problem, we follow the ideas of Engquist and Majda in formulating approximate b.c.'s. It can be shown that the left eigenvector v^L is a function of $\phi = l/\omega$, and so can be written in a Taylor series expansion about $\phi = 0$. Keeping just the first two terms gives

$$\left(v_n^L \Big|_{\phi=0} + \frac{l}{\omega} \frac{dv_n^L}{d\phi} \Big|_{\phi=0} \right) U = 0. \qquad (3.8)$$

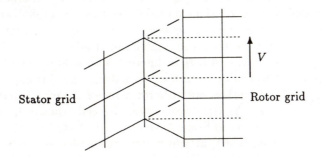

Fig. 2: Shearing cells at unsteady stator/rotor interface

Multiplying by ω, and replacing ω and l by $i\dfrac{\partial}{\partial t}$ and $-i\dfrac{\partial}{\partial y}$ respectively gives,

$$\boldsymbol{v}_n^L\bigg|_{\phi=0}\frac{\partial \boldsymbol{U}}{\partial t} - \frac{d\boldsymbol{v}_n^L}{d\phi}\bigg|_{\phi=0}\frac{\partial \boldsymbol{U}}{\partial y} = 0. \tag{3.9}$$

This is now a local boundary condition and so can be implemented without difficulty. It is important to check the wellposedness of this boundary condition using the theory of Kreiss. It is quite complicated but it is based on proving the non-existence of any generalized incoming modes which satisfy the homogeneous boundary conditions. The analysis shows that the outflow boundary condition is wellposed, but the inflow conditions need to be modified slightly to become wellposed. Analysis also shows that the inflow boundary conditions are fourth order and the outflow boundary condition is second order, meaning that an outgoing pressure wave of unit amplitude produces an incoming pressure wave whose amplitude is $O(\phi^4)$ and $O(\phi^2)$ in the two cases, respectively.

In some applications one wishes to specify an incoming disturbance, due to either wakes or pressure waves. This can still be accomplished using non-reflecting boundary conditions, by applying the theory to the perturbations of the unsteady flow from some specified flow which includes the desired wakes or pressure waves. See (Giles 1988c) for additional details.

4.   Stator/rotor interface

The extension from the calculation of single blade rows, to the calculation of combined stator and rotor rows is surprisingly simple, at least at an algorithmic level. The computational grid is composed of two parts, one part fixed to the stator row and the other fixed to the rotor row. Using relative flow variables on each grid the Euler equations are used to calculate the unsteady flow. The only problem is how to handle the interface between the stator and rotor grids. The approach used by Rai (1985) was to have a sliding interface with no gap between the two halves, and interpolation to obtain flow variables along the common interface.

The approach developed here is different and perhaps slightly simpler. A small gap is retained between the two halves, and this gap is filled by connecting nodes on either side to form a set of shearing interface cells, as shown in Fig. 2. As time progresses, the cells change from State 1 (shown with solid lines) to State 2 (shown with dotted lines), to State 3 (shown with dashed lines), at which time the nodes are reconnected to form cells in State 1. The basic algorithm for solving the Euler equations has to be modified to include the extra flux terms due to the motion of the computational cell. Apart from this the only difficulty is taking care to transform from rotor-relative variables to stator-relative variables and vice-versa, as appropriate when calculating fluxes and for numerical smoothing and when updating the flow field. Further details will be available in (Giles 1988a).

5.   Time-inclined computational planes

An interesting problem arises in the formulation of the periodic boundary condition. Consider the wake/rotor interaction problem shown schematically in Fig. 3a. In this figure the pitch of the wake, P_s, which is the pitch of the upstream stator row which generated it, happens to be equal to the pitch of the rotor row, P_r. The wake is steady in the stator frame of reference, but in the frame of reference of the rotor which is moving upwards with wheel speed V the wakes sweep downwards along the inlet boundary with speed V. Thus, the flow in the rotor passage satisfies the following periodic boundary condition

$$U(x,y,t) = U(x,y+P_r,t). \tag{5.1}$$

When P_s is different from P_r this has to be changed. Taking the usual case in which the stator pitch is larger than the rotor pitch, then an incoming wake crosses the inlet boundary/upper periodic boundary junction a small time ΔT after the neighboring wake crosses the inlet/lower periodic junction (see Fig. 3b). The time lag is equal to the difference in pitches divided by the rotor wheel speed

$$\Delta T = (P_s - P_r)/V. \tag{5.2}$$

Thus the inlet boundary conditions satisfy the lagged periodic condition,

$$U(x,y,t) = U(x,y+P_r,t+\Delta T). \tag{5.3}$$

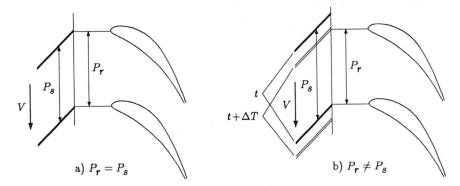

a) $P_r = P_s$ b) $P_r \neq P_s$

Fig. 3: Wake/rotor interaction

The next step is to apply this lagged periodic condition to the upper and lower periodic lines. Strictly speaking this involves an assumption that the periodic inflow does not generate a subharmonic flow behavior, but this possibility is ignored at present.

So far we have discussed only the mathematical formulation of the problem. Now we consider the problem of how to calculate the flow numerically, and in particular how to handle the periodic boundary condition. If the flow is steady or the pitch ratio P_s/P_r is unity, then one simply enforces periodicity by letting the computational grid "wrap round" so that the points on the periodic line are treated exactly the same as any other points in the grid.

When the pitch ratio is not unity, we adopt the technique of "time-inclined" computational planes. If, as illustrated in Fig. 4, one inclines the computational grid in time by an amount $\Delta T/P_r$ then one automatically satisfies the lagged periodic boundary condition by simply wrapping round on the computational grid as before. This eliminates the problem of the lagged periodic boundary condition at the expense of introducing complexity into the solution of the interior flow.

The effect of the inclination of the time-like faces on the equations of motion can be found in two different ways, both of which will be presented as each is helpful in understanding the changes. The mathematical approach is to consider a coordinate transformation from (x, y, t), the physical coordinates, to (x', y', t'), the computational coordinates in which t' is constant on each computational time level. The equations defining the transformation are,

$$x' = x$$

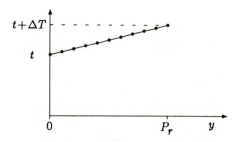

Fig. 4: Time-inclined computational plane

$$y' = y \tag{5.4}$$
$$t' = t - \lambda y,$$

and the inverse transformation is

$$x = x'$$
$$y = y' \tag{5.5}$$
$$t = t' + \lambda y',$$

where

$$\lambda = \frac{\Delta T}{P_r} = \frac{P_s - P_r}{V P_r}. \tag{5.6}$$

When one transforms the Euler equations into the new computational coordinate system the resultant equations are

$$\frac{\partial}{\partial t'}(U - \lambda G) + \frac{\partial F}{\partial x'} + \frac{\partial G}{\partial y'} = 0. \tag{5.7}$$

Thus the conservation state variables have changed from U to $Q = U - \lambda G$.

The alternative way of arriving at the same conclusion is to consider a one-dimensional conservation cell shown in Fig. 5 in the original (y,t) plane. The integral form of the Euler equations, integrated over the space-time volume Ω, is

$$\oint_{\partial \Omega} (-U \, dy + G \, dt) = 0. \tag{5.8}$$

When this equation is then approximated on the conservation cell the discrete flux through the time-like face is $U \Delta y - G \Delta t = (U - \lambda G) \Delta y = Q \Delta y$.

The change in the conservation variables requires just minor changes to numerical algorithms because fortunately one can calculate U from Q in closed form for a perfect gas (Giles 1987). Only two modifications are required in the standard Ni implementation of the Lax-Wendroff method. The first step is to calculate the total flux out of each cell. This determines

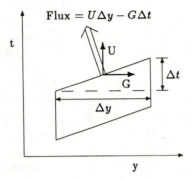

Fig. 5: Inclined conservation cell

the average change ΔU in each cell in the standard method, but in the modified algorithm this gives ΔQ which must then be converted to ΔU. The next steps are to calculate the second-order changes due to the flux terms ΔF and ΔG and distribute these and the first-order changes to the nodes. These remain the same, except that the distributed changes give ΔQ at the nodes, and so these have to be converted again to ΔU before being added to the original values of U at the nodes to obtain the new flow solution. The total increase in CPU per iteration is approximately 15%, and there is no increase in memory requirements.

The use of time-inclined computational planes changes the usual time-step stability limits for explicit methods. More fundamentally, there is a limit on the magnitude of λ which can be used. This limit is due to the restriction that the computational plane cannot be inclined in time more steeply than the fastest propagating characteristic wave. If it were more steeply inclined, it would violate causality, and effectively be computing backwards in time instead of forwards. This limitation and the change in the explicit time-step stability limit are discussed in full detail in an earlier paper (Giles 1987).

6. Results
6.1 Subsonic and transonic turbine
To verify the effectiveness of the steady-state non-reflecting boundary conditions, Figs. 6 and 7 show results for a high-turning turbine cascade. The first figure shows results for subsonic outflow conditions, with two different locations of the far-field boundaries. The results are almost identical. The second figure shows the corresponding results for a supersonic outflow condition which has two weak, oblique shocks extending from the trailing edge. The agreement in this case in not quite as good, but under the

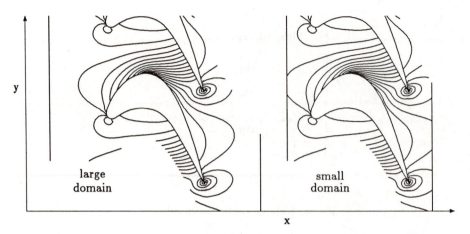

Fig. 6: Pressure contours for turbine with subsonic outflow, $M_{exit} = 0.75$

standard boundary conditions, which impose uniform exit pressure, the outgoing shocks are artificially reflected and greatly contaminate the solution on the blade. Thus the non-reflecting boundary conditions give a major improvement in accuracy.

6.2   Flat plate cascade test case

The first test case is a relatively simple linear test case, the addition of a low amplitude sinusoidal wake to a steady uniform flow past an unloaded flat plate cascade. This case was chosen because the results can be compared to those obtained using LINSUB, a program developed by Whitehead (1986) based upon the linear singularity theory of Smith (1971). The sinusoidal wake is clearly not a good approximation to an actual physical wake, but a real wake can be decomposed into its Fourier components and with this linear theory each component can be analyzed separately and then the results summed to obtain the unsteady lift and moment.

The steady flow has a Mach number of 0.7 and a flow angle of 30°, parallel to the flat plates, which have a pitch/chord ratio of 0.577. The unsteady wakes have a pitch which is a factor 0.9 smaller, and an angle of $-30°$ which corresponds to the outflow angle relative to the upstream blade row. The magnitude of the wake velocity defect was chosen to be 5% which was found to be large enough to avoid difficulties with machine accuracy (which would be a problem if a value such as 0.01% had been used) and small enough that the solution remained linear. Computational results using UNSFLO were obtained on a grid of size 400x50, with sufficently many periods being calculated to ensure that the solution had converged

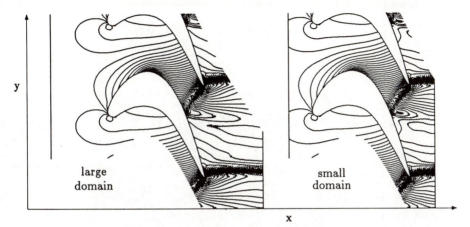

Fig. 7: Pressure contours for turbine with supersonic outflow, $M_{exit} = 1.1$

to a periodic form in which the unsteady lift and moment were converged to within 1%.

Fig. 8 shows contour plots of the entropy and pressure at one instant in time. These were obtained by interpolation from the time-inclined computational planes. The entropy plot shows that to first order the entropy simply convects through the blade passage. The shearing of the entropy lines behind the blade is an indication of the unsteady vortex sheet which extends from the trailing edge. The pressure plot shows that there are plane pressure waves radiating upstream of the blade row. These represent the noise generated by the unsteady wake/rotor interaction. Downstream of the blade there is a mixture of different pressure modes. There is some indication of peculiarities in the neighborhood of the outflow boundary. This is due to the trailing vortex sheet passing through the boundary. This excites the full spectrum of pressure waves at the boundary. However most of the modes decay exponentially away from the boundary, and the non-reflecting boundary conditions perform an excellent job of preventing any reflection in the incoming acoustic modes.

To obtain a quantitative comparison the unsteady pressures computed by UNSFLO were Fourier transformed, and then non-dimensionalized in exactly the same manner as in LINSUB, using the product of the mean density, the mean speed, and the maximum wake velocity normal to the blade. To be able to compare phase results the wake definitions at the inflow are specified such that the centerline of the wake passes through the blade leading edge at the beginning of the period. Fig. 9 shows the real and imaginary components of the complex amplitude of the first Fourier

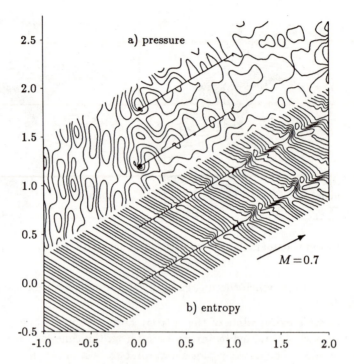

Fig. 8: Flat plate pressure and entropy contours

mode of the pressure jump across the blade. The agreement between the UNSFLO computation and the LINSUB theory is good except at the leading edge where the $x^{-\frac{1}{2}}$ singularity causes some minor oscillations. The integrated lift and moment also agree to within 5%. This test case shows that the computational method is capable of correctly predicting the unsteady forces due to a wake/vortex interaction, and validates the use of the time-inclined computational plane to solve the problem posed by the differing pitches of the wake and blade row.

6.3   Transonic turbine stator/rotor interaction

The final case is an example of the capability of the method to analyze unsteady stator/rotor interactions. The calculation is for a high pressure ratio turbine stage, in which the flow in both the stator and the rotor passages is choked, causing oblique shocks to be generated at the trailing edges of all blades. The stator/rotor pitch ratio is approximately 1.7, and the stator exit Mach number and the rotor relative exit Mach number are both approximately 1.1. A point of note is that this calculation is quasi-three-dimensional, including the effects of variations in the streamtube

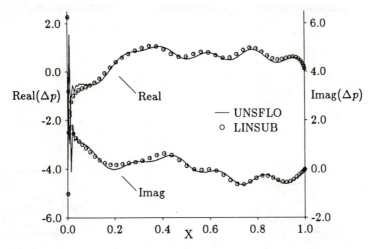

Fig. 9: Complex amplitude of flat plate pressure jump

thickness in the third z-direction: this involves only minor changes to the basic flow algorithm. In this application there is a 20% decrease in streamtube thickness from the stator leading edge to the stator trailing edge, and a corresponding 20% increase from the rotor leading edge to the rotor trailing edge. These significantly change the flow field and need to be included in the computational model.

The principal point of interest in the results is the shock dynamics generated by the interaction. Fig. 10 shows unsteady pressure contours at eight equally-spaced instants during a rotor blade passing period. The calculation was performed on a grid with approximately 18,000 nodes, and took about 12 CPU hours on a three-processor Alliant FX/8 to converge to a periodic state. At the beginning of the period ($t = 0$), the oblique shock from the lower stator is making contact on the crown of the suction surface of a rotor blade. As time progresses the rotor blade moves upward and so the shock moves forward to the rotor leading edge and produces a strong reflected shock which moves upstream. At $t = 0.5$ the oblique shock begins to diffract around the rotor leading edge. At around $t = 0.7$ the reflected shock impinges on the rearward portion of the stator suction surface, and reflects again, which can be distinctly seen at $t = 0.875$ and $t = 0.0$ (which is equivalent to $t = 1$). Another interesting feature to note is

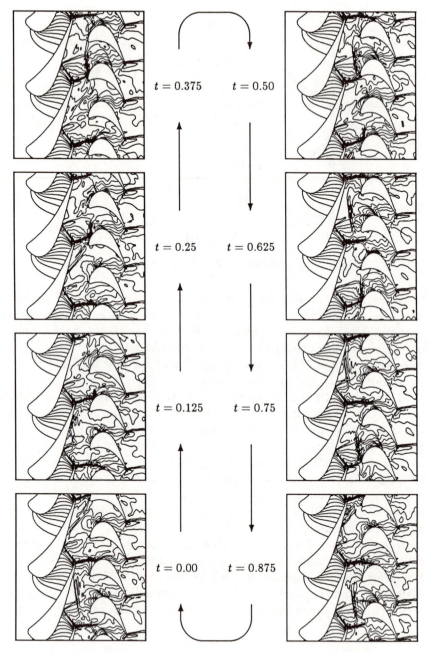

Fig. 10: Pressure contours in unsteady stator/rotor interaction

a compression wave which starts near the rearward portion of the upper rotor pressure surface at $t = 0.25$, and moves upstream forming a shock on the suction surface of the adjacent rotor blade at $t = 0.375$. The shock continues moving upstream around the suction surface of the rotor until it is no longer discernible at $t = 0.875$. These predicted shock motions, for both the primary reflection and the self-steepening compression wave, are in substantial agreement with experimental results obtained at Oxford University (Johnson 1988). Unfortunately it is not possible at present to show comparisons.

7. Conclusions

This paper has discussed three different advances in the calculation of unsteady flows in turbomachinery. The first is the development of a unified approach to non-reflecting boundary conditions for the Euler equations. These are required to minimize the non-physical reflections at the computational far-field boundary. The theory was described briefly for three different classes of problems, unsteady flows with one single frequency, steady-state flows and unsteady flows with multiple frequencies. Numerical examples demonstrate the effectiveness of the steady-state boundary conditions for both subsonic and supersonic conditions.

The second development is the concept of "time-inclined" computational planes. This is introduced as a computational technique to solve the problem associated with the lagged periodic boundary condition in unsteady flow situations where the stator and rotor pitches are not equal. The approach is validated by a simple test case of sinusoidal wakes interacting with an unloaded flat plate cascade, for which an analytic solution exists.

The third innovation is a new approach to the calculation of unsteady stator/rotor interactions through the simple use of shearing cells in an interface region between one grid fixed to the stator and another fixed to the rotor. Results show the capability to calculate complex shock dynamics in a highly unsteady transonic turbine stage.

Acknowledgements

This research was supported by Rolls-Royce PLC and the Air Force Office of Scientific Research under contract number F49620-78C-0084.

References

Dannenhoffer III, J. F. and Baron, J. R. (1986), *Robust grid adaptation for complex transonic flows*, AIAA Paper 86-0495.

Donea, J. (1984), A Taylor-Galerkin method for convective transport problems, *International Journal of Numerical Methods in Engineering*, Vol. 20, pp. 101–119.

Engquist, B., and Majda, A. (1977), Absorbing boundary conditions for the numerical simulation of waves, *Mathematics of Computation*, Vol. 31, July 1977, pp. 629–651.

Giles, M. B. (1987), *Generalized conservation cells for finite volume calculations*, AIAA Paper 87-1118-CP.

Giles, M. B. (1988a), *UNSFLO: A Numerical Method for Calculating Unsteady Stator/Rotor Interaction; Second Edition*, Technical Report (currently in preparation), MIT Computational Fluid Dynamics Laboratory, 1988.

Giles, M. B. (1988b), *Non-Reflecting boundary conditions for the Euler equations*, Technical Report TR-88-1, MIT Computational Fluid Dynamics Laboratory, 1988.

Giles, M. B. (1988c), Calculation of unsteady wake rotor interaction, *AIAA Journal of Propulsion and Power*, Vol. 4, No. 3, May/June 1988.

Johnson, A. B., Rigby, M. J., Oldfield, M. L. G., Ainsworth, R. W., and Oliver, M. J. (1988), *Surface heat transfer fluctuations on a turbine rotor blade due to upstream shock wave passing*, ASME Paper 88-GT-72.

Kreiss, H.-O. (1970), Initial boundary value problems for hyperbolic systems, *Communications on Pure and Applied Mathematics*, Vol. 23, pp. 277–298.

Lindquist, D. R. (1988), *A Comparison of Numerical Schemes on Triangular and Quadrilateral Meshes*, Master's thesis, M.I.T., May 1988.

Lohner, R., Morgan, K., Peraire, J., and Zienkiewicz, O. C. (1985), *Finite Element Methods for High Speed Flows*, AIAA Paper 85-1531.

Morton, K. W. (1987), *Finite volume and finite element methods for the steady Euler equations of gas dynamics*, presented at the MAFELAP Conference.

Ni, R.-H. (1981), A Multiple grid scheme for solving the Euler equations, *AIAA Journal*, Vol. 20, Nov 1981, pp. 1565–1571.

Rai, M. M. (1985), *Navier–Stokes simulations of rotor-stator interaction using patched and overlaid grids*, AIAA Paper 85-1519.

Smith, S. N. (1971), *Discrete frequency sound generation in axial flow turbomachines*, University of Cambridge, Department of Engineering Report CUED/A-Turbo/TR 29.

Stow, P. (1988), CFD applications to the aero-thermodynamics of turbomachinery, *These proceedings*.

Whitehead, D. S. (1986), *LINSUB User's Guide*, Personal communication.

Efficient solution of the steady Euler equations with a centered implicit method

A. Lerat[1]   J. Sidès

ENSAM, Paris, France   *ONERA, Chatillon, France*

1. Introduction

A large number of finite-difference, finite-volume or finite-element methods have been developed for the solution of the Euler equations in aerodynamics. The most classical methods of second-order accuracy are based on a space-centered approximation. Such an approach is economical but it generally suffers from the necessity of adding some artificial viscosity to damp spurious oscillations in the numerical solution, or even to stabilize the method especially in the several space-dimensions. Recently, upwind methods have become very popular since they can produce non-oscillatory solutions without artificial viscosity. However, the price to pay for this improvement is a significant increase in the complexity of the algorithm and also difficulties in the extension to Navier–Stokes equations. In the present paper, a centered Euler solver will be described. It is based on an implicit method of second-order accuracy which can approximate steady weak solutions without artificial viscosity, since its internal dissipation is just sufficient to ensure the success of the calculation when the CFL number is large enough. For a multidimesional problem, that requires a careful treatment of the space differencing and the use of a local time-step.

The next section presents the method in one-space dimension. Then, Section 3 gives the 2-D extension compared to a more classical approach. Section 4 discusses briefly the accuracy obtained at steady state and finally, Section 5 describes various applications to transonic aerodynamics, namely the internal flow in a channel with a bump and several external flows over an airfoil at low and high angles of attack. Numerical results illustrate the efficiency and the remarkable shock-capturing capabilities of the method.

2. 1-D method

Let us first consider the hyperbolic system of m conservation laws:

$$w_t + f(w)_x = 0 \tag{2.1}$$

---

[1]Consultant at ONERA.

where $w(x, t)$ is a m-component vector. Let $A(w)$ be the Jacobian matrix of the flux function, that is:

$$A(w) = \frac{df}{dw}(w).$$

We approximate System (2.1) with the implicit schemes introduced by Lerat (1979) — see also (Lerat 1981 and 1983). Their time-differencing can be expressed as:

$$\Delta w + \alpha \Delta t (A \Delta w)_x + \beta \frac{\Delta t^2}{2}[A^2(\Delta w)_x]_x + \gamma \frac{\Delta x^2}{2}(\Delta w)_{xx}$$
$$= -\Delta t f_x + (1 - 2\alpha)\frac{\Delta t^2}{2}(A f_x)_x \qquad (2.2)$$

where

$$f = f(w^n), \quad A = A(w^n), \quad \Delta w = w^{n+1} - w^n,$$

w^n being the numerical solution at time level $t = n\Delta t$, and α, β, γ are three parameters (real numbers). The space derivatives in equation (2.2) are approximated by using 3-point centered formulae. The full-discrete schemes are conservative and second-order accurate in time and space, for any value of approximations derived by Beam and Warming (1976).

It has been proved (Lerat 1979 and 1981) that under the following necessary and sufficient condition on the parameters:

$$\alpha < \frac{1}{2}, \quad \beta \leq \alpha - \frac{1}{2}, \quad \gamma < \frac{1}{2}, \qquad (2.3)$$

and for a pure initial-value problem, the above schemes are *always* (i.e. for any Δt):

- linearly solvable

- linearly stable in L_2

- linearly dissipative in the sense of Kreiss (1964) unless A is singular.

Furthermore, if in addition:

$$\beta < -\frac{\alpha^2}{4(1 - \gamma)}, \qquad (2.4)$$

the linear algebraic system to be solved (to obtain Δw) *always* satisfies some criterion of strict diagonal dominance.

The class of schemes includes approximations which are third and fourth-order accurate for a linear hyperbolic system (i.e. when A is a constant matrix). These approximations are defined by

$$\beta = \alpha - \frac{1}{3}, \quad \gamma = \frac{1}{3}. \tag{2.5}$$

Unfortunately, these relations are not compatible with the constraints (2.3). The schemes associated with (2.5) can only be conditionally stable.

When the implicit schemes (2.2) are used a time-dependent method to reach a steady state, it is important also to take into account the convergence rate towards the steady solution. A linear analysis of this convergence rate (Daru and Lerat 1985) has shown that the best choice is:

$$\beta = 2\alpha - 1. \tag{2.6}$$

For this choice, the efficiency increases monotonic with the mesh ratio $\Delta t / \Delta x$.

Clearly, the simplest scheme meeting the requirements (2.3), (2.4), and (2.6) is given by:

$$\alpha = 0, \quad \beta = -1, \quad \text{and} \quad \gamma = 0,$$

that is

- Explicit stage [R.H.S. of (2.2)]

$$\Delta w^{expl} = -\Delta t f_x + \frac{\Delta t^2}{2} (A f_x)_x. \tag{2.7}$$

- Implicit stage [L.H.S. of (2.2)]

$$\Delta w - \frac{\Delta t^2}{2} [A^2 (\Delta w)_x]_x = \Delta w^{expl}. \tag{2.8}$$

This is precisely the basic method that we use in the present work. It can be easily interpreted. The explicit stage (2.7) is nothing but the approximation of Lax and Wendroff (1960). Concerning the implicit stage (2.8), let us note that

$$A^2 (\Delta w)_x = A^2 w_x^{n+1} - A^2 w_x^n = A^2 w_x^{n+1} - A f_x.$$

Therefore, the implicit stage (2.8) introduces a linearly implicit counterpart of the higher-order term in the Lax–Wendroff approximation.

In order to write down the fully-discrete form of the method, we introduce the following operators:

$$(\delta \Psi)_j = \Psi_{j+\frac{1}{2}} - \Psi_{j-\frac{1}{2}},$$

$$(\mu \Psi)_j = \frac{1}{2}(\Psi_{j+\frac{1}{2}} + \Psi_{j-\frac{1}{2}}).$$

where Ψ_j is a mesh function defined at $x = j\Delta x$ for integer values of $2j$. With these notations, the implicit method can be expressed as:

- Explicit stage

$$\Delta w_j^{expl} = -\frac{\Delta t}{\Delta x}\delta(\mu f)_j + \frac{1}{2}(\frac{\Delta t}{\Delta x})^2 \delta[(\mu A)\delta f]_j$$

or else

$$\Delta \tilde{w}_{j+\frac{1}{2}} = -\Delta t \left(\frac{\delta f}{\Delta x}\right)_{j+\frac{1}{2}} \tag{2.9}$$

$$\tilde{f}_{j+\frac{1}{2}} = \left[(\mu f) + \frac{1}{2}(\mu A)\Delta \tilde{w}\right]_{j+\frac{1}{2}} \tag{2.10}$$

$$\Delta w_j^{expl} = -\Delta t \left(\frac{\delta \tilde{f}}{\Delta x}\right)_j \tag{2.11}$$

- Implicit stage

$$\Delta w_j - \frac{1}{2}\left(\frac{\Delta t}{\Delta x}\right)^2 \delta[(\mu A)^2 \delta(\Delta w)]_j = \Delta w_j^{expl}. \tag{2.12}$$

Let us note that (2.12) leads to the solution of a block-tridiagonal linear system.

3.   2-D Method

We consider now the hyperbolic system

$$w_t + f(w)_x + g(w)_y = 0 \tag{3.1}$$

with the Jacobian matrices

$$A(w) = \frac{df}{dw}(w) \text{ and } B(w) = \frac{dg}{dw}.$$

By extending directly the time-differencing (2.7) and (2.8), system (2.12) can be discretized in time as

- Explicit stage

$$\Delta w^{expl} = -\Delta t(f_x + g_y) + \frac{\Delta t^2}{2}[A(f_x + g_y)]_x + \frac{\Delta t^2}{2}[B(f_x + g_y)]_y, \tag{3.2}$$

- Implicit stage

$$\Delta w - \frac{\Delta t^2}{2}[A^2(\Delta w)_x + AB(\Delta w)_y]_x$$

$$-\frac{\Delta t^2}{2}[BA(\Delta w)_x + B^2(\Delta w)_y]_y = \Delta w^{expl}, \quad (3.3)$$

where

$$f = f(f(w^n)), \quad g = g(w^n), \quad \Delta w = w^{n+1} - w^n,$$

and

$$A = A(w^n), \quad B = B(w^n).$$

In the present calculations, the implicit stage (3.3) is simplified by suppressing the cross derivatives and by using an approximate factorization. The actual implicit stage can be expressed as

- Implicit stage presently used:

$$\Delta w^* - \frac{\Delta t^2}{2}[A^2(\Delta w^*)_x]_x = \Delta w^{expl}, \quad (3.4)$$

$$\Delta w - \frac{\Delta t^2}{2}[B^2(\Delta w^*)_y]_y = \Delta w^*, \quad (3.5)$$

It is worth noting that the time-differencing (3.2), (3.4) and (3.5) is still second-order accurate.

Let us now define a centered space-differencing which involves 3×3 mesh points. Several choices are possible for this discretization and, contrary to the one-dimensional case, one can obtain distinct approximations even when system (3.1) is linear (i.e. when A and B are constant matrices). We shall consider two space-discretizations of the explicit stage (3.2), both being implemented in two steps as in the 1-D explicit stage (2.9)-(2.11). The first discretization of the explicit stage is equivalent to the formulation of the Lax–Wendroff method used by Ni (1981) in his multigrid scheme. It makes use of one predictor-step at mesh points $j + 1/2, \ k + 1/2$. The second discretization involves two predictor-steps located at $j + 1/2, \ k$ and $j, \ k + 1/2$, which is known to be a favourable feature to obtain a dissipative approximation (Lerat 1981). Similarly as in the previous section, we introduce the following spatial-operator

$$(\delta_1 \Psi)_{j,k} = \Psi_{j+\frac{1}{2},k} - \Psi_{j-\frac{1}{2},k}, \quad (\delta_2 \Psi)_{j,k} = \Psi_{j,k+\frac{1}{2}} - \Psi_{j,k-\frac{1}{2}},$$

$$(\mu_1 \Psi)_{j,k} = \frac{1}{2}(\Psi_{j+\frac{1}{2},k} + \Psi_{j-\frac{1}{2},k}), \quad (\mu_2 \Psi)_{j,k} = \frac{1}{2}(\Psi_{j,k+\frac{1}{2}} + \Psi_{j,k-\frac{1}{2}})$$

where $\Psi_{j,k}$ is defined at $x = j\Delta x$, $y = k\Delta y$ for integer values of $2j$ and $2k$. Using these notations, the first formulation of the explicit stage can be written as

$$\Delta\tilde{w}_{j+1,k+\frac{1}{2}} = -\Delta t \left(\mu_2 \frac{\delta_1 f}{\Delta x} + \mu_1 \frac{\delta_2 g}{\Delta y} \right)_{j+\frac{1}{2},k+\frac{1}{2}}, \tag{3.6}$$

$$\tilde{f}_{j+\frac{1}{2},k+\frac{1}{2}} = [(\mu_1\mu_2 f) + \frac{1}{2}(\mu_1\mu_2 A)\Delta\tilde{w}]_{j+\frac{1}{2},k+\frac{1}{2}}, \tag{3.7}$$

$$\tilde{g}_{j+\frac{1}{2},k+\frac{1}{2}} = [(\mu_1\mu_2 g) + \frac{1}{2}(\mu_1\mu_2 b)\Delta\tilde{w}]_{j+\frac{1}{2},k+\frac{1}{2}}, \tag{3.8}$$

$$\Delta w_{j,k}^{expl} = -\Delta t \left(\mu_2 \frac{\delta_1 \tilde{f}}{\Delta x} + \mu_1 \frac{\delta_2 \tilde{g}}{\Delta y} \right)_{j,k}. \tag{3.9}$$

The second formulation of the explicit stage is

$$\Delta\tilde{w}_{j+1,k} = -\Delta t \left(\frac{\delta_1 f}{\Delta x} + \mu_1\mu_2 \frac{\delta_2 g}{\Delta y} \right)_{j+\frac{1}{2},k}, \tag{3.10}$$

$$\tilde{f}_{j+\frac{1}{2},k} = [(\mu_1 f) + \frac{1}{2}(\mu_1 A)\Delta\tilde{w}]_{j+\frac{1}{2},k}, \tag{3.11}$$

$$\tilde{\tilde{w}}_{j,k+\frac{1}{2}} = -\Delta t \left(\mu_1\mu_2 \frac{\delta_1 f}{\Delta x} + \frac{\delta_2 g}{\Delta y} \right)_{j,k+\frac{1}{2}}, \tag{3.12}$$

$$\tilde{\tilde{g}}_{j,k+\frac{1}{2}} = [(\mu_2 g) + \frac{1}{2}(\mu_2 B)\Delta\tilde{\tilde{w}}]_{j,k+\frac{1}{2}}, \tag{3.13}$$

$$\Delta w_{j,k}^{expl} = -\Delta t \left(\frac{\delta_1 \tilde{f}}{\Delta x} + \frac{\delta_2 \tilde{g}}{\Delta y} \right)_{j,k}. \tag{3.14}$$

For the implicit stage, we use the straightforward discretization

$$\Delta w_{j,k}^* - \frac{1}{2}\left(\frac{\Delta t}{\Delta x}\right)^2 \delta_1[(\mu_1 A)^2\delta_1(\Delta w^*)]_{j,k} = \Delta w_{j,k}^{expl}, \tag{3.15}$$

$$\Delta w_{j,k} - \frac{1}{2}\left(\frac{\Delta t}{\Delta y}\right)^2 \delta_2[(\mu_2 B)^2\delta_2(\Delta w)]_{j,k} = \Delta w_{j,k}^* \tag{3.16}$$

which leads to solve a block-tridiagonal linear system on each mesh line. The following results have been proven.

Assume that the matrices A and B commute, then

- both implicit schemes (3.6)-(3.9), (3.15)- (3.16) and (3.10)-(3.14), (3.15)- (3.16) are always linearly stable in L_2

- the scheme (3.6)-(3.9), (3.15)- (3.16) is never linearly dissipative (in the sense of Kreiss), but the scheme (3.10)-(3.14), (3.15)-(3.16) is linearly dissipative for any Δt, except for some special states (for which A and B have a zero eigenvalue corresponding to the same eigenvector).

When using the scheme (3.6)–(3.16), the shortest waves are not damped in the direction of the two mesh-diagonals and furthermore, in another direction depending on the velocity vector, all waves are undamped. As a consequence, this scheme cannot be implemented without adding some artificial viscosity. In order to avoid this ingredient, we use the formulation (3.10)-(3.14) and (3.15)-(3.16).

4. Accuracy at steady-state

When the implicit method has reached a steady-state, the following system is solved

$$\frac{\Delta w^{expl}}{\Delta t} = 0$$

that is, in semi-discrete form

$$(f_x + g_y) - \frac{\Delta t}{2}Q = 0 \tag{4.1}$$

where

$$Q = [A(f_x + g_y)]_x + [B(f_x + g_y)]_y$$

The internal dissipation at steady-state is due to the term Q in (4.1). An original feature of this internal dissipation is to be connected to the exact system because

$$w_t = -(f_x + g_y) \quad \text{yields} \quad w_{tt} = -(Aw_t)_x - (Bw_t)_y = Q.$$

However, (4.1) shows that the steady numerical solution depends on the time-step Δt. Of course for small Δt, (4.1) implies

$$f_x + g_y = O(\Delta t^2).$$

In fact when the CFL number is smaller than unity, one obtains the same steady solutions as in the Lax–Wendroff method. It is well known that these solutions are oscillatory, especially when CFL<< 1. As the time step increases, one observes that the numerical solution improves and looks monotonic for some CFL> 1, large enough.

Let us study the behaviour of the steady numerical solution in the limit of large Δt. Dividing (4.1) by Δt and letting Δt go to infinity, one obtains

$$Q = 0. \tag{4.2}$$

It is easy to analyse the solutions of this equation in one space-dimension and also for a model problem in two space-dimensions.

First for a one-dimensional problem, (4.2) reduces to

$$(Af_x)_x = 0, \qquad (4.3)$$

that is

$$Af_x = \text{const.}$$

By choosing the numerical boundary conditions so that the constant be null, one can obtain the correct solution. On the discrete level, this gives the approximation

$$\frac{1}{2}(A_j + A_{j+1})(f_{j+1} - f_j) = 0$$

which is second-order accurate at $x = (j + 1/2)\Delta x$. Moreover this two-point discretization is non-oscillatory.

We consider now the scalar equation

$$w_t + aw_x + bw_y = 0.$$

where a and b are constants. In this case, (4.2) becomes

$$a(aw_x + bw_y)_x + b(aw_x + bw_y)_y = 0. \qquad (4.4)$$

This is a parabolic equation in space, equivalent to

$$aw_x + bw_y = \text{const.}$$

on the lines defined by $dx/a = dy/b$.

Again, by prescribing correctly the conditions on the outflow part of the boundary, one can ensure that

$$aw_x + bw_y = 0, \quad \text{everywhere.}$$

5. Application to transonic flows

5.1 General features

The implicit method (3.10)-(3.16) has been used to solve the Euler equations for the internal flow in a channel with a bump and for external flows over an airfoil at low and high angles of attack, in transonic conditions. The calculations have been performed by replacing the unsteady energy equation by the condition of constant total-enthalpy which holds at steady state. Thus, a hyperbolic system of the form (3.1) has been solved with

$$w = \begin{pmatrix} \rho \\ \rho u \\ \rho v \end{pmatrix}, \quad f(w) = \begin{pmatrix} \rho u \\ \rho u^2 + p \\ \rho uv \end{pmatrix}, \quad g(w) = \begin{pmatrix} \rho v \\ \rho uv \\ \rho v^2 + p \end{pmatrix}$$

and

$$p = \frac{\gamma - 1}{\gamma} \rho \left(H_\infty - \frac{u^2 + v^2}{2} \right)$$

where ρ, p, u and v denote the density, pressure, and cartesian velocity-components, γ is the ratio of specific heats ($\gamma = 1.4$) and H_∞ is the total enthalpy (constant value). The implicit method is applied on a structured mesh by using a finite-volume formulation.

On a rigid wall, the slip condition is prescribed and the pressure is obtained from a linear combination of the discrete form of the x and y-momentum equations in order to obtain a conservative approximation of the normal momentum equation. On an external boundary, we prescribe the freestream direction and the entropy for a subsonic inflow, or the pressure for a subsonic outflow. We have also investigated the asymptotic correction given by Thomas and Salas (1985). All the boundary condition are treated implicitly by using a linearization technique in the implicit stage.

The time evolution starts always from a freestream uniform flow. In the present method, it is very important to use a local time-step, not only to enhance the convergence rate but also to avoid spurious oscillations which might happen if the CFL number were small somewhere in the mesh. Here, the local time step $\Delta t_{j,k}^n$ is such that the *local CFL number* is *uniform* all over the spatial mesh and *does not vary in time* throughout the pseudo-unsteady evolution. Naturally, the time step is calculated by using the characteristic velocities of the present hyperbolic system and not those of the complete Euler equations which are slightly different. *All the computations* are run without adding any artificial viscosity, so that there is no parameter to adjust.

5.2 Channel with a bump

In order to discuss the efficiency and accuracy of the implicit method in terms of the CFL number, we first consider a classical test case defined by the GAMM Workshop on transonic flows organized by Rizzi and Viviand (1981). This is the internal flow through a parallel channel having a 4.2% thick circular arc bump on its lower wall. The ratio of static downstream pressure to total upstream pressure corresponds to a Mach number of 0.85 in isentropic flow and the distance between the walls is 2.07 times the chord length of the bump. The computational mesh is made of 71×20 cells. Fig. 1 shows the whole field in the channel for CFL= 18 and Fig. 2 presents the Mach number and the entropy distributions on the lower wall for various CFL numbers. Here

$$\sum = \frac{p}{p_\infty} \left(\frac{\rho_\infty}{\rho} \right)^\gamma - 1.$$

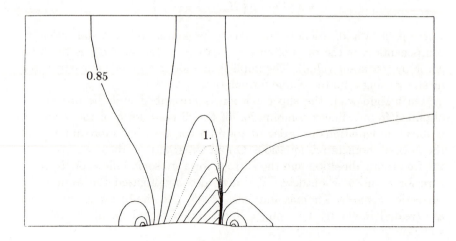

Fig. 1: Mach number contours in the channel ($\Delta M = 0.05$) for CFL=18

For CFL=1, the numerical shock is oscillatory (but the method does converge). The classical remedy is to add a nonlinear artificial viscosity. For comparison, Fig. 2 shows the effect of a second-order artificial viscosity of Lax–Wendroff type: the oscillations are damped but the numerical shock structure is spread out. Without artificial viscosity, the spurious oscillations weakens as the CFL number grows up while the shock profile remains very sharp. The Mach number distribution is monotonic in the shock structure when the CFL is larger than about 8. For CFL= 18 though the mesh is rather coarse, the shock wave is very well captured on the body and in the flowfield.

In the expansion wave upstream of the shock, no anomaly happens on the sonic line. However, a small oscillation appears in the entropy distribution at the origin and the end of the bump. This is due to the discontinuity of the wall slope (no special effort having been made to treat this singularity).

The convergence history is presented on Fig. 3. More precisely the root mean-square R_2 of the residual for the mass equation over the whole mesh is plotted versus the number of time iterations, for CFL= 1, 3, 6, 9, 12, 15 and 18. In this range, one observes that the convergence rate increases monotonically with the CFL number. For greater CFL numbers, the method fails to converge (starting from an uniform flow with a constant

CFL number). For any CFL number between 1 and 18, the convergence is quite sound, i.e. one can always reach the machine zero (10^{-13} presently). In practice, a high level of convergence is obtained when $R_2 = 10^{-5}$. Using CFL=18, that needs 400 time-iterations.

Let us now consider the real efficiency of the method. Fig. 4 shows the residual in terms of the CPU-time on a Cray XMP 18 computer for the present method (with CFL=18) compared to other methods used in their optimal conditions, namely the present explicit stage (CFL=1) alone or with artificial viscosity, or else with an implicit *residual smoothing* (CFL=9) as in Lerat–Sidès–Daru (1982). This residual smoothing comes from replacing the matrices $(\mu_1 A)^2$ and $(\mu_2 B)^2$ by their spectral radii in the implicit stage (3.15)-(3.16). The present implicit method appears to be much more efficient than the other three. To reach a residual R_2 equal to 10^{-5}, it requires only 6 seconds CPU time.

5.3 A classical but difficult external test case

We now consider the external flow over the NACA 0012 airfoil at Mach number $M_\infty = 0.85$ with an angle of attack of 1 deg. This flow configuration involves two shock waves and a slip line downstream of the trailing edge. We have first calculated the flowfield in a 256×32 C-mesh. The convergence history is shown on Fig. 5 and steady Mach contours are displayed on Fig. 6. This calculation requires 1min 30s to reach $R_2 = 10^{-5}$. More details on this test case can be found in Lerat and Sidès (1987) for a similar grid (same number of mesh lines). In both calculations, the lift coefficient is $C_l = 0.370$.

Since various values of C_l have been obtained by other authors even in very fine meshes, we have also applied the present method on a very fine C- grid composed of 500×50 cells with an attempt to align the mesh with the shocks (mesh generation due to Le Balleur and Henry). Results are shown in Figs. 7 and 8. On the airfoil, the upper shock is captured over one mesh cell and the lower shock over two. The Rankine–Hugoniot relations across the shocks are perfectly satisfied. The entropy distribution on the airfoil shows that the dissipative error is peculiarly low. The influence of the location of the remote boundary has been studied. The present mesh stretches to 20 chord lengths from the airfoil. No significant difference has been found on the airfoil by removing the exernal boundary further away or else by applying the correction proposed by Thomas and Salas (1985).

The lift coefficient C_l is given in Table 1, comparison being made with other contributions on similar meshes. The drag coefficient is $C_d = 0.0580$. Finally, we give the foot locations of the upper and the lower shock (x-

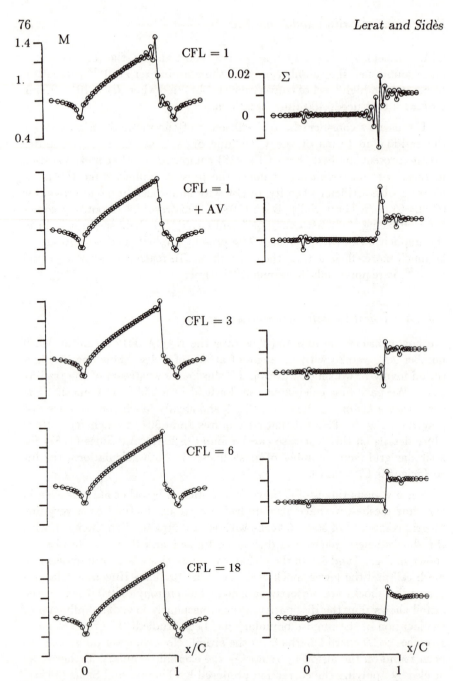

Fig. 2: Mach number and entropy distributions on the lower wall of the channel

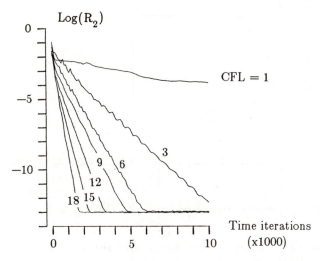

Fig. 3: Convergence history in terms of time-iterations for various CFL numbers

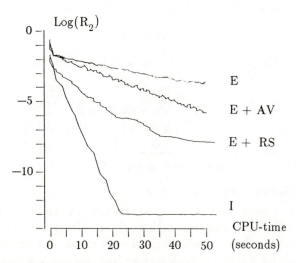

E: Present explicit stage alone (CFL=1).
E+AV: Present explicit stage with artificial viscosity (CFL=1).
E+RS: Present explicit stage with implicit residual smoothing (CFL=9).
I: Present implicit method (CFL=18).

Fig. 4: Convergence history in terms of CPU-time for various methods used in their optimal conditions

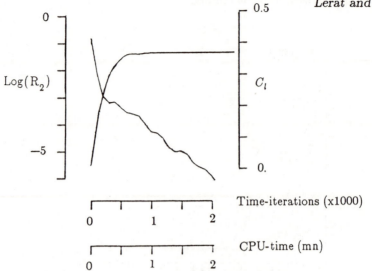

Fig. 5: Convergence history for the NACA 0012 airfoil at $M_\infty = 0.85$ and $i = 1$ deg (256×32 C-mesh)

coordinate of the sonic point, based on the chord):

$$x_u = 0.866, \qquad x_l = 0.629.$$

One can reasonably hope that the present solution improves the results obtained in previous works.

Reference	Mesh	C_l
Schmidt and Jamenson (1985)[2]	320×64 O-mesh	0.3584
Pulliam and Barton (1985)	560×65 C-mesh	0.3938
Present contribution	500×50 C-mesh	0.3762

Table 1: Lift coefficient for the NACA 0012 airfoil at $M_\infty = 0.85$, $i = 1$ deg.

5.4 Airfoil calculation at high angles of attack

To check also the robustness of the method, we have calculated transonic flows over the NACA 0012 airfoil at high angles of attack. In the paper by Barton and Pulliam (1984), we have considered the flow at Mach number 0.301 and 13.5 deg angle of attack and we have obtained an unsteady solution too. The results show a self-induced oscillation with a separation near the trailing edge, though the present calculation is not really consistent with the unsteady Euler equations. For lower angles of attack (12.5 and 13 deg) we have obtained steady solutions without separation. Fig. 9 shows the lift coefficient in terms of the number of iterations for the three angles of attack in a very fine mesh.

---

[2]See Viviand (1985)

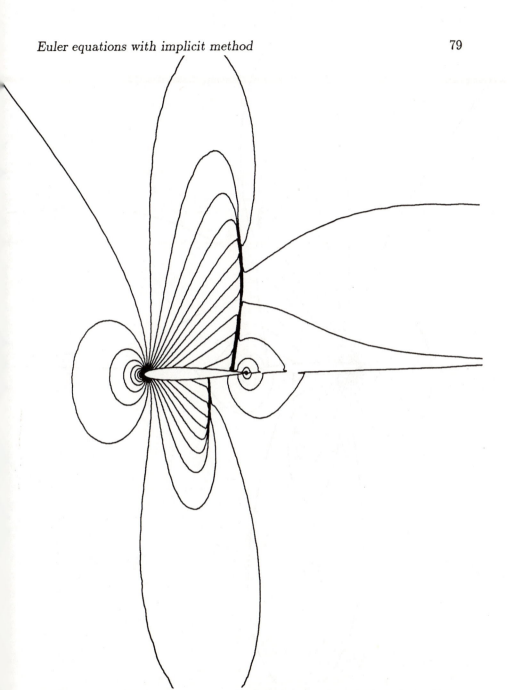

Fig. 6: Mach number countours ($\Delta M = 0.05$) around the NACA 0012 airfoil at $M_\infty = 0.85$ and $i = 1$ deg (256×32 C-mesh)

Lerat and Sidès

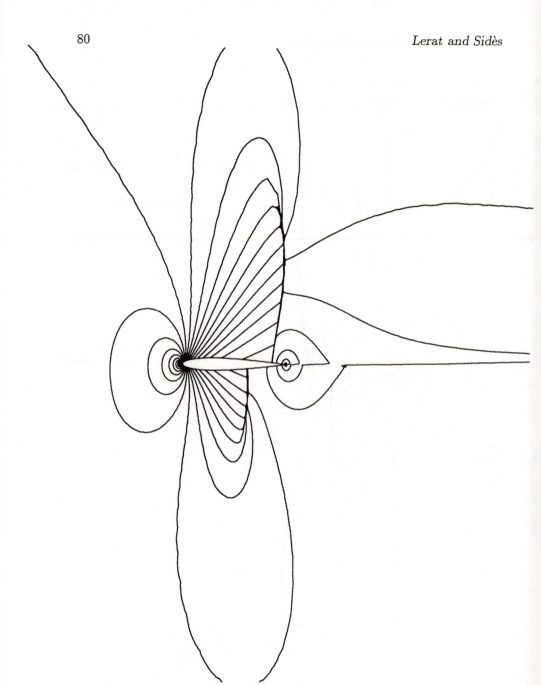

Fig. 7: Mach number countours ($\Delta M = 0.05$) around the NACA 0012 airfoil at $M_\infty = 0.85$ and $i = 1$ deg (500×50 C-mesh)

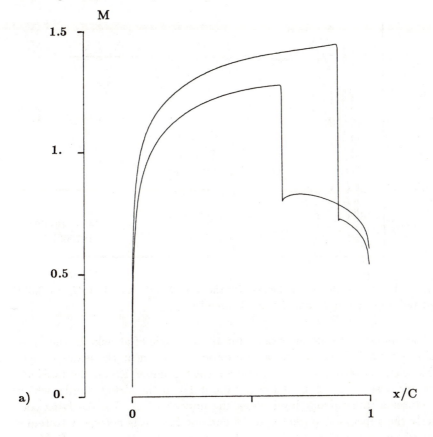

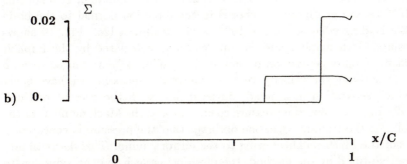

Fig. 8: a) Mach number and b) Entropy on the NACA 0012 airfoil 0012 airfoil at $M_\infty = 0.85$ and $i = 1$ deg (500 × 50 C-mesh)

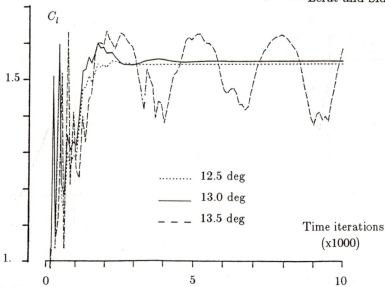

Fig. 9: History of the lift coefficient for the NACA 0012 airfoil at $M_\infty = 0.301$ and various angles of attack (500 × 50 C-mesh)

Let us now present the results for 13 deg angle of attack. In the flow-field, a curved shock occurs on the upper surface near the leading edge. This shock is rather strong (upstream Mach number 1.5 on the body) it extends over a very short distance (about 1% of the airfoil chord), which produces a thin entropy layer along the upper surface. For resolving correctly the supersonic pocket and the entropy layer it is necessary to refine the grid near the upper surface as already noticed by Barton and Pulliam.

With a view to obtaining a reference solution, we have used a very fine mesh composed of 500 × 50 cells. This mesh has a minimum grid spacing of 10^{-3} (based on the airfoil chord) in the direction normal to the body at the leading edge and 1.6×10^{-3} at the trailing edge. Fig. 10 shows pressure, Mach number and entropy contours calculated by the implicit method. No flow separation occurs at the trailing edge as can be seen on Fig. 11 showing the velocity vectors. Generally speaking, we have never obtained separation in a steady inviscid flow with the present method. Finally Fig. 12 shows the pressure coefficient and the Mach number on the airfoil. At the trailing edge, one observes that the pressure is continuous, but not the Mach number owing to the entropy jump. This discontinuity is well captured by the method, the mesh being such that no value has to be computed on the actual trailing edge.

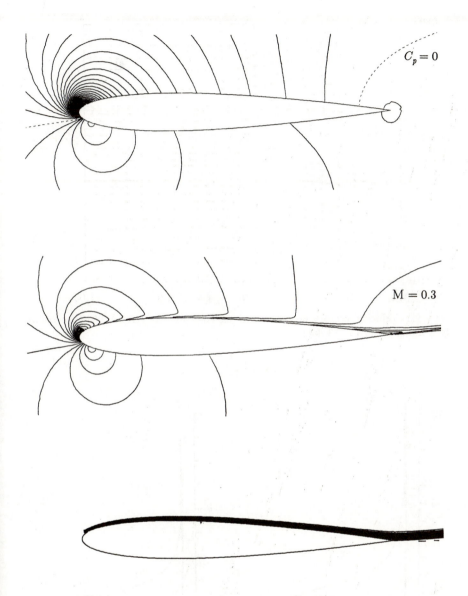

Fig. 10: Pressure, Mach number and entropy contours ($\Delta C_p = 0.2$, $\Delta M = 0.05$, $\Delta \sum = 0.001$) for the NACA 0012 airfoil at $M_\infty = 0.301$ and $i = 13$ deg.

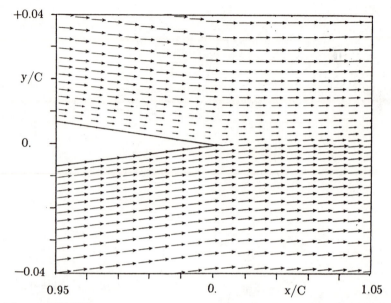

Fig. 11: Velocity vectors near the trailing edge for the NACA 0012 airfoil at $M_\infty = 0.301$ and $i = 13$ deg.

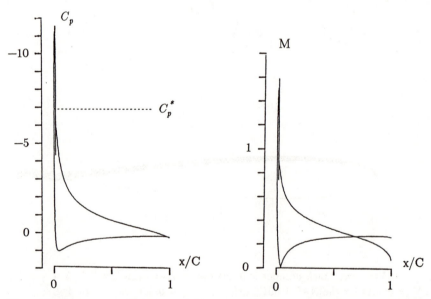

Fig. 12: Pressure coefficient and Mach number on the NACA 0012 airfoil at $M_\infty = 0.301$ and $i = 13$ deg.

References

Barton, J. T. and Pulliam, T. H. (1984) Airfoil computation at high angles of attack, inviscid and viscous phenomena, AIAA paper 84-0524, also: *AIAA Journal* (1986), 24 : 705–712.

Beam, R. and Warming, R. F. (1976) An implicit finite-difference algorithm for hyperbolic systems in conservation-law form, *Journal of Computational Physics*, 22 : 87–110.

Daru, V. and Lerat, A. (1985) Analysis of an implicit Euler solver, *Numerical Methods for the Euler Equations of Fluid Dynamics*, Angrand, F. et al. eds, SIAM Publ., 246–280.

Kreiss, H. O. (1964) On difference approximations of the dissipative type for hyperbolic differential equations. *Comm. Pure and Appl. Math.*, 17 : 335–353.

Lax, P.D. and Wendroff, B. (1960) Systems of conservation laws. *Comm. Pure. Appl. Math.* 13 : 217–237.

Lerat, A. (1979) Une classe de schémas aux différences implicites pour les systèmes hyperboliques de lois de conservation, *Comptes-Rendus de l'Académie des Sciences*, 288A : 1033–1036.

Lerat, A. (1981) Sur le calcul des solutions faibles des systèmes hyperboliques de lois de conservation à l'aide de schémas aux différences, *Publication ONERA*, 1981-1.

Lerat, A. (1983) Implicit methods of second-order accuracy for the Euler equations, AIAA Paper 83-1925; also: *AIAA Journal*, 1985, 23 : 33–40.

Lerat, A. and Sidès, J. (1987) Implicit transonic calculations without artificial viscosity or upwinding, *GAMM Workshop on Numerical Simulation of Compressible Euler Flows*, June 1986, to appear. Also: *TP ONERA*, 1987-195.

Lerat, A. Sidès, J. and Daru, V. (1982) An implicit finite-volume method for solving the Euler equations, *Lecture Notes in Physics*, 170 : 343–349.

Ni, R. H. (1981) A multiple-grid scheme for solving the Euler equations. AIAA Paper 81-1025. Also: *AIAA Journal*, 1982, 20 : 1565-1571.

Pulliam, T. H. and Barton, J. T. (1985) Euler computations of AGARD Working Group 07 airfoil test cases, AIAA Paper 85-0018.

Rizzi, A. and Viviand, H. (1981) Eds. Numerical methods for the computation of inviscid transonic flows with shock waves, *Notes on Numerical Fluid Mechanics*, **3**.

Thomas, J. L. and Salas, M. D. (1985) Far-field boundary conditions for transonic lifting solutions to the Euler equations, AIAA Paper 85-0020.

Viviand, H. (1985) Numerical solutions of two-dimensional reference test cases, *AGARD Advisory Report* 221, 6.1–6.68.

Some current trends in numerical grid generation

J. F. Thompson

Mississippi State University

1. Introduction

With the advent of supercomputers with very large storage and high speed, it has become possible to treat physical field problems on very complex regions by the numerical solution of systems of partial differential equations. Computational fluid dynamics, for instance, has progressed to the point of finite-difference (or finite-volume) flow solutions about full aircraft configurations. Finite element solutions for solid mechanics are even more developed. Similar advances are emerging in hydrodynamics, electromagnetics, magnetohydrodynamics, heat and mass transfer and, to some degree, in all field problems.

An essential element of these solutions on general regions is the construction of a mesh on which to represent the partial differential equations in finite form. This mesh may be structured, i.e., formed by intersections of the three coordinate surfaces of a curvilinear coordinate system fitted to the boundaries of the region, or may be unstructured, e.g. composed of tetrahedrons connecting a random distribution of points. In either case the mesh must be generated for the region of interest, and this is far from being a trivial problem.

In fact, it may at present take orders of magnitude more man-hours to construct the mesh than it does to construct and analyze the physical solution on the mesh (Thompson and Steger). This is especially true in some areas where solution codes of wide applicability are becoming available but which require that a mesh be generated on which to run the code. Again computational fluid dynamics is a prime example, and mesh generation has been cited repeatedly as being a major pacing item in the treatment of realistic aircraft configurations (National Research Council 1986), (Kutler 1986). The flow codes now available require much less esoteric expertise of the knowledgeable user than do the mesh generation codes. Very general mesh generation codes are now becoming available, and the theory of the generation systems has matured to the point of considerable standardization, but the construction of a mesh for very complex regions with these codes requires considerable ingenuity and experience.

The present paper summarizes some current techniques for the generation of structured meshes for general configurations.

2. Mesh types

Each type of mesh has its advantages and disadvantages. For instance, the structured mesh provides a more natural representation of normal derivative boundary conditions and allows more straightforward approximations based on prevailing directions, e.g. parallel to a boundary or flow direction. The structure also leads to a much more simple data set construction, and allows the use of directional time splitting and flux representations. On the other hand, the unstructured meshes can be much more readily imagined for complicated boundary configurations.

General configurations can be treated conceivably with either type of mesh, and (Flores and Gundy 1987), (Jameson and Baker 1987) give examples of structured and unstructured meshes applied to full aircraft flow simulations. Combinations are also possible, of course, using perhaps individual structured meshes near boundaries with these sub-regions being connected by an unstructured mesh (Nakahashi and Obayashi 1987). Structured grids are particularly attractive near solid boundaries in solutions involving very strong gradients normal to the boundary, e.g. boundary layers. Dynamically adaptive meshes, i.e., with the mesh coupled with the physical solution in order to reduce the error in the solution, can also be constructed on both types, and (Dannenhoffer III and Baron 1985), (Oden and Devloo 1987) provide recent examples of such adaption on structured an unstructured meshes.

3. Composite structured grids

The efficiency of the computation is greatly enhanced if there is some organization to the mesh. This organization can be provided by having the discretization defined by the nodes of a curvilinear coordinate system filling the physical field. Such systems are readily available from handbooks for certain simple configurations such as regions that are cylindrical, spherical, elliptical, etc. For general regions of arbitrary shape, numerical grid generation provides the curvilinear system. The techniques of numerical grid generation, and its application to the numerical solution of partial differential equations, are covered in detail in a recent text on the subject (Thompson and Mastin 1985), and in a chapter of (Thompson 1988). Several surveys of the field have also been given (Thompson and Mastin 1982), (Thompson 1984), (Thompson 1985), (Eiseman 1985), and four conference proceedings dedicated to the area have appeared (Thompson 1982), (Smith 1980), (Ghia and Ghia 1983), (Hauser and Taylor 1986).

Although in principle it is possible to establish a correspondence between any physical region and a single empty rectangular block for general three-dimensional configurations, the resulting grid is likely to be much too skewed and irregular to be usable when the boundary geometry is complicated. A better approach with complicated physical boundaries is to segment the physical region into contiguous sub-regions, each bounded by six curved sides (four in 2D) and each of which transforms to a rectangular block in the computational region, with a grid generated in each sub-region, (Thompson 1986), (Thomas 1982). Each sub-region has its own curvilinear coordinate system irrespective of that in the adjacent sub-regions.

This then allows both the grid generation and numerical solutions on the grid to be constructed to operate in a rectangular computational region, regardless of the shape or complexity of the full physical region. the full region is treated by performing the solution operation in all of the rectangular computational blocks. With the composite framework, partial differential equation solution procedures written to operate on rectangular regions can be incorporated into a code for general configurations in a straightforward manner, since the code only needs to treat a rectangular block. The entire physical field then can be treated in a loop over all the blocks.

The generally curved surfaces bounding the sub-regions in the physical region form internal interfaces across which information must be transferred, i.e., from the sides of one rectangular computational block to those of another. These interfaces occur in pairs, an interface on one block being paired with another on the same, or another, block, since both correspond to the same physical surface. Grid lines at the interfaces may meet with complete continuity, with or without slope continuity, or may not even meet.

Complete continuity of grid lines across the interface requires that the interface be treated as a branch cut on which the generation system is solved just as it is on the interior of blocks. The interface locations are then not fixed, but are determined by the generation system (Thompson 1986). This is most easily handled in coding by providing an extra layer of points surrounding each block. Here the grid points on an interface of one block are coincident in physical space with those on another interface of the same or another block, and also the grid points on the surrounding layer outside the first interface are coincident with those just inside the second, and vice versa. This coincidence can be maintained during the course of an iterative solution of an elliptic generation system by setting the values on the surrounding layers equal to those at the corresponding interior points after each iteration. All the blocks are thus iterated to convergence together, so that the entire composite grid is generated at once.

The construction of flow codes for the solution of partial differential equations (PDE) in complicated regions is greatly simplified by the composite grid structure since, withe the use of the surrounding layer of points on each block, a PDE code is only required basically to operate on a rectangular computational region. The necessary correspondence of points on the surrounding layers (image points) with interior points (object points) is set up by the grid code and made available to the PDE solution code. The entire physical field then can be treated in a loop over all the blocks.

Such a composite structure has been incorporated in several recent grid codes, e.g. (Thomas 1982), (Coleman 1985), (Soni 1985), (Miki and Takagi 1984), (Weatherill and Forsey 1984), (Sorenson and Steiger 1983), (Thompson 1987a), (Woan 1987), (Holcomb 1987), (Sorenson 1986) of various degrees of generality; cf. also (Hauser and Taylor 1986) and (Thompson 1982). The curved surfaces bounding the sub-regions in the physical region form internal interfaces across which information must be transferred, i.e., from the sides of one rectangular computation block to those of another. regardless of whether the composite grid is formed using contiguous sub-grids (i.e. a blocked grid) or from overset (or overlapped) grids, these interface boundaries occur in pairs. For a blocked grid an interface of one block is paired with another on the same, or different, block, since both correspond to the same physical surface. Grid lines at the interfaces may meet with complete continuity, with or without slope continuity, or may not meet at all. The codes of (Coleman 1985), (Miki and Takagi 1984), (Weatherill and Forsey 1984), (Thompson 1987a), (Woan 1987) and (Holcomb 1987) provide complete continuity, while those of references (Sorenson and Steiger 1983) and (Sorenson 1986) are based on slope continuity.

4.   Orthogonality

Coordinate systems that are orthogonal, or at least nearly orthogonal, near the boundary make the application of boundary conditions most straightforward. Although strict orthogonality is not necessary, the accuracy deteriorates if the departure from orthogonality is too large (i.e. grid lines crossing at angles less than $45°$, (Thompson and Mastin 1985). The implementation of algebraic turbulence models is more reliable with nearorthogonality at the boundary, since information on local boundary normals is usually required in such models. The formulation of boundary-layer equations is also more straightforward and unambiguous in such systems. It is thus better in general, other considerations being equal, for grid lines to be nearly normal to boundaries.

5. Grid generation schemes

The generation procedures for curvilinear grids are of two general types, cf. (Thompson and Mastin 1985): (1) by numerical solution of partial differential equations, and (2) construction by algebraic interpolation. In the former, the PDE system may be elliptic, parabolic, or hyperbolic.

The relative merits of the various types of grids and generation procedures have been discussed in the various surveys noted above, as well as in the works cited therein. Basically, the algebraic generation systems are faster, but the grids generated from partial differential equations are smoother. The hyperbolic and parabolic generation systems are faster than the elliptic systems, but are more limited in the configurations that can be treated. The elliptic systems are the most generally applicable with complicated boundary configurations, but transfinite interpolation is also effective in the composite grid framework.

5.1 Algebraic grid generation

Algebraic grid generation consists of the determination of the interior values in the rectangular array $r_{i,j,k}$ from the set values on the sides by interpolation. A number of different forms of interpolation are discussed in (Thompson and Mastin 1985). Such generation systems are surveyed in (Thompson and Mastin 1982), (Thompson 1984), and (Eiseman 1985) as well as in (Smith 1982) and (Smith 1983).

A generally effective grid generation procedure is provided by the transfinite interpolation technique (Gordon and Thiel 1982), (Rizzi and Eriksson 1981), in which all of the boundary values are matched the the interpolation function. Transfinite interpolation in multiple dimensions can be built up of one-dimensional interpolations, cf. (Thompson and Mastin 1985), (Thompson 1987a).

General algebraic grid generation codes have been reported in (Soni 1985), (Thompson 1987a), and (Eiseman 1982).

5.2 Elliptic grid generation

Since elliptic partial differential systems determine a function in terms of its values on the entire closed boundary of a region, such a system can be used to generate the interior values in the array $r_{i,j,k}$ from the values set on the sides. The properties of elliptic grid generation systems are discussed in (Thompson and Mastin 1985). The extremum principles that are exhibited by some elliptic systems serve to prevent the grid overlap that can occur with algebraic grid generation in some configurations. Grids generated from some elliptic systems also generally tend to be smoother than those from algebraic systems. In fact, it can be shown by the calculus of variations that a grid generated as the solution of Laplace equations is the smoothest possible grid. The lines of such a grid tend to concentrate over convex portions of the physical boundary and to be more widely spaced over concave portions, however.

Control over the spacing of the grid lines can be exercised by incorporating non-zero Laplacians into the generation system. The most common form at present is the following system; cf. (Thompson 1987a), (Thompson 1987b):

$$\sum_{m=1}^{3} \sum_{n=1}^{3} g^{mn} \mathbf{r}_{\xi^m \xi^n} + \sum_{n=1}^{3} g^{nn} P_n \mathbf{r}_{\xi^n} = 0 \qquad (5.1)$$

where the g^{mn} are the elements of the contravariant metric tensor, and the P_n are the "control functions" which serve to control the spacing and orientation of the grid lines in the field.

Control functions can be evaluated on the boundaries using the specified boundary point distribution in the generation system, with certain necessary assumptions to eliminate some terms, and then can be interpolated from the boundaries into the field. Earlier approaches (Thomas 1982) interpolated the entire functions from the boundaries in this manner. More general regions can, however, be treated by interpolating elements of the control functions separately, cf. (Thompson 1987a), (Thompson 1987b), (Thomas 1984). Alternatively, the three components of the elliptic grid generation system, Eq. (5.1), provide a set of three equations, that can be solved simultaneously at each point for the three control functions, P_k ($k = 1, 2, 3$) when $\mathbf{r}$ is given by an algebraic grid. The derivatives here are represented by central differences. This produces control functions which will reproduce the algebraic grid from the elliptic system solution in a single iteration. Thus evaluation of the control functions in this manner would be of trivial interest except when these control functions are smoothed before being used in the elliptic generation system. This smoothing is done by replacing the control function at each point with the average of the four neighbors in the two curvilinear directions (one in 2D) other than that of the function. No smoothing is done in the direction of the function because to do so would smooth the spacing distribution (Thompson 1987a).

A code normally generates an algebraic grid by transfinite interpolation from the boundary point distribution to serve as the starting solution for the iterative solution for the elliptic system. With the boundary point distribution set from the hyperbolic sine or tangent functions (Vinokur 1983), (Thompson and Mastin 1985), which have been shown to give reduced truncation error, this algebraic grid has a good spacing distribution but may have slope breaks propagated from corners into the field. The use of smoothed control functions evaluated from the algebraic grid produces a smooth grid that retains essentially the spacing of the algebraic grid.

A second-order elliptic generation system allows either the point locations on the boundary or the coordinate line slope at the boundary to be specified, but not both. It is possible, however, to iteratively adjust the control functions in the generation system of the Poisson type discussed above until not only a specified line slope but also the spacing of the first coordinate surface off the boundary is achieved, with the point locations on the boundary specified (Sorenson 1986).

In three dimensions the specification of the coordinate line slope at the boundary requires the specification of two quantities, e.g., the direction cosines of the line with two tangent to the boundary. The specification of the spacing of the first coordinate surface off the boundary requires one more quantity, and therefore the three control functions in the system Eq. (5.1) are exactly sufficient to allow these three specified quantities to be achieved, while one boundary condition allowed by the second order system provides for the point locations on the boundary to be specified.

An iterative solution procedure for the determination of the three control functions for the general three-dimensional case can be constructed as detailed in (Sorenson 1986), (Thompson 1987a) and (Thompson 1987b).

6. Surface grids

The specification of the boundary point distribution is a two-dimensional grid problem in its own right, which can also be done either by interpolation or a PDE solution. In general, this is a two-dimensional boundary value problem on a curved surface, i.e., the determination of the locations of the points on the surface from specified distributions of points on the four edges of the surface. This is best approached through the use of surface parametric coordinates, cf. (Warsi 1986), (Jones 1988), whereby the surface is first defined by a 2D array of points, $r_{i,j}$, e.g. a set of cross-sections. The surface is then splined, and the spline coordinates (u,v surface parametric coordinates) are then made the dependent variables for the interpolation or PDE generation system. The generation of the surface grid can then be accomplished by first specifying the boundary points in the array $r_{i,j}$ on the four edges of the surface grid, converting these Cartesian coordinate values to spline coordinate values (u_{ij}, v_{ij}) on the edges, then determining the interior values in the arrays u_{ij} and v_{ij} from the edge values by interpolation or PDE solution, and finaly converting these spline values to Cartesian coordinates $r_{i,j}$.

7. Adaptive grid schemes

Finally, dynamically-adaptive grid continually adapt to follow developing gradients in the physical solution. This adaption can reduce oscillations associated with inadequate resolution of large gradients, allowing sharper shocks and better representation of boundary layers. Another advantageous feature is the fact that in the viscous regions where real diffusion effects must not be swamped, the numerical dissipation from upwind biasing is reduced by the adaption. Dynamic adaption is at the frontier of numerical grid generation and may well prove to be one of its most important aspects, along with the treatment of real three-dimensional configurations through the composite grid structure. There are three basic strategies that mat be employed in dynamically adaptive grids, cf. (Thompson 1985) coupled with the partial differential equations of the physical problem. Combinations are also possible, of course:

7.1   Redistribution of a fixed number of points

In this approach, points are moved from regions of relatively small error or solution gradient to regions of large error or gradient As long as the redistribution of points does not seriously deplete the number of points in other regions of possible significant gradients, this is a viable approach. The increase in spacing that must occur somewhere is not of practical consequence if it occurs in regions of small error or gradient, even though in a formal mathematical sense the global error is not improved. The redistribution approach has the advantage of not increasing the computer time and storage during the solution, and of being straightforward in coding and data structure. The disadvantages are the possible deletrious depletion of points in certain regions and the possibility of the grid becoming too skewed.

Recent examples of this adaptive approach in CFD are (Brackbill and Saltzman 1982) in 2D and (Kim and Thompson 1988) in 3D.

7.2   Local refinement of a fixed set of points

In this approach, points are added (or removed) locally in a fixed point structure in regions of relatively large error or solution gradient. Here there is, of course, no depletion of points in other regions and therefore no formal increase of error occurs. Since the error is locally reduced in the areas of refinement, the global error does formally decrease. The practical advantage of this approach is that the original point structure is preserved. The disadvantages are that the computer time and storage increase with the refinement, and that the coding and data structures are difficult, especially for implicit flow solvers.

Recent examples of this adaptive approach in CFD are (Dannenhoffer III and Baron 1985) and (Oden and Devloo 1987), both in 2D.

7.3   Local increase in algorithm order

In this approach, the solution method is changed locally to a higher order approximation in regions or relatively large error or solution gradient without changing the point distribution. This again increases the formal global accuracy since a local increase is achieved without an attendant decrease in formal accuracy elsewhere. The advantage is that the point distribution is not changed at all. The disadvantage is the great complexity of implementation in implicit flow solvers.

This adaptive approach has not had any significant application in CFD in multiple dimensions.

Adaptive redistribution of points traces its roots to the principle of equidistribution of error, cf. (Thompson and Mastin 1985), (Thompson 1985), by which a point distribution is set so as to make the product of the spacing and a weight function constant over the points. With the point distribution defined by a function $x(\xi)$, where ξ varies by a unit increment between points, the equidistribution principle can be expressed as

$$wx_\xi = \text{constant} \tag{7.1}$$

This one-dimensional equation can be applied in each direction in an alternating fashion, but a direct extension to multiple dimensions can be made in either of two ways as follows:

From the calculus of variations, equation (7.1) can be shown, cf. (Thompson and Mastin 1985), to be the Euler variational equation for the function $x(\xi)$ which minimizes the integral

$$\int w(\xi)x_\xi^2 d\xi \tag{7.2}$$

Generalizing this, a competitive enhancement of grid smoothness, orthogonality, and concentration can be accomplished by representing each of these features by integral measures over the grid and minimizing a weighted average of the three. This approach was put forward in (Brackbill and Saltzman 1982) and discussed in detail in (Thompson and Mastin 1985).

The second approach is to note the correspondence between (7.1) and the one-dimensional form of the commonly-used elliptic grid generation system, equation (5.1).

It is logical then to represent the control functions in 3D as

$$P_n = \frac{w_{\xi^n}}{w} \qquad n = 1, 2, 3 \tag{7.3}$$

This approach was put forward in (Anderson 1986) and has been applied in 3D in (Kim and Thompson 1988)

This control function adaptive approach has the significant advantage of being based on the same elliptic generation equations that are in common use in grid generation codes, and the adaptive control functions given by equation (7.3) can be added to those already evaluated from the configuration geometry.

The complete generalization of (7.3) is

$$P_i = \sum_j \frac{g^{ij}}{g^{ii}} \frac{(w_i)_{\xi^j}}{w_i} \qquad (7.4)$$

involving three weight functions, $w_i (i = 1, 2, 3)$, as given in (Eisman 1987). In some applications the availability of three separate control functions could be a definite advantage, with perhaps the velocity gradient used in the weight function in one direction and the pressure gradient in another.

8. Conclusion

Three-dimensional grid codes should hopefully become suitable for general use in the near future. Further development is now needed in automation of the field segmentation decisions and refinement of the geometric properties of the physical boundaries. The incorporation of dynamically-adaptive grids in the composite framework is only just emerging and should prove to be of considerable importance to general flow solutions.

References

Anderson, D. A. (1986). Generating adaptive grids with a conventional grid scheme. AIAA–86–0427, AIAA 24th Aerospace sciences meeting, Reno.

Brackbill, J. U. and Saltzman, J. S. (1982). Adaptive zoning for singular problems in two dimensions. *Journal of Computational Physics*, 46:342.

Coleman, R. M. (1985). INMESH: An interactive program for numerical grid generation. DTNSRDC–85–054, David W. Taylor naval ship reseach and development center, Bethesda, MD.

Dannenhoffer III, J. and Baron, J. R. (1985). Grid adaption for the 2-D Euler equations. AIAA–85–0484, AIAA 23rd Aerospace sciences meeting, Reno.

Eiseman, P. R. (1982). Automatic algebraic coordinate generation. In *Numerical grid generation*, Thompson, J. F., editor. Noth-Holland.

Eiseman, P. R. (1985). Grid generation for fluid mechanics computations. *Annual review of fluid mechanics*, 17.

Eisman, P. R. (1987). Adaptive grid generation. *Computer Methods in Applied Mechanics and Engineering*, 64(321).

Ghia, K. N. and Ghia, U., editors (1983). *Advances in grid generation*. ASME applied mechanics, bioengineering, and fluis engineering conference, Houston.

Gordon, J. W. and Thiel, L. C. (1982). Transfinite mappings and their applications to grid generation. In *Numerical grid generation*, Thompson, J. F., editor. Noth-Holland.

Hauser, J. and Taylor, C., editors (1986). *Numerical Grid generation in computational fluid dynamics*. Pineridge Press.

Holcomb, J. E. (1987). Development of a grid generator to support 3-D multizone Navier-Stokes analysis. AIAA–87–0203, AIAA 25th Aerospace sciences meeting, Reno.

J. F. Thompson, Z. U. A. W. and Mastin, C. W. (1982). Boundary fitted coordinates systems for numerical solution of partial differential equations — a review. *Journal of Computational Physics*, 47:1.

J. F. Thompson, Z. U. A. W. and Mastin, C. W. (1985). *Numerical grid generation: foundations and applications*. North-Holland.

J. Flores, S. G. Reznick, T. L. H. and Gundy, K. (1987). Transonic Navier-Stokes solutions for a fighter-like configuration. AIAA–87–0032, AIAA 25th Aerospace sciences meeting, Reno.

J. T. Oden, T. S. and Devloo, P. (1987). An adaptive finite element strategy for complex flow problems. AIAA–87–0557, AIAA 25th Aerospace sciences meeting, Reno.

Jameson, A. and Baker, T. G. (1987). Improvements to the aircraft Euler method. AIAA–87–0452, AIAA 25th Aerospace sciences meeting, Reno.

Jones, G. A. (1988). *Surface grid generation for composite block grids*. PhD thesis, Mississippi State University.

Kim, H. J. and Thompson, J. F. (1988). Three dimensional adaptive grid generation on a composite block grid. AIAA–88–0311, AIAA 26th Aerospace sciences meeting, Reno.

Kutler, P. (1986). A perspective of computational fluid dynamics. In *Numerical grid generation in computational fluid dynamics*, Hauser, J. and Taylor, C., editors, page 547. Pineridge Press.

Miki, K. and Takagi, T. (1984). A domain decomposition and overlapping method for the generation of three-dimensional boundary-fitted coordinate systems. *Journal of Computational Physics*, 53:319.

Nakahashi, K. and Obayashi, S. (1987). FDM-FEM zonal approach for viscous flow computations over multiple bodies. AIAA–87–0604, AIAA 25th Aerospace sciences meeting, Reno.

National Research Council (1986). *Current capabilities and future directions in computational fluid dynamics*. National Academy Press.

Rizzi, A. and Eriksson, L. E. (1981). Transfinite mesh generation and damped Euler equation algorithm for transonic flow around wing-body configurations. AIAA–81–0999, AIAA 5th conputational fluid dynamics conference, Palo Alto, California.

Smith, R. E., editor (1980). *Numerical grid generation techniques*, NASA Langley research center. NASA conference publication 2166.

Smith, R. E. (1982). Alegbraic grid generation. In *Numerical grid generation*, Thompson, J. F., editor. Noth-Holland.

Smith, R. E. (1983). Three-dimensional algebraic grid generation. AIAA–83–1904, AIAA computational fluid dynamics conference, Danvers, Mass.

Soni, B. K. (1985). Two– and Three– dimensional grid generation for internal flow applications of computational fluid dynamics. AIAA–85–1526, AIAA 7th computational fluid dynamics conference, Cincinnati, Ohio.

Sorenson, R. L. (1986). Three-dimensional elliptic grid generation about fighter aircraft for zonal finite-difference computations. AIAA–86–0429, AIAA 24th Aerospace sciences meeting, Reno.

Sorenson, R. L. and Steiger, J. L. (1983). Grid generation in three dimensions by Poisson equations with control of cell size and skewness at boundary surfaces. In *Advances in grid generation*, Ghia, K. N. and Ghia, U., editors. ASME Applied mechanics, bioengineering, and fluids engineering conference.

Thomas, P. D. (1982). Composite three-dimensional grids generated by elliptic systems. *AIAA Journal*, 20:1195.

Thomas, P. D. (1984). Stationary interior grids and computation of moving interfaces. In *Advances in computational methods for boundary and interior layers*, Miller, J. J. H., editor. Boole Press, Dublin.

Thompson, J. F., editor (1982). *Numerical grid generation*. North-Holland. (Also published as Vol. 10 and 11 of Applied Mathematics and Computation, 1982).

Thompson, J. F. (1985). A survey of dynamically-adaptive grids in the numerical solution of partial differential equations. *Applied Numerical Mathematics*, 1:3.

Thompson, J. F., editor (1986). *A survey of composite grid generation for general three-dimensional regions*. AIAA.

Thompson, J. F. (1987a). A composite grid generation code for general 3-D regions. AIAA–87–0275, AIAA 25th Aerospace sciences meeting, Reno.

Thompson, J. F. (1987b). A general three-dimensional elliptic grid generation system on a composite block structure. *Computer methods in applied mechanics and engineering*, 64:377.

Thompson, J. F. and Mastin, C. W. (1985). Order of difference expressions in curvilinear coordinate systems. *Journal of Fluids Engineering*, 107:241.

Thompson, J. F. and Steger, J. L. Three-dimensional grid generation for complex configurations – recent progress. AGARDOGRAPH to be released 1988. ed. H. Yoshihara.

Thompson, J. T. (1984). Grid generation techniques in computational fluid dynamics. *AIAA Journal*, 22:1505.

Thompson, J. T. (1988). Grid generation. In *Handbook of numerical heat transfer*, W. J. Minkowycz, E. M. Sparrow, G. E. S. and Pletcher, R. H., editors. John Wiley.

Vinokur, M. (1983). On one-dimensional stetching functions for finite-difference calculations. *Journal of Computational Physics*, 50:215.

Warsi, Z. U. A. (1986). Numerical grid generation in arbitrary surfaces though a second-order differential geometric model. *Journal of Computational Physics*, 64:82.

Weatherill, N. P. and Forsey, C. R. (1984). Grid generation and flow calculations for complex aircraft geometries using a multi-block scheme. AIAA–84–1665, AIAA 17th fluid dynamics, plasma dynamics, and laser conference, Snowmass, Colorado.

Woan, C. J. (1987). Three-dimensional elliptic grid generations using a multi-block method. AIAA–87–0278, AIAA 25th Aerospace sciences meeting, Reno.

A strategy for the use of hybrid structured–unstructured meshes in computational fluid dynamics

N. P. Weatherill

Institute for Numerical Methods in Engineering
University College of Swansea

1. Introduction

In recent years there has been much interest in the development of techniques which are capable of automatically generating mesh points around complex aeronautical configurations. This effort has concentrated on two contrasting philosophies; structured and unstructured mesh generation. In the former, sets of curvilinear coordinates are derived which can be mapped to a Cartesian network in computational space. For realistic applications and complex geometries this technique is implemented within a multiblock environment (Weatherill 1985; Thompson 1988; Boerstoel 1988). The method has proved successful with finite difference and finite volume techniques to solve the flow equations. In contrast, the second approach allows mesh points to be connected such that they can possess different numbers of direct neighbouring points. Because of the flexibility inherent to this approach it is necessary to define a so-called connectivity matrix which details the connections of every point to its adjacent points. Unstructured meshes have been traditionally used with the finite element method (Jameson 1986; Peraire 1987). Both these approaches have produced some impressive demonstrations of their capabilities. These results have stimulated a general discussion on the apparent advantages of one method over the other. It is true to say that the advantages of the structured approach are

1. the flexibility of implementation of all classes of flow algorithm;
2. efficient utilisation of vector architecture of computers;
3. efficiency in CPU time and computer memory;
4. good environment for the multigrid technique.

The relative disadvantages include

1. lack of total flexibility for very complex regions;

2. not generally amenable to mesh point enrichment.

The unstructured method is

1. flexible for very complicated geometrical regions;

2. is a natural environment for mesh/flow adaption.

Some disadvantages would include

1. not necessarily amenable to all classes of flow algorithms, and to the implementation of multigrid;

2. relatively inefficient in computer memory requirements.

It is interesting to note that the real advantages of one approach are the disadvantages of the other. This observation is at the heart of the investigation described in this paper. We sought to investigate if the advantages of the unstructured method could be combined with the advantages of the structured method.

Here we describe some initial work into this unification of structured–unstructured mesh generation. Two strategies are presented. One, to encompass mesh enrichment/flow adaptivity; and one to utilise the suitability of unstructured meshes to produce a discretisation of complex geometrical regions. Both these strategies, in general, result in meshes with regions of triangular and quadrilateral cells. It is, therefore, necessary to construct flow algorithms which can be used on mixed structured–unstructured meshes.

2.   Flow algorithm on hybrid meshes

The time dependent Euler equations for inviscid compressible flow can be written

$$\frac{d}{dt}\int_V \mathbf{W}\,d\mathbf{V} + \int_C (\mathbf{F}\,dy - \mathbf{G}\,dx) = 0, \qquad (2.1)$$

where

$$\mathbf{W} = (\rho,\ \rho u,\ \rho v,\ \rho E)^T,$$

$$\mathbf{F} = (\rho u,\ \rho u^2 + p,\ \rho uv,\ \rho uH)^T,$$

$$\mathbf{G} = (\rho v,\ \rho uv,\ \rho v^2 + p,\ \rho vH)^T,$$

$$E = \frac{p}{(\gamma - 1)\rho} + \frac{1}{2}(u^2 + v^2), \quad H = E + \frac{p}{\rho}$$

in which, ρ, u, v, E, H, denote the density, x and y components of velocity, E the total energy and H the total enthalpy, γ is the ratio of specific

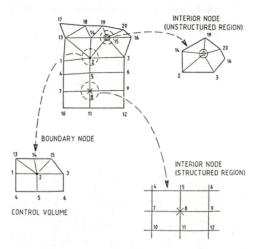

Fig. 1: Evaluation of the flux matrix for a cell vertex scheme

heats and **V** and C the volume and boundary of the flow domain, respectively.

The solution technique, used to solve this system of equations follows the work of Jameson (Jameson 1986; Jameson 1983). Discretisation of the Euler equations results in a set of coupled ordinary differential equations which are integrated in time. Artificial dissipation, $D(\mathbf{W})$ is added to eliminate the occurence of undamped waves and oscillations near shock waves.

The resulting descritized equation set takes the form

$$S_i\frac{dW_i}{dt} + [Q(W_i) - D(W_i)] = 0, \quad i = 1, 2, \cdots. \qquad (2.2)$$

where $Q(W_i)$ is the flux associated with the control volume of area S_i. These equations, once $S_i, Q(W_i)$ and $D(W_i)$ have been determined at an iteration level, are integrated using a multi–stage time–stepping procedure.

To implement such a scheme on a mesh of mixed triangles and quadrilaterals it is necessary to pay particular attention to the control volume associated with each node. Fig. 1 shows a typical region of mesh which contains nodes either, i) interior to a region of unstructured triangles, ii) on a boundary between triangles and quadrilaterals and iii) in the interior of a structured quadrilateral region. In each case the appropriate control volume is indicated. The area of each control volume and the artificial

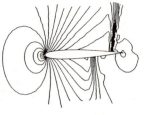

CONTOURS OF PRESSURE   CONTOURS OF DENSITY

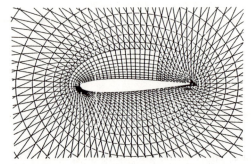

Fig. 2: Flow field simulation on a hybrid mesh

dissipation associated with each node are evaluated using similar computational stencils.

The data structure used to implement the evaluations of the flux vector etc. utilises a connectivity matrix in the regions of triangles but in the domains of quadrilaterals the explicit structured relationship between points is used. Edges between regions are treated as part of the unstructured data set and are thus included within the connectivity matrix.

Fig. 1 is illustrated for a cell vertex scheme. However, it is possible to construct a similar scheme which utilises a cell centre philosophy. In addition, it is possible to construct analogous schemes to those discussed here but with the equations augmented with the stress tensor and heat flux vector. Such schemes for solution of the Reynolds Averaged Navier–Stokes equations have been developed. Further details of the flow algorithms for mixed triangles and quadrilaterals can be found elsewhere (Weatherill 1988a).

As an illustration of the flow simulation on a hybrid grid we present in Fig. 2 the flow field contours for density and pressure for the flow over a NACA0012 at $M = 0.85$ and incidence 1.5^0. The contours are free from spurious oscillations on the boundaries between triangles and quadrilaterals.

3. Strategies for hybrid meshes

In order to incorporate the advantages of unstructured meshes within the structured mesh framework two simple strategies are presented. Fig. 3 and Fig. 4 show the procedures for mesh generation and mesh/flow adaptivity, respectively.

The details contained in Fig. 3 provide a basis by which a poor quality mesh can be transformed into a high quality mesh or alternatively, a method of further enhancing an already good quality structured mesh. The strategy presented in Fig. 4 is designed to provide the flexibility of local mesh adaption or refinement whilst maintaining many of the features of a structured method.

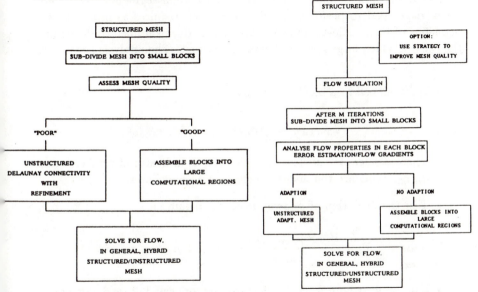

Fig. 3: Strategy for mesh generation   Fig. 4: Strategy for mesh adaption

Common to both strategies is an initial structured mesh which is subdivided into a set of small rectangular blocks. This enables either mesh quality or flow features to be assessed locally. If mesh quality is poor it is assumed that a refinement of some kind is required and thus points within such blocks are relabelled as requiring triangulation. Similarly, if gradients in the flow field are detected as being significantly steep the associated block is nominated as requiring mesh refinement and thus the points within are connected to form a triangulation. The subdivision of the structured mesh into small blocks is a convenient way of assessing local regions of the domain. However, blocks with dimensions small compared with optimum vector lengths of computers inhibit the efficiency of vector machines. It is necessary, therefore, after the sub–division and assessment processes to reassemble the blocks, which are to remain structured, into large rectangular computational regions. Both strategies usually result in hybrid meshes in which local regions of the domain are triangulated whilst the majority remain structured quadrilaterals.

4.   Mesh adaption to flow field

To illustrate the strategy for mesh adaption, assume that a mesh around a NACA0012 aerofoil is given and that no improvements in its quality are required. The strategy suggests that the Euler algorithm is used on the quadrilateral mesh to establish the flow features. This having been achieved, the flow domain is subdivided into small segments. The algorithm developed to perform this decomposition requires as input the dimensions of the smallest segment. The flow field is examined to determine the cells in which large flow gradients are present. The analysis performed is

- For each cell in a block determine

$$\phi_{\text{average}} = \frac{1}{4}(\phi_{n1} + \phi_{n2} + \phi_{n3} + \phi_{n4}),$$

  where ϕ refers to the flow variable used as indicator of gradients and n_1, n_2, n_3, n_4 are the nodes which form a quadrilateral.

- Test the deviation of ϕ_i away from the mean, i.e.

$$\Gamma_i = \frac{100 \times |\phi_{\text{average}} - \phi_i|}{\phi_{\text{average}}}, \quad i = n_1, n_2, n_3, n_4.$$

- If $\Gamma_i \geq \Phi$, where Φ is the deviator factor, say $1, 2$ or 3, for any i, then $D_c = 1$, where the subscript refers to the cell. If $\Gamma_i < \Phi$, then $D_c = 0$.

- If $\frac{1}{N_c} \sum_{c=1}^{N_c} D_c \geq 0.5$, where N_c are the total number of quadrilateral cells, then the block contains high gradients in ϕ and further point resolution within the block is rquired. The block is nominated unstructured, points are added and a triangulation of the total set of points is performed.

  If $\frac{1}{N_c} \sum_{c=1}^{N_c} D_c < 0.5$, the gradients are assumed to be small and no enrichment is required. The block remains fully quadrilateral.

Having established all blocks for triangulation the assembly of all blocks with quadrilateral cells is performed. For an initial polar mesh of 80×25

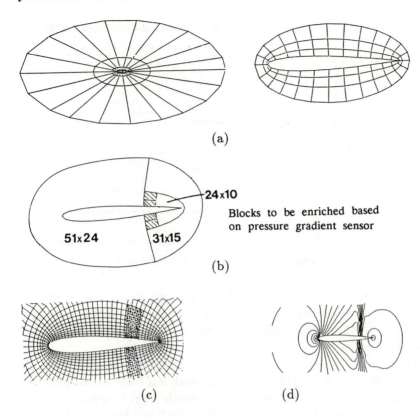

(a)

24×10

Blocks to be enriched based
on pressure gradient sensor

51×24   31×15

(b)

(c)   (d)

Fig. 5: Illustration of process for mesh adaption

and with subdivision size of 4 × 4, 104 blocks are produced (see Fig. 5a).
Mesh adaptivity outlined above with ϕ taken to be the pressure for the flow
field of $M = 0.85$ and $\alpha = 0^0$, results in 4 blocks for mesh refinement and
3 computational blocks with dimensions (51×24), (24×10) and (31×15)
after assembly (see Fig. 5b). The pressure contours obtained from the flow
simulation on the hybrid mesh together with the mesh are shown in Fig. 5c
and Fig. 5d.

Consider now the efficiency of this procedure. The flow results presented for the hybrid process were obtained a factor of 2.5 times faster
than the calculation performed on the mesh when the unstructured un-
vectorised triangle data structure was used throughout the domain. This
factor has been achieved by minimising the computations required by utilising quadrilaterals in the majority of the flow field and capitalising on the
relatively long vector lengths achievable in the three structured rectangular
regions. Clearly the vectorisation factor is heavily dependent on how the

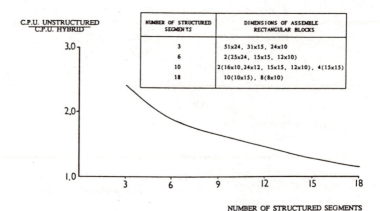

Fig. 6: Computational efficiency of hybrid flow code

small segments are reassembled and Fig. 6 shows a plot of efficiency versus number of assembled structured regions. More rectangles, for a given number of mesh points, implies small vector lengths and with this lower computational speed.

Mesh enrichment on structured meshes has been previously studied (Berger 1985; Dannenhoffer 1985). Such work has used a refinement process which embeds quadrilaterals within quadrilaterals resulting in a requirement for special formulations within the flow algorithms for coarse grid-fine grid interface. The approach advocated here requires flow algorithms which are constructed for hybrid meshes. However, not only can such algorithms be used for mesh enrichment but as will now be demonstrated such algorithms also provide considerable flexibility for new mesh generation techniques.

5. Mesh generation

The potential flexibility of the use of triangular and quadrilateral cells to discretise a domain affords several possibilities for the mesh generation process and some ideas will be highlighted.

5.1 *Local refinement*

This application is a straight forward implementation of the strategy outlined in Fig. 3 and is best illustrated with reference to Fig. 7. Assume that a mesh has been generated around an aerofoil. In this case a structured mesh generated from a system of elliptic partial differential equations has been constructed around a NACA0012 aerofoil. The global structure is an H–mesh topology and the grid is deficient in quality at the leading and trailing edges (see Fig. 7a). A flow simulation performed on the mesh of quadrilateral cells for $M = 0.5$, incidence $= 0^0$ results in a poor prediction of the flow field (see Fig. 7b).

Following the proposed strategy the mesh is subdivided into a set of blocks of dimensions approximately 4×4. Of the 56 blocks produced 8 blocks are declared of inadequate mesh quality and the regions contained within are refined using a triangulation procedure. As shown in Fig. 7c, 4 blocks at the trailing and leading edges, respectively, are combined and a refinement implemented in each of the two regions. The refinement algorithm adds points local to the aerofoil surface in a manner which produces triangles which are approximately equilateral (Weatherill 1988b). The triangulation of the set of points is performed using an algorithm which constructs the Delaunay triangulation (Weatherill 1988c).

The resulting mesh in the vicinity of the refined regions is shown in Fig. 7d. The pressure coefficients on the surface of the aerofoil for two flow parameters are shown in Fig. 7e. In both cases the cell vertex Euler algorithm has been used. The flow simulations obtained on the refined hybrid mesh are in sharp contrast to the set of results on the quadrilateral mesh. The effective computational cost has been the addition of 159 new points.

5.2 Local mesh regeneration

Fig. 8 shows a structured mesh with H–topology around a two element aerofoil configuration. The mesh quality is very poor in the vicinity of the leading edge of the main element and around the flap. In fact there is mesh crossover around the flap. Clearly, no reasonable flow solution is possible on this mesh and in most circumstances the mesh would be rejected. Undoubtedly a better structured mesh can be constructed for this configuration but it is a suitable case to demonstrate the flexibility of hybrid mesh generation.

Following the basic strategy the mesh is subdivided into 90 blocks, of which the mesh in 4 blocks around the leading edge of the main element and 16 blocks around the flap are required to be refined. It would be possible to use the idea of local refinement outlined in Section 5.1 but in this case a demonstration is given of mesh point regeneration. The mesh points within the nominated blocks are deleted and a new set of points is defined. A number of alternative techniques are available for generating the new points but in this case an auxiliary mesh generator using a conformal mapping has been used. The new points are added and triangulated using the Delaunay criterion. An illustration of this is given in Fig. 9 where it is clear that the new points added in the two regions of triangulation are generated from polar meshes centred around the aerofoil

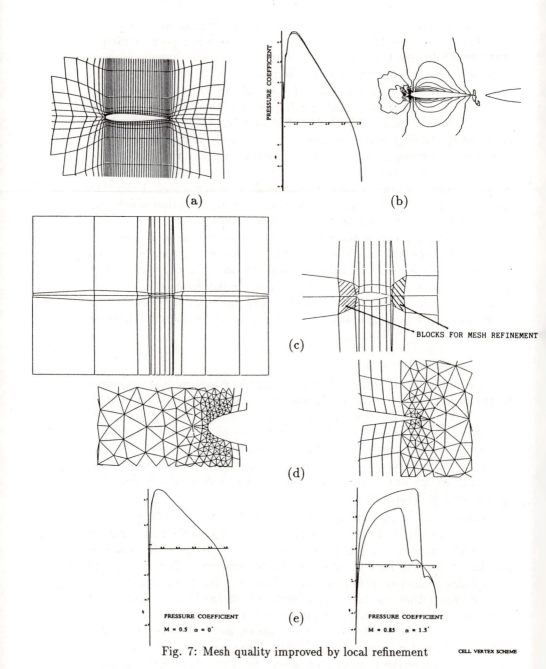

Fig. 7: Mesh quality improved by local refinement   CELL VERTEX SCHEME

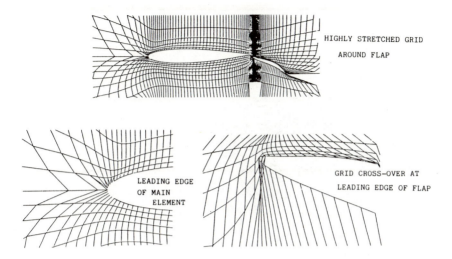

Fig. 8: Structured mesh for a two element configuration

and flap, respectively.

5.3 Non-aligned mesh

In Section 5.2 the initial structured mesh possessed mesh folding i.e. cells of negative area, at the leading edge of the flap. The technique of mesh point regeneration readily alleviated this problem. A natural extension of the ideas outlined in Section 5.2 is to initially ignore the presence of any auxiliary components and generate a global mesh around the main component. The additional components are added but, in general, are not aligned with any co-ordinate line. After the block subdivision process the points contained in the blocks local to each auxiliary component are deleted and new points particular to the added component are introduced. These points are connected by a Delaunay triangulation and subsequently the triangulated region is reconnected to the global mesh. The general strategy, which has some commonality with other techniques (Jameson 1986; Weatherill 1988c; Steger 1983), is shown in Fig. 10. A typical application of the idea to a two body interference configurations is shown in Fig. 11. Here we see a global polar mesh in which a segment has been deleted to allow for a second body section. The subsequent flow simulation on the hybrid mesh with second body present is shown in Fig. 12 and was computed using three structured regions and one unstructured region.

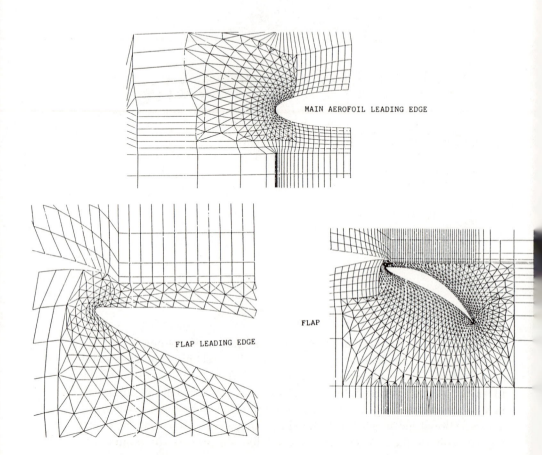

Fig. 9: Hybrid mesh after local regeneration of mesh

IN GENERAL

- TAKE ANY GOOD MESH

- ADD NEW COMPONENTS

- SUB-DIVIDE MESH INTO SMALL BLOCKS

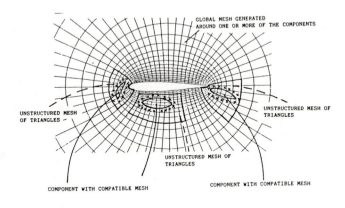

GLOBAL MESH GENERATED
AROUND ONE OR MORE OF THE COMPONENTS

UNSTRUCTURED MESH
OF TRIANGLES

UNSTRUCTURED MESH OF
TRIANGLES

UNSTRUCTURED MESH OF
TRIANGLES

COMPONENT WITH COMPATIBLE MESH

COMPONENT WITH COMPATIBLE MESH

- DEFINE BLOCKS FOR REGENERATION

- ADD NEW POINTS AND TRIANGULATE

- RECONNECT WITH MAIN MESH

Fig. 10: General strategy for use of non-body conforming mesh generation

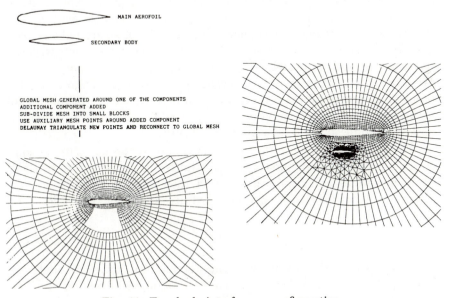

MAIN AEROFOIL

SECONDARY BODY

GLOBAL MESH GENERATED AROUND ONE OF THE COMPONENTS
ADDITIONAL COMPONENT ADDED
SUB-DIVIDE MESH INTO SMALL BLOCKS
USE AUXILIARY MESH POINTS AROUND ADDED COMPONENT
DELAUNAY TRIANGULATE NEW POINTS AND RECONNECT TO GLOBAL MESH

Fig. 11: Two body interference configuration

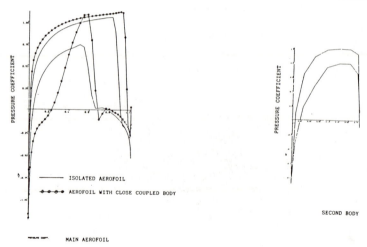

Fig. 12: Flow simulation on hybrid mesh for the two body configuration

6. Discussion

Here preliminary results obtained from using structured-unstructured meshes have been presented. The investigation has highlighted some of the potential benefits of such an approach. In reality, it is unlikely that hybrid meshes will be used for any configurations other than the most complex. A more likely role will be to further enhance existing mesh generation techniques. For example, the non-aligned procedure demonstrated in Section 5.3 could be of value when , having generated a mesh around a configuration, it is necessary to add one or more additional components without the requirement for a complete reconfiguration of the mesh topology. Experience shows that as more components are added to a configuration the problem of structured mesh generation can become over constrained and with this a rapid deterioration in mesh quality. The use of local regions of unstructured mesh could alleviate this difficulty.

The penalty to pay for this added flexibility is the necessity to construct algorithms for the flow equations which can utilise hybrid meshes. Some algorithms will require more effort than others to convert to this environment but for the scheme presented here no major difficulties were encountered.

7. Conclusion

The findings from this investigation into the use of mixed structured- unstructured meshes have been sufficiently encouraging to warrant further study of this technique. The flow algorithms constructed to work on hybrid meshes appear to work well and the approach opens new avenues for exploration in mesh generation techniques.

8. Acknowledgements

The author is grateful to the Aircraft Research Association, Bedford and the Procurement Executive, Ministry of Defence, Royal Aircraft Establishment Farnborough for their support of this work.

References

Berger, M. J. and Jameson, A. (1985) Automatic adaptive grid refinement for the Euler equation. AIAA Journal, **23**, No. 4, 561-568.

Boerstoel, J. F., (1988) Numerical grid generation in 3-D Euler flow simulation. *These proceedings.*

Dannenhoffer, J. F. and Baron, J. R. (1985) Grid adaption for the 2-D Euler equations. AIAA paper 85-0484.

Jameson, A., Baker, T. J. and Weatherill, N. P. (1986) Calculation of inviscid transonic flow over a complete aircraft. AIAA paper 86-0103.

Jameson, A. and Baker, T. J. (1983) Solution of the Euler equations for complex configuration. AIAA paper 83-1929.

Peraire, J., Peiro, J., Formaggio, L., Morgan, K. and Zienkiewicz, O. C. (1987) Finite element Euler computations in three dimensions. AIAA paper 87-0032.

Steger, J. L., Dougherty, F. C. and Benek, J. A. (1983) A chimera grid scheme. In *Advances in Grid Generation*, Ghia, K. N. and Ghia, V., Editors, pages 59–70.

Thompson, J. F., (1988) Three-dimensional adaptive grid generation on a composite block grid. *These proceedings.*

Weatherill, N. P. and Forsey, C. R. (1985) Grid generation and flow calculations for aircraft geometries. *Journal of Aircraft* **22**, No. 10.

Weatherill, N. P. (1988a) An investigation into the use of hybrid structured-unstructured meshes in computational aerodynamics–Part I Flow simulation and mesh adaptivity. *University College of Swansea, Institute for Numerical Methods in Engineering*, Report No. C/R/604/88.

Weatherill, N. P. (1988b) An investigation into the use of hybrid structured- unstructured meshes in computational aerodynamics–Part II Mesh generation. *University College of Swansea, Institute for Numerical Methods in Engineering*, Report No. C/R/605/88.

Weatherill, N. P. (1988c) A method for generating irregular computational grids in multiply connected planar domains. *International Journal for Numerical Methods in Fluids* 8:181–197.

Implicit methods in CFD

Thomas H. Pulliam

NASA Ames Research Center, USA

1. Introduction

1.1 Discussion

Implicit finite difference techniques have taken the dominant role in the computation of fluid dynamics and aerodynamics. Over a decade ago one could have made the observation that a substantial portion of all computational results were obtained using explicit techniques such as Lax-Wendroff or most likely MacCormack's scheme (MacCormack 1969), (the remainder came under the classification of analytic, spectral or perturbation schemes). At that time the limitations of computer speed and algorithm development made the explicit schemes the most practical. Explicit schemes are easy to program, requiring no additional numerical algorithms. In contrast an implicit scheme invariably leads to the need to solve either linear or nonlinear systems of equations. Linear algebra plays a major role in the solution process, where large sparse (but usually structured) matrices are inverted. Until recently, Gaussian elimination (or LU decompositions such as the Thomas algorithm) was commonly used to invert the matrices. With the advent of parallel and vector computer architectures, sparse matrix solvers may be the optimal way to proceed. Methods such as cyclic reduction, nested dissection, and sparse factorization techniques may take over especially if they can be more efficient on the new generation of high speed processors.

For a given time accuracy, explicit methods can require significantly less computational operations then implicit schemes and at least at first glance explicit methods seem to be easier to vectorize. Implicit methods have the advantage of being typically unconditionally stable (in terms of linear analysis) as compared with limited stability for explicit schemes. In cases where time accuracy restricts one to time steps on the order of or less than the stability bounds of an explicit method, such methods may be more efficient than a corresponding implicit scheme. In general though, even within the limit of time accuracy, explicit schemes can be more restrictive in terms of the allowable time step than is warranted by the time scales of the solution. For instance, boundary layer mesh scales may be more restrictive than the spatial scales of a plunging body.

Implicit schemes have been used more extensively for steady state problems where the restriction of a time step of integration commensurate with the physics of a problem is ignored in the hope of converging the solution more rapidly. In fact, both explicit and implicit schemes commonly employ time steps which vary across the physical domain (local time steps) so that time accuracy is completely lost. In these cases, implicit schemes have an advantage over explicit schemes. The one exception where explicit based schemes have performed as well or better than implicit schemes for steady state is in the case of multigrid acceleration of multi-stage integration, see (Jameson 1981; Jespersen 1984). Multigrid has been applied to both inviscid (Jameson 1981) and viscous cases (Martinelli 1988) and has even been analyzed for unsteady applications (Jespersen 1985). It should be noted that even in those schemes some form of an implicit operator is used (residual averaging).

1.2   Equations

The methods by which schemes are made implicit can take a wide variety of forms in terms of efficiency, consistency, accuracy and resulting stability and convergence. In this paper we shall examine various forms of implicit algorithms which can be applied to the compressible Euler and Navier–Stokes equations. The starting point is a set of partial differential equations, (in this case the Euler equations in conservation law form)

$$\partial_t Q + \partial_x E + \partial_y F + \partial_z G = 0 \tag{1.1}$$

where

$$Q = \begin{bmatrix} \rho \\ \rho u \\ \rho v \\ \rho w \\ e \end{bmatrix}, \tag{1.2}$$

$$E = \begin{bmatrix} \rho u \\ \rho u^2 + p \\ \rho u v \\ \rho u w \\ u(e + p) \end{bmatrix}, \quad F = \begin{bmatrix} \rho v \\ \rho v u \\ \rho v^2 + p \\ \rho v w \\ v(e + p) \end{bmatrix}, \quad G = \begin{bmatrix} \rho w \\ \rho w u \\ \rho w v \\ \rho w^2 + p \\ w(e + p) \end{bmatrix}.$$

Pressure is related to the conservative flow variables, Q, by the equation of state

$$p = (\gamma - 1) \left(e - \frac{1}{2}\rho(u^2 + v^2 + w^2) \right)$$

where γ is the ratio of specific heats, generally taken as 1.4. The speed of sound is a which for ideal fluids is given by, $a^2 = \gamma p/\rho$.

In the presentation below we will periodically be switching between the full nonlinear system given above and various reductions and simplifications. In some cases we will just consider the one-dimensional ($v = w = 0$) and two-dimensional forms ($w = 0$). We shall also employ a representative scalar form where we have

$$u_t + f(u) = 0$$

with either $f(u) = au_x + bu_y + cu_z$ the scalar wave equation, or $f(u)$ a general nonlinear function of u, which could be multi-dimensional, having similar definitions as needed.

If we examine the $1 - D$ wave equation

$$u_t + u_x = 0$$

and apply Fourier analysis $u(x, t) = w(t)e^{i\theta x}$, for spatial difference operators, such as central differencing we have

$$u_t + u_x = u_t + \delta_x u_j = u_t + \frac{u_{j+1} - u_{j-1}}{2\Delta x} \rightarrow w_t + \frac{i \sin \theta \Delta x}{\Delta x} w = w_t + \lambda_x^c w = 0$$

or for 1st order backward

$$u_t + u_x = u_t + \nabla_x u_j = u_t + \frac{u_j - u_{j-1}}{\Delta x} \rightarrow$$

$$w_t + \frac{1 - \cos \theta \Delta x + i \sin \theta \Delta x}{\Delta x} w = w_t + \lambda_x^b w = 0.$$

When we talk of the differencing signature we will be referring to the eigenvalue λ_x associated with the choice of spatial differencing. This is generally termed the semi-discrete eigenvalue of the partial differential equation. In the above, λ_x can be either pure imaginary or complex, which indicates that we should consider complex $f(u)$ in our analysis since spatial differences will produce complex eigensystems.

There are a number of issues to be examined when developing an implicit algorithm for (1.1). First, a spatial differencing scheme must be chosen. Within the framework of finite difference schemes we shall consider central and upwind differencing of the flux terms, $\partial_x E$ and $\partial_y F$. In Section 2, we shall motivate the choices, discuss various splittings of the flux vectors E and F which lead to flux vector and flux difference schemes and provide the framework for the implicit methods.

As the second component in developing an implicit scheme, we adopt a Newton like approach to the development, assessment and analysis of implicit operators. In the next Section we discuss the equivalence of an implicit operator to an approximate Newton scheme. Various approximations are analyzed using an eigensystem analysis technique introduced in (Jespersen 1983), which is a periodic Neumann-like approach where the system coupling is maintained leading to generalized eigensolutions.

2.   Newton's method

Newton's method can be used to understand the approximations employed in implicit schemes. If we start out looking at Newton's method for the fixed point problem

$$F(Q) = 0 \qquad (2.1)$$

we first do a Taylor series expansion with remainder of the function $F(Q)$ about some iteration level Q^n which leads to

$$F(Q) = F(Q^n) + \frac{\partial F^n}{\partial Q}(Q - Q^n) + \frac{\partial^2 F^*}{\partial Q^2}(Q - Q^n)^2 \qquad (2.2)$$

where Q^* lies in the solution interval between Q and Q^n. Note that by $\frac{\partial F^n}{\partial Q}$ we mean the matrix Jacobian of $F(Q)$ with respect to Q evaluated at Q^n.

Dropping the remainder term (leading to a first order approximation) and evaluating (2.2) at $n + 1$, assuming $F(Q^{n+1}) \approx 0$ gives the iterative scheme

$$Q^{n+1} = Q^n - \left[\frac{\partial F^n}{\partial Q}\right]^{-1} F(Q^n)$$

which is Newton method for a nonlinear system.

It is easy to show that Newton's method is quadratically convergent for any initial solution in the domain of attraction. Using the scalar form of (2.1), (2.2)

$$f(u) = 0: \quad u^{n+1} = u^n - \frac{f(u^n)}{f'(u^n)} = g(u^n) \qquad (2.3)$$

let

$$f(v) = 0: \quad v \quad \text{the root.}$$

The iterative scheme (2.3) is convergent if $|g'(u)| < 1$. We have

$$g'(u) = \frac{f''(u)f(u)}{[f'(u)]^2}.$$

Expanding $g(u^n)$ in a Taylor series as above, with ξ between u^n and v, we have

$$g(u^n) = g(v) + g'(v)(u^n - v) + \frac{g''(\xi)}{2}(u^n - v)^2. \qquad (2.4)$$

Letting the error be defined as $e^n = u^n - v$ and using $g(v) = v$, and $g'(v) = 0$ we have

$$e^{n+1} = u^{n+1} - v = g(u^n) - v = \frac{g''(\xi)}{2}(e^n)^2$$

which shows quadratic convergence.

One of the most commonly used implicit schemes is the 1st order Euler implicit method. Writing (1.1) for now as

$$Q_t + F(Q) = 0$$

we have

$$\frac{(Q^{n+1} - Q^n)}{h} = -F(Q^{n+1}) \tag{2.5}$$

with $h = \Delta t$.

The function $F(Q)$ is nonlinear in Q and therefore we linearize (2.5) about time level $n + 1$ which results in

$$\left[\frac{1}{h}I + \frac{\partial F^n}{\partial Q}\right]\Delta Q^n = -F(Q^n)$$

with $\Delta Q^n = (Q^{n+1} - Q^n)$, the so called "delta" form of the implicit scheme. Rearranging terms we get

$$Q^{n+1} = Q^n - \left[\frac{1}{h}I + \frac{\partial F^n}{\partial Q}\right]^{-1} F(Q^n)$$

which in the limit as $h \to \infty$ reduces to (2.2), Newton's method. For a finite h we have an approximate Newton scheme.

Analyzing the scalar form of the Euler implicit scheme as we did for full Newton, we have

$$u_t + f(u) = 0: \quad u^{n+1} = u^n - \frac{hf(u^n)}{1 + hf'(u^n)} = g(u^n)$$

again letting

$$f(v) = 0: \quad v \quad \text{the root.}$$

Continuing as before

$$g'(u) = 1 - \frac{hf'(u^n)}{1 + hf'(u^n)} + h^2 \frac{f''(u)f(u)}{[1 + hf'(u)]^2} \tag{2.6}$$

Using (2.4), $f(v) = 0, g(v) = v$ and (2.6) evaluated at v we have

$$g'(v) = \frac{1}{1 + hf'(v)} = \sigma(h)$$

which gives

$$e^{n+1} = \sigma(h)e^n + \frac{g''(\xi)}{2}(e^n)^2 .$$

We will consider $f'(u)$ to be either real or complex (it could be complex if $f(u)$ is derived from a differential operator). Note then that for small $h : \sigma(h) \approx 1$ and convergence is linear, but as $h \to \infty : \sigma(h) \to 0$ and we approach quadratic convergence.

As the above example demonstrates, approximations to Newton's method can lead to slower convergence, and in fact sometimes instability (although we shall restrict ourselves to examining schemes which are stable). In general, an implicit scheme has some equivalence to a modified Newton's method and we shall be analyzing schemes from that point of view.

As we have seen the Euler implicit scheme is a good approximation to Newton's method producing good convergence characteristics for large time steps. Other approximations usually inhibit the convergence and we shall be examining various forms below. There is also a class of problems where we are interested in time accurate computations. In those cases, though, we usually start with an initial solution which generates a transient which is hopefully eliminated at some time and then an accurate resolution of the time variations is obtained. One candidate would be the 2nd order accurate trapezoidal scheme, here written for the scalar equation $u_t + f(u) = 0$ as

$$\frac{u^{n+1} - u^n}{h} + \frac{f(u^{n+1}) + f(u^n)}{2} = 0$$

which after expanding $f(u^{n+1}) = f(u^n) + f'(u^n)(u^{n+1} - u^n) + O(h^2)$ leads to the 2nd order scheme

$$u^{n+1} = u^n - \frac{2hf(u^n)}{2 + hf'(u^n)} = g(u^n).$$

For this we have

$$g'(u) = 1 - \frac{2hf'(u)}{2 + hf'(u)} + \frac{2hf''(u)f(u)}{[2 + hf'(u)]^2}$$

where using (2.4), $f(v) = 0$ and $g(v) = v$ gives

$$e^{n+1} = -e^n + \frac{g''(\xi)}{2}(e^n)^2$$

which is nonconvergence, i.e. there is no reduction in the error with iteration. In general the trapezoidal scheme is unsuitable for steady-state and usually not desirable for time accurate problems because of this lack of error reduction.

3.   Factorizations, splitting and approximations

Applying the Euler implicit time differencing to the two-dimensional form of the Euler equations (1.1) we have

$$\frac{(Q^{n+1} - Q^n)}{h} + E_x^{n+1} + F_y^{n+1} = 0.$$

Linearizing the fluxes gives

$$
\begin{aligned}
E^{n+1} &= E^n + A^n \left(Q^{n+1} - Q^n\right) + O\left((Q^{n+1} - Q^n)^2\right), \\
F^{n+1} &= F^n + B^n \left(Q^{n+1} - Q^n\right) + O\left((Q^{n+1} - Q^n)^2\right)
\end{aligned}
$$

where the Jacobian matrices $A = \frac{\partial E}{\partial Q}^n$ or $B = \frac{\partial F}{\partial Q}^n$ are given by

$$
\begin{bmatrix}
0 & \kappa_x & \kappa_y & 0 \\
-u\theta + \kappa_x\phi^2 & \theta - (\gamma-2)\kappa_x u & \kappa_y u - (\gamma-1)\kappa_x v & (\gamma-1)\kappa_x \\
-v\theta + \kappa_y\phi^2 & \kappa_x v - (\gamma-1)\kappa_y u & \theta - (\gamma-2)\kappa_y v & (\gamma-1)\kappa_y \\
\theta[\phi^2 - a_1] & \kappa_x a_1 - (\gamma-1)u\theta & \kappa_y a_1 - (\gamma-1)v\theta & \gamma\theta
\end{bmatrix}
$$

with $a_1 = \gamma(e/\rho) - \phi^2$, $\theta = \kappa_x u + \kappa_y v$, $\phi^2 = \frac{1}{2}(\gamma-1)(u^2 + v^2)$, and $\kappa_x = 1$, $\kappa_y = 0$ for A or $\kappa_x = 0$, $\kappa_y = 1$ for B. These are 3×3 matrices in one-dimension (1-D), 4×4 in 2-D, and 5×5 in 3-D. Their exact form for the Euler equations can be found in numerous papers, e.g. (Pulliam 1985a).

Using the "delta" form leads to

$$[I + h\delta_x A^n + h\delta_y B^n]\left(Q^{n+1} - Q^n\right) = -h\left(\delta_x E^n + \delta_y F^n\right) \tag{3.1}$$

where δ represents a derivative operator which can be either analytic or numerical.

Assuming the use of 2nd order central differences for δ, the solution of (3.1) requires the inversion of a large sparse banded matrix.

3.1   *Matrix form of unfactored algorithm*

If central differences are used in (3.1) it is easy to show that the implicit algorithm produces a large banded system of algebraic equations. Let the mesh size in x be $Jmax$ and in y by $Kmax$. We choose an ordering of the data with the j index running first and the k index second, other orderings being permutations of the data. Then the banded matrix is a

$(Jmax \cdot Kmax \cdot 4) \times (Jmax \cdot Kmax \cdot 4)$ rectangular matrix of the form

$$
\begin{bmatrix}
\square & \square & & & & \square & & & & & \\
\square & \square & \square & & & & \square & & & & \\
& \square & \square & \square & & & & \square & & & \\
& & \square & \square & & & & & \square & & \\
\square & & & & \square & & \square & \square & & & \square \\
& \ddots & & & & \ddots & \ddots & \ddots & & & \ddots \\
& & -B & & & & -A & I & A & & & B \\
& & & \ddots & & & & \ddots & \ddots & \ddots & & & \ddots \\
& & & \square & & & & \square & \square & & & & \square \\
& & & & \square & & & & & \square & \square & \\
& & & & & \square & & & & \square & \square & \square \\
& & & & & & \square & & & & \square & \square & \square \\
& & & & & & & \square & & & & \square & \square \\
\end{bmatrix}.
$$

The matrix is sparse but it would be very expensive (computation-ally) to solve the algebraic system. For instance, for a reasonable two-dimensional calculation of transonic flow past an airfoil we could use approximately 80 points in the x direction and 40 points in the y direction. The resulting algebraic system is a 12,800 × 12,800 matrix problem to be solved and although we could take advantage of its banded sparse structure it would still be very costly in both CPU time and storage.

3.2   Approximate factorization

As we have seen, the integration of the full two-dimensional operator is too expensive. One way to simplify the solution process is to introduce an approximate factorization of the two-dimensional operator into two one-dimensional operators. The implicit side (left hand) of (3.1) can be written as

$$
[I + h\delta_x A^n + h\delta_y B^n]\Delta Q^n =
$$
$$
[I + h\delta_x A^n][I + h\delta_y B^n]\Delta Q^n - h^2 \delta_x A^n \delta_y B^n \Delta Q^n. \tag{3.2}
$$

The cross term (h^2 term) is second order in time since ΔQ^n is $O(h)$. It can therefore be neglected without degrading the time accuracy of any second order scheme which we may choose.

The resulting factored form of the algorithm is

$$
[I + h\delta_x A^n][I + h\delta_y B^n]\Delta Q^n = -h[\delta_x E^n + \delta_y F^n]. \tag{3.3}
$$

We now have two implicit operators each of which is block tridiagonal. The structure of the block tridiagonal matrix is

$$
\begin{bmatrix}
\square & \square & & & & & & \\
\square & \square & \square & & & & & \\
& \square & \square & & \square & & & \\
& & \ddots & \ddots & \ddots & & & \\
& & -A & I & A & & & \\
& & & \square & \square & \square & & \\
& & & & \square & \square & \square & \\
& & & & & \square & \square &
\end{bmatrix}.
$$

The solution algorithm now consists of two one-dimensional sweeps, one in the x and one in the y direction. The block matrix size is now at most $(\max[Jmax, Kmax] \cdot 4) \times (\max[Jmax, Kmax] \cdot 4)$. Each step requires the solution of a linear system involving a block tridiagonal which is solved by block LUD (lower-upper decomposition). The resulting solution process is much more economical than the unfactored algorithm in terms of computer storage and CPU time.

3.3 Newton scalar analysis of factored implicit scheme

The use of factorization produces an efficient and practical algorithm, but as we shall see the use of the factorization leads to a reduction in the convergence properties and sometimes stability of the resulting algorithm. First we shall use the scalar Newton analysis as above and then introduce a model system analysis which retains the structure of the Euler equations. A representative scalar model of the factored algorithm is

$$
u_t + f_1(u) + f_2(u) = u_t + r(u) = 0
$$

which leads to

$$
[1 + hf_1'(u^n)] [1 + hf_2'(u^n)] \left(u^{n+1} - u^n\right) = -hr(u^n) \tag{3.4}
$$

where we are seeking solutions to $r(v) = 0$. Note that the "delta form" guarantees that the steady solution $r(v) = 0$ is satisfied independent of h (Δt). We recast (3.4) as

$$
u^{n+1} = u^n - \frac{hr(u^n)}{[1 + hf_1'(u^n)][1 + hf_2'(u^n)]} = g(u^n)
$$

$$
g'(u) = 1 - \frac{hr'(u)}{[1 + hf_1'(u)][1 + hf_2'(u)]}
$$
$$
+ \frac{h^2 r(u) f_1''(u)}{[1 + hf_1'(u)]^2 [1 + hf_2'(u)]} + \frac{h^2 r(u) f_2''(u)}{[1 + hf_1'(u)][1 + hf_2'(u)]^2}.
$$

Using the definitions and (2.4) of Section 2 we have

$$e^{n+1} = \sigma(h)e^n + \frac{g''(\xi)}{2}(e^n)^2$$

with

$$\sigma(h) = \frac{1 + h^2 f_1'(v) f_2'(v)}{[1 + h f_1'(v)][1 + h f_2'(v)]}.$$

Now for real or complex $f_1'(v)$ and $f_2'(v)$ the coefficient $\sigma(h)$ of e^n is bounded by 1 for all h and as $h \to \infty$, $\sigma(h) \to 1$ showing that for large h we approach nonconvergence. For small h it can be shown that $\sigma(h) < 1$ and the scheme does converge and in fact there is an optimal h for maximum convergence.

We can next look at the equivalent of a three-dimensional factorization using

$$u_t + f_1(u) + f_2(u) + f_3(u) = u_t + r(u) = 0$$

Forming the iterative scheme,

$$u^{n+1} = u^n - \frac{hr(u^n)}{[1 + h f_1'(u^n)][1 + h f_2'(u^n)][1 + h f_3'(u^n)]} = g(u^n)$$

which gives

$$e^{n+1} = \sigma(h)e^n + \frac{g''(\xi)}{2}(e^n)^2$$

with

$$\sigma(h) = \frac{1 + h^2 f_1'(v) f_2'(v) + h^2 f_2'(v) f_3'(v) + h^2 f_1'(v) f_3'(v) + h^3 f_1'(v) f_2'(v) f_3'(v)}{[1 + h f_1'(v)][1 + h f_2'(v)][1 + h f_3'(v)]}.$$

Now again in this case we have $\sigma(h) \to 1$ as $h \to \infty$, but in the case of complex f_i we also have that for some $h > h_c$: $\sigma(h) > 1$ which implies that the three-dimensional factored algorithm is at best conditionally stable. In fact, consider the three-dimensional wave equation solved using second order central differences. In that case, the differencing signature ($\lambda's$) produces pure imaginary f_i and the scheme would then be unconditionally unstable. Any real part occurring in the f_i would be associated with dissipative differencing or the addition of a dissipative term, see discussion on artificial dissipation below.

3.4   Newton system analysis of factored implicit scheme

The scalar analysis presented above provides guidelines for algorithm choices and gives useful information about the characteristics of the resulting schemes. We can go one step further in understanding the application to systems, such as the Euler equations, by employing an analysis technique which maintains the system structure, see (Jespersen 1983). As an example of the usefulness of such an analysis, consider (3.1). The left hand side of that equation could be reduced if we could simultaneously diagonalize A and B, thereby reducing the system to 4 scalar operators instead of a 4×4 block operator. This can not be done since A and B do not commute (they don't have a common set of eigenvectors). For the unfactored algorithm this complicated the computational work, but in the case of the factored algorithm it also has an impact on the stability and accuracy since the order of factorization, see (3.2), can produce a cross term of either $h^2 \delta_x A^n \delta_y B^n (Q^{n+1} - Q^n)$ or $h^2 \delta_y B^n \delta_x A^n (Q^{n+1} - Q^n)$ which can be significantly different. A linear scalar analysis of a model equation could not detect the effect of such a term.

The Newton system analysis as described below can provide a very useful tool to help in the development and choice of numerical algorithms for the Euler and Navier–Stokes equations. Jespersen and Pulliam (Jespersen 1983) examined the effect of using approximate Jacobians in the implicit operator for flux split schemes, see Section 4, and showed limited stability bounds for approximate forms. Anderson (Anderson 1986) employed the analysis to examine various approximate factorizations for stability. Barth (Barth 1987; Barth 1988) examined various approximate Jacobians for Riemann solvers and MUSCL differencing schemes and in (Barth and Steger 1985) he employed the analysis to examine new splittings of the Euler fluxes to produce a class of efficient implicit solvers.

The system analysis as outlined in (Jespersen 1983) retains the system matrix form of the equations while applying Fourier analysis. Recast (3.3) as

$$M(Q^n)\left(Q^{n+1} - Q^n\right): \; = \; [I + h\delta_x A^n][I + h\delta_y B^n]\left(Q^{n+1} - Q^n\right)$$
$$= \; -h(\delta_x E^n + \delta_y F^n) := P(Q^n).$$

Rearranging terms

$$Q^{n+1} = Q^n + M(Q^n)^{-1}P(Q^n) = G(Q^n)$$

and we can examine the stability and convergence by looking at $G'(Q^n)$. At steady-state $P(Q^*) = 0$ and $G'(Q^*) = I + M(Q^*)^{-1}P'(Q^*)$ so that we have

$$\lambda(G)\,[M(Q^*) + P(Q^*)]\,\overline{v} \; = \; \lambda M(Q^*)\overline{v}$$
$$L\overline{v} \; = \; \lambda K\overline{v}$$

The analysis now includes the full system with $L = M + P$ and $K = M$. For the unfactored scheme $L = M + P = I$, for the factored scheme

$L = M + P = I +$ cross term error and in other cases $M + P$ may involve
approximate flux Jacobians, different spatial operators on the implicit M
and explicit P sides and other approximations to the implicit operator. In
the best case $L = M + P = I$ and as $h \to \infty$ we have full Newton.

Considering constant in space and time A, B, and C, let $\overline{v} = e^{i(j\theta+k\phi+l\alpha)}\overline{w}$,
for Fourier analysis and replace derivatives by their Fourier signature, e.g.
$\delta_x \overline{v} = i(\sin\theta\Delta x/\Delta x)\,\overline{w} = \lambda_x \overline{w}$.

Combining terms we have

$$\hat{L}\overline{w} = \lambda \widehat{K}\overline{w}$$

where $\hat{L}$ and $\widehat{K}$ are functions of θ, ϕ and α. A generalized eigenvalue
problem is then solved over a spectrum of wave numbers θ, ϕ, α. It should
be stressed that this analysis is by no means the most general possible.
One has to choose a finite spectrum of wave numbers, the assumption
of constant A, B, and C in space and time, and periodicity is another
limiting factor. The analysis, though, does provide a next step past linear
scalar analysis since the effect of the coupled system enters in through the
eigensystem (eigenvalues and eigenvectors) of A, B, and C.

3.5   Artificial dissipation

In solving practical problems with central difference schemes artificial dissipation (Pulliam 1985b) must be added to ensure stability and enhance
robustness. There are many forms of artificial dissipation and these is an
equivalence between upwind schemes and forms of artificial dissipation, see
(Pulliam 1985b). We consider artificial dissipation here in reference to the
way it modifies the λ's associated with the spatial differencing. Artificial
dissipation applied to both the explicit and implicit operators can have the
form

$$[I + h\delta_x A^n - hD_x^i][I + h\delta_y B^n - hD_y^i]\left(Q^{n+1} - Q^n\right) = $$
$$-h[\delta_x E^n + \delta_y F^n] - h[D_x^e + D_y^e]$$

where for our purposes here we choose

$$D_x^e = -\epsilon_e(\nabla_x \Delta_x)^2, \quad D_x^i = -\epsilon_i \nabla_x \Delta_x$$

with, for example,

$$\nabla_x\, u_{j,k} = \frac{u_{j,k} - u_{j-1,k}}{\Delta x} \quad \text{and} \quad \Delta_x\, u_{j,k} = \frac{u_{j+1,k} - u_{j,k}}{\Delta x}. \tag{3.5}$$

The effect of the artificial dissipation is to add a real part to the Fourier
signature of the differencing eigenvalues. The coefficients ϵ_e and ϵ_i are usually chosen to be order 1. An example of Newton system analysis for the

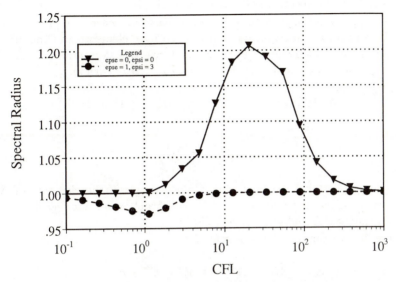

Fig. 1: Newton system analysis for 3D factored scheme

3D factored 1st order Euler implicit schemes is given in Fig. 1. A typical result without artificial dissipation and with artificial dissipation is shown. The addition of the dissipation produces a stability region, while the case of no dissipation is unconditionally unstable, as expected. The linear analysis result of instability for the three-dimensional factored algorithm is well known and has often been pointed out as a weakness of the factored schemes. In practice, though, for nonlinear problems in the presence of some form of dissipation (whether it is added artificial or inherent numerical, such as in upwind schemes, or physical resulting from resolved viscous terms) the algorithm has never behaved any more restrictedly than its two-dimensional linearly unconditionally stable counterpart. As we shall see below, there are alternative schemes which avoid the unconditional stability, but in practical problems they behave no differently than the three-dimensional factored scheme.

4. Flux-vector splitting

The concept of splitting operators or fluxes based on the eigensystem of the Euler equations leads naturally to the flux vector or flux difference split schemes which are currently the most popular in terms of research and application to shock capturing methods. One of the earliest schemes is due to Steger and Warming (Steger 1981) where they constructed new fluxes F^+, F^- which were upwind differenced in a stable fashion based on the sign of the eigenvalues of the Jacobians $\tilde{A}^+, \tilde{A}^-$.

The approach taken is to split the eigenvalue matrix Λ of the flux Jacobians into two matrices, one with all positive elements and the other with all negative elements. Then the similarity transformations T_x or T_y are used to form new matrices A^+, A^- and B^+, B^-. Formally,

$$A = T_x \Lambda_x T_x^{-1} = T_x(\Lambda_x^+ + \Lambda_x^-)T_x^{-1} = A^+ + A^-$$

with

$$\Lambda_x^{\pm} = \frac{\Lambda_x \pm |\Lambda_x|}{2}.$$

Here, $|\Lambda|$ implies that we take the absolute values of the elements of Λ. The two matrices, A^+ and A^- have by construction all nonnegative and all nonpositive eigenvalues, respectively.

New flux vectors can be constructed as

$$\begin{aligned} E &= AQ = (A^+ + A^-)Q = E^+ + E^- \\ F &= BQ = (B^+ + B^-)Q = F^+ + F^-. \end{aligned}$$

A general form of the flux vector can be written as $\hat{F} =$

$$\frac{\rho}{2\gamma} \begin{bmatrix} 2(\gamma-1)\lambda_1 + \lambda_3 + \lambda_4 \\ 2(\gamma-1)\lambda_1 u + \lambda_3(u+c) + \lambda_4(u-c) \\ 2(\gamma-1)\lambda_1 v + \lambda_3 v + \lambda_4 v \\ (\gamma-1)\lambda_1(u^2+v^2) + (\lambda_3/2)[(u+c)^2 + v^2] + (\lambda_4/2)[(u-c)^2 + v^2] + w \end{bmatrix}$$

where $w = \frac{(3-\gamma)(\lambda_3+\lambda_4)c^2}{2(\gamma-1)}$ with $\lambda_1 = u, \lambda_3 = u+c$, and $\lambda_4 = u-c$ recovering the flux vector E of (1.3).

The flux vectors F^+, F^- are formed by inserting $\lambda_i = \lambda_i^+$ and $\lambda_i = \lambda_i^-$, respectively, where for example $\lambda_i^{\pm} = (\lambda_i \pm |\lambda_i|)/2$.

The Steger-Warming flux splitting suffers from discontinuous derivatives of the fluxes at zeros of the eigenvalues (i.e. stagnation or sonic points) and one can smooth out the discontinuities by the modification $\lambda_i^{\pm} = (\lambda_i \pm (\lambda_i^2 + \epsilon^2)^{\frac{1}{2}})/2$ for small ϵ. An alternate flux splitting is proposed by Van Leer (van Leer 1983) where $F^{\pm}$ are given in terms of a local one-dimensional Mach number $M = \frac{u}{c}$ where

$$\begin{aligned} F^+ &= F, \quad F^- = 0 \quad \text{for } M \leq 1 \\ F^+ &= 0, \quad F^- = F \quad \text{for } M \leq -1 \end{aligned}$$

and for $|M| < 1$

$$
F^\pm = \begin{bmatrix} f^\pm \\ f^\pm[(\gamma - 1)u \pm 2c]/\gamma \\ f^\pm v \\ f^\pm[((\gamma - 1)u \pm 2c)^2/2(\gamma^2 - 1) + v^2/2] \end{bmatrix}
$$

with $f^\pm = \pm\rho c\,[(M \pm 1)/2]^2$. This flux is continuously differentiable at sonic and stagnation points.

The splitting of the fluxes into terms with Jacobians with either all positive or all negative characteristic speeds allows us to choose upwind differences for each operator. For the positive terms a backward difference can be used and for the negative terms a forward difference is applied. Different type of spatial differencing can now be used for each of the new flux vectors. The one-sided difference operators are usually either first order accurate, (3.5), or second order accurate

$$
\delta_x^b\, u_{j,k} = \frac{\frac{3}{2}\,u_{j,k} - 2\,u_{j-1,k} + \frac{1}{2}\,u_{j-2,k}}{\Delta x}, \quad \delta_x^f u_{j,k} = \frac{-\frac{3}{2}\,u_{j,k} + 2\,u_{j+1,k} - \frac{1}{2}\,u_{j+2,k}}{\Delta x}.
$$

An unfactored 1st order Euler implicit algorithm can be written as

$$
\left[I + h\delta_x^b\tilde{A}^+ + \delta_x^f\tilde{A}^- + \delta_y^b\tilde{B}^+ + \delta_y^f\tilde{B}^- + \delta_z^b\tilde{C}^+ + \delta_z^f\tilde{C}^-\right]\left(Q^{n+1} - Q^n\right) =
$$
$$
-h\left(\delta_x^b E^+ + \delta_x^f E^- + \delta_y^b F^+ + \delta_y^f F^- + \delta_z^b G^+ + \delta_z^f G^-\right) = R^n
$$

where, e.g., $\tilde{A}^\pm$ is the flux Jacobian of $E^\pm$. Note that $A^\pm \neq \tilde{A}^\pm$ and in fact if $A^\pm$ were used, instability could could result, see (Steger 1981) and (Jespersen 1983). The full unfactored algorithm can be reduced by factoring as we did before. One possibility is a three-factor scheme, $R^n =$

$$
\left[I + h\delta_x^b\tilde{A}^+ + \delta_x^f\tilde{A}^-\right]\left[I + \delta_y^b\tilde{B}^+ + \delta_y^f\tilde{B}^-\right]\left[I + \delta_z^b\tilde{C}^+ + \delta_z^f\tilde{C}^-\right]\left(Q^{n+1} - Q^n\right)
$$

which requires three block tridiagonal inversions, and is equivalent to central differencing and added artificial dissipation, (Pulliam 1985b). An alternative is a two factor scheme where all the positive terms and all the negative terms are lumped together,

$$
\left[I + h\delta_x^b\tilde{A}^+ + \delta_y^b\tilde{B}^+ + \delta_z^b\tilde{C}^+\right]\left[I + \delta_x^f\tilde{A}^- + \delta_y^f\tilde{B}^- + \delta_z^f\tilde{C}^-\right]\left(Q^{n+1} - Q^n\right) = R^n.
$$

The two factor scheme produces a purely upper and lower triangular matrix system and can be solved as lower/upper sweeps. The disadvantage of the two factor scheme is that it is harder to vectorize, the recursive nature of the sweeps making the identification of a vectorizable direction

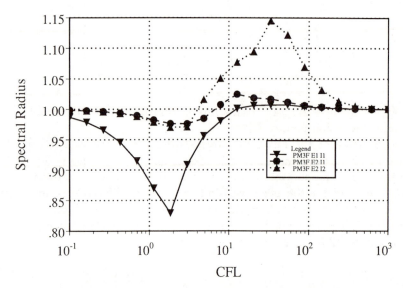

Fig. 2: Newton system analysis for 3 factor ± scheme

difficult. In contrast, for the three factor scheme, each operator is one-dimensional and can be vectorized over one of the other directions (see (Anderson 1986) for more discussion).

Newton system analysis of these schemes is presented in Fig. 2 and Fig. 3. In Figure 2, the three-factor scheme shows conditional stability at low CFL. In these schemes, one can choose either purely first order differences, which results in an implicit block tridiagonal system, or second order differences. The latter requires an implicit block pentadiagonal solver or one can reduce the computational work of the implicit operator but maintain second order steady-state accuracy by using first order differences in the implicit operator and second order for the explicit operator. This mixing of the order of differencing also affects the stability characteristics as shown in Fig. 2. (The label E1 I1 refers to 1st order explicit and 1st order implicit differences, etc.) Figure 3 shows results for the two factor scheme, where all forms are unconditionally stable with the purely 1st order scheme showing the best characteristics.

5.   F3D ± flux split scheme

The three-factor implicit central difference scheme suffers from a bad reputation resulting from the linear instability as shown above. In general, for practical problems this doesn't seem to be a real restriction. Nevertheless, one would like to employ schemes which are at least stable in the

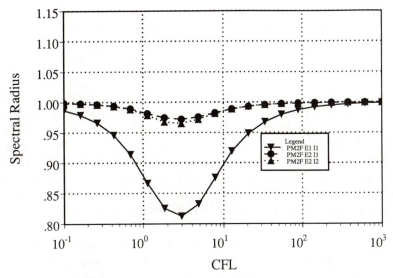

Fig. 3: Newton system analysis for 2 factor $\pm$ scheme

linear sense. In that regard, Ying (Ying 1986), developed the Factored Three-Dimensional algorithm (F3D) which employs flux splitting in one coordinate direction and central differences in the other 2 directions. This produces a two factor implicit scheme which has central difference characteristics in two coordinate directions (usually the near normal and some other cross flow direction) and an upwind nature in one direction (usually chosen in the major flow direction). The advantages of this scheme is the two factor operator which can be shown to be unconditionally stable in the linear constant coefficient case and the upwind nature of one of the operators which is usually chosen in the direction perpendicular to a shock or flow disconinuity. Consider, for example, a blunt cone at angle of attack. The F3D scheme would use flux splitting in the axial direction, and central differences in the circumferential and body normal directions. Note that in inviscid supersonic axial flow, the scheme could reduce to pure supersonic marching, which can be very efficient.

Ying (Ying 1986) choose the x direction to flux split and the resulting equations can be written as

$$Q_t + \delta_x^b E^+ + \delta_x^f E^- + \delta_y F + \delta_z G$$

where first or second order differences can be employed for δ_x^b and δ_x^f and second order central differences for δ_y and δ_z. The "delta" form of Euler implicit time differencing is given as

$$\left[I + h\delta_x^b \tilde{A}^+ + \delta_y B\right] \left[I + \delta_x^f \tilde{A}^- + \delta_z C\right] \left(Q^{n+1} - Q^n\right) =$$
$$-h\left(\delta_x^b E^+ + \delta_x^f E^- + \delta_y F + \delta_z G\right) = 0.$$

where the $\delta_x^b \tilde{A}^+$ implicit operator is placed with the y operator and the $\delta_x^f \tilde{A}^-$ operator is placed with the z operator. This produces a two-factor scheme which is lower block diagonal in x coupled with block tridiagonal in y for the first operator, which can be solved by sweeping in x within an LU decomposition in y. The z operator is handled similarly except that it is upper block diagonal in x and block tridiagonal in z. This produces an efficient algorithm which does not suffer directly from the three-factor linear instability. The first operator can be vectorized in z while the second is vectorized in y.

Figure 4 shows Newton system analysis for the F3D scheme. Results are shown for 1st order upwind in x and all 2nd order differencing. To enhance the efficiency, i.e. reduce the band width of the implicit x component of the algorithm, 1st order upwind differences can be used on the implicit side and 2nd order on the implicit side, resulting in a 2nd order steady-state. Unfortunately, this produces limited stability, although the stability range may still be in the useful region. The above results are shown for no added artificial dissipation in the central y and z directions. Adding artificial dissipation improves the stability in much the same way as for the fully central three-factor algorithm.

6. Summary

A class of implicit approximate factorization schemes has been examined for stability and convergence characteristics. In general, all the schemes suffer from some limited stability or asymptotic convergence restriction. Practical schemes will almost always fall within this class. The unconditional instability of the three-dimensional factored scheme is one end of the spectrum, where conditional stability can be achieved with added artificial dissipation. The F3D scheme avoids the unconditional instability, but in the end has similar convergence characteristics. Full Newton schemes are currently being pursued (see (Beam 1988) and (Venkatakrishnan 1988)) for the Euler and Navier–Stokes equations with some success. These efforts are more restricted by the computer resources than by any numerical analysis considerations, such as stability or consistency. At present, though, schemes such as F3D and the 3D factored method are the most useful and practical.

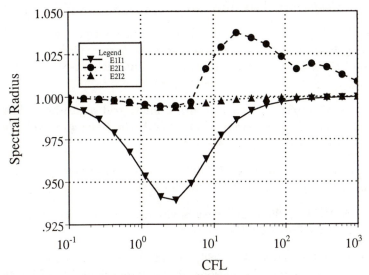

Fig. 4: Newton system analysis for F3D scheme

References

Anderson, W. K. (August 1986). *Implicit Multigrid Algorithms for the Three-Dimensional Flux Split Euler Equations*, Ph.D. Thesis, Mississippi State University.

Barth, T. J. (1987). Analysis of implicit local linearization techniques for upwind and TVD algorithms, AIAA Paper 87-0595, *AIAA 25th Aerospace Sciences Meeting, Reno, NV.*

Barth, T. J. (1988). Implicit linearization procedures for upwind algorithms. *Proceedings of the International Conference on Computational Engineering Science, Atlanta, GA*, Vol. 2.

Barth, T. J. and Steger, J. L. (1985). A fast efficient implicit scheme for the gasdynamics equations using a matrix reduction technique. AIAA Paper 85-0439, *AIAA 23rd Aerospace Sciences Meeting, Reno, Nevada.*

Beam, R. M. and Bailey, H. E. (1988). Newton's method for the Navier–Stokes equations. *Proceedings of the International Conference on Computational Engineering Science, Atlanta, GA*, Vol. 2.

Jameson, A., Schmidt, W. and Turkel, E. (1981). Numerical solutions of the Euler equations by finite volume methods using Runge–Kutta time-stepping schemes. AIAA Paper 81-1259, *AIAA 14th Fluid and Plasma Dynamics Conference, Palo Alto.*

Jespersen, D. C. (1984). Recent developments in multigrid methods for the steady Euler equations. *Lecture Notes for Lecture Series on Computational Fluid Dynamics*, von Karman Institute, Rhode-Saint-Genèse, Belgium.

Jespersen, D. C. (1985). A time-accurate multi-grid algorithm. AIAA Paper 85-1493-CP, *Proceedings of the AIAA 7th Computational Fluid Dynamics Conference, Cincinnati, OH.*

Jespersen, D. C. and Pulliam, T. H. (1983). Approximate Newton methods and flux vector splitting. AIAA Paper 83-1899.

MacCormack, R. W., (1969). The effect of viscosity in hypervelocity impact cratering. AIAA Paper 69-354.

Martinelli, L. and Jameson, A. (1988). Validation of a multigrid method for the Reynolds averaged equations. AIAA Paper 88-0414, *AIAA 26th Aerospace Sciences Meeting, Reno, NV.*

Pulliam, T. H. (1985). Efficient solution methods for the Navier–Stokes equations. *Lecture Notes for the von Karman Institute For Fluid Dynamics Lecture Series : Numerical Techniques for Viscous Flow Computation In Turbomachinery Bladings*, von Karman Institute, Rhode-Saint-Genèse, Belgium.

Pulliam, T. H. (1985). Artificial dissipation models for the Euler equations. *AIAA Journal* **24** No. 12.

Steger, J. L. and Warming, R. F. (1981). Flux vector splitting of the inviscid gas dynamic equations with applications to finite difference methods. *Journal of Computational Physics* 40:263–293.

van Leer, B. (1983). Flux-vector splitting for the Euler equations. *Eighth International Conference on Numerical Methods in Fluid Dynamics*, Springer Lecture Notes in Physics no. 170, E. Krause, editor.

Venkatakrishnan, V. (1988). Newton Solution of Inviscid and Viscous Problems. AIAA Paper 88-0413, *AIAA 26th Aerospace Sciences Meeting, Reno, NV.*

Ying, S. X. (1986). *Three-Dimensional Approximately Factored Schemes for Equations in Gasdynamics*, Ph.D. Thesis, Stanford University.

The cell vertex method for steady compressible flow

K. W. Morton, P. N. Childs, and M. A. Rudgyard

ICFD, Oxford University Computing Laboratory, Oxford

1. Introduction

Following the pioneering work of Denton (1981), Jameson (1981) and Ni (1981), some form of finite volume method has become by far the most widely used approach to the solution of the compressible flow equations. On a quadrilateral mesh in two dimensions there are two natural ways of arranging the discretisation: *the cell centre approach,* in which the unknowns are associated with the centre of the quadrilateral over which the integral of the conservations laws is carried out; and the *cell vertex approach,* in which the unknowns are associated with the vertices of this quadrilateral. In Morton and Paisley (1986,1988) arguments are given for the cell vertex approach to be preferred for inviscid computations. These stem from the greater compactness of the grid stencil defining the cell residual, which in steady flows one endeavours to set to zero, and include the capability of capturing properly aligned shocks over one mesh interval without artificial damping: its disadvantages largely arise from the greater difficulty one has in devising effective iterative procedures to accomplish the task of driving the residuals to zero.

These residuals should not be set to zero for those cells which are crossed obliquely by shocks. Morton and Paisley therefore introduced techniques for detecting the presence of a shock and adjusting the mesh in its neighbourhood so that the shock could be fitted and therefore treated as an internal boundary. In combination with the cell vertex method this gives very accurate results on quite coarse meshes. However, its applicability is limited because of the need for local mesh adaption.

In the present paper we have two objectives. Firstly we shall give preliminary results from an analysis of the key features of the cell vertex method and the associated Lax–Wendroff or Taylor–Galerkin iteration procedures which have so far been used to solve the equations. Secondly, we shall describe shock recovery and fitting procedures which do not need any adaption of the underlying mesh, and give some numerical results obtained from their use. The analysis leads one to procedures which are just as effective as the earlier ones, while having the capacity for wider application:

it also indicates the need to move away from the Lax-Wendroff procedure if the full potential of the scheme is to be realised.

The preference for quadrilateral meshes perhaps needs some comment in view of the present interest in unstructured triangular meshes. On an infinite mesh of the latter type, there are twice as many triangles as vertices so that the cell vertex method is not feasible on such a mesh, in the sense that we cannot set to zero the residuals over individual triangles. What one has to do is calculate the residuals over the Voronoi cells around each vertex, which gives a scheme much closer to the cell centre approach. Boundary conditions for outgoing characteristics will lead to some uncoupling of the residuals, but to nothing like the extent found with quadrilaterals (see Morton and Paisley, loc. cit.). There is no problem however in admitting a few triangles as degenerate quadrilaterals at sharp corners of the boundary of a generally quadrilateral mesh.

2.   The cell vertex approach with the Taylor-Galerkin update

The system of conservation laws for the vector of unknowns $\mathbf{w}$ in two cartesian dimensions,

$$\frac{\partial \mathbf{f}(\mathbf{w})}{\partial x} + \frac{\partial \mathbf{g}(\mathbf{w})}{\partial y} = \mathbf{0}, \tag{2.1}$$

results from the integral relation for the fluxes through the boundary

$$\int_{\Omega} div(\mathbf{f}, \mathbf{g}) d\Omega = \int_{\partial\Omega} \mathbf{f} dy - \mathbf{g} dx = \mathbf{0}, \tag{2.2}$$

applied to an arbitary region Ω with boundary $\partial\Omega$. Applied to a quadrilateral with vertices (x_i, y_i), $i = 1 \to 4$, and using the trapezoidal rule to approximate the line integrals, this gives the delightfully simple and compact approximation to (2.1) and (2.2):

$$(\mathbf{f}_1 - \mathbf{f}_3)(y_2 - y_4) + (\mathbf{f}_2 - \mathbf{f}_4)(y_3 - y_1) -$$
$$(\mathbf{g}_1 - \mathbf{g}_3)(x_2 - x_4) - (\mathbf{g}_2 - \mathbf{g}_4)(x_3 - x_1) = \mathbf{0}. \tag{2.3}$$

Here the unknowns are associated with the vertices so that $\mathbf{f}_i := \mathbf{f}(\mathbf{W}_i)$ and $\mathbf{g}_i := \mathbf{g}(\mathbf{W}_i)$: and we wish to solve the system of nonlinear equations resulting from (2.3) applied to all the quadrilateral cells of some region Ω^h, together with appropriate boundary conditions, for the complete set of unknowns $\{\mathbf{W}_i\}$, where i runs over all the vertices of Ω^h.

Iterative methods based on time-stepping the unsteady form of (2.1) are normally used to solve these equations, with various modifications which are valid only for steady flows used to accelerate the convergence. We will use the second order Taylor–Galerkin scheme described in Morton

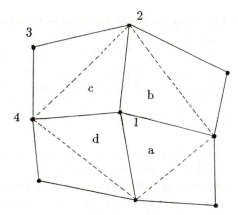

Fig. 1: Layout of quadrilaterals around an updated vertex

(1987). It results, like the familiar Lax–Wendroff scheme, from expanding $\mathbf{w}(t + \triangle t)$ in a Taylor series and using the differential equation $\mathbf{w}_t + \mathbf{f}_x + \mathbf{g}_y = 0$ to replace time derivatives by space derivatives: then, assuming the approximation $\mathbf{W}$ has a bilinear form on each quadrilateral, taking the Galerkin projection of the expansion, using one point quadrature to approximate the inner products and applying mass-lumping because we wish to use the maximal timesteps, we obtain the final update procedure. Referring to Fig. 1 this can be written in the form

$$\mathbf{W}_1^{n+1} = \mathbf{W}_1^n - \triangle t[D_a\mathbf{R}_a + D_b\mathbf{R}_b + D_c\mathbf{R}_c + D_d\mathbf{R}_d], \qquad (2.4)$$

where the residuals are defined by setting $\mathbf{R}_c$ to the left hand side of (2.3) divided by twice the cell measure V_c, and the matrices giving their relative 'weights' are given by

$$D_c := \frac{1}{V_a + V_b + V_c + V_d}\{V_cI - \triangle t[(y_4 - y_2)A_c - (x_4 - x_2)B_c]\} \qquad (2.5)$$

in terms of the Jacobian matrices $A := \partial\mathbf{f}/\partial\mathbf{w}$, $B := \partial\mathbf{g}/\partial\mathbf{w}$. Note that the second order update can be regarded as resulting from integrating $div(A\mathbf{R}, B\mathbf{R})$ over the region bounded by the diagonals of the four cells (shown by the dashed lines in Fig. 1): the boundary integral is then approximated using the mid-point rule with A and B evaluated from cell averages of $\mathbf{W}$.

In this update procedure for interior nodes, $\triangle t$ occurs in two places and we have not indicated any dependence that it may have on local conditions, either relating to the vertices or the cells. In practice there are many possible choices, including distinguishing between the occurrence of $\triangle t$ in

(2.4) and that in (2.5). But we defer discussion of this issue until later
when we consider attempts to optimise the convergence rate. However, we
do note here that one can program the update algorithm either vertex-by-
vertex as in (2.4) or cell-by-cell. In the latter case we would wish to absorb
the sum of cell measures occurring in the denominator of (2.5) into the
choices of Δt: then the main step of the algorithm consists of distributing
multiples of the cell residual $\mathbf{R}_c$ to update the values at the cell's four
vertices.

At a boundary vertex, the update procedure is first applied using only
the residuals corresponding to cells contained in the region Ω^h: thus at
a non-reentrant corner, for example, only one cell residual is used. Then
boundary conditions are used to modify or overwrite these updated values.
The number and choice of boundary conditions will be dealt with in numerical examples below: but generally they are determined by the ingoing
characteristics of the Jacobian corresponding to the normal flux function
at the boundary point.

In the first stage of a computation, the above procedure is applied in
a shock-capturing mode to identify and locate any shocks. A dissipation
operator needs to be added to (2.4) to compensate for the fact that the
residuals for shocked cells should not be driven to zero, but its form is
not particularly critical. The important point is that the shock profiles
at this stage should be such as to be recognisable by the shock detection
algorithm: thus they should not have spurious oscillations but also not be
too smeared. At the next stage the detected shocks are taken account of
in the update procedure and the dissipative mechanisms removed, at least
largely so. Details of these algorithms will be given below. One of the
main objectives is to use as little detailed information about the shock as
possible in order to devise a robust algorithm of wide applicability.

3. Analysis of the update algorithm

Many choices have to be made in the design and application of the algorithm and some analysis of simple model situations gives valuable guidance
in these choices. Such analyses are the subject of this section, since a thorough analysis of the complete algorithm has not yet proved feasible.

First of all, though, we should consider the accuracy that can be expected when the update algorithm has done its job and all the residuals
are set to zero. Morton and Paisley (1988) have shown that the truncation
error for the cell vertex scheme is $O(h^2)$ so long as each cell is within $O(h)$
of being a parallelogram: indeed, this robustness against grid distortion is
one of the key advantages over the cell centre approach. However, there
is considerable experimental evidence of a supraconvergence phenomenon:
that is, the global error in the solution may continue to be $O(h^2)$ even

when the grid is so non-uniform that the truncation error is of lower order. Giles (1988) has given theoretical arguments for the occurrence of this phenomenon, which has been well documented in one dimension, but the analysis for the undamped cell vertex approximation is as yet incomplete.

3.1 Fourier analysis of the Lax–Wendroff and Taylor–Galerkin updates

Consider the linear hyperbolic system

$$\mathbf{w}_t + A\mathbf{w}_x + B\mathbf{w}_y = 0, \tag{3.1}$$

where A and B are constant $p \times p$ matrices. On a uniform rectangular mesh a cell residual has the difference operator form

$$\mathbf{R}_c = \left(\frac{1}{\Delta x}\right) A\mu_y \delta_x \mathbf{W} + \left(\frac{1}{\Delta y}\right) B\mu_x \delta_y \mathbf{W}, \tag{3.2}$$

where $\delta_x W_j := W_{j+\frac{1}{2}} - W_{j-\frac{1}{2}}$, $\mu_x W_j := \frac{1}{2}(W_{j+\frac{1}{2}} + W_{j-\frac{1}{2}})$ and δ_y, μ_y have similar definitions in the y-direction. Both the Lax–Wendroff update and the Taylor–Galerkin update (2.4), (2.5) then take the form at mesh point (j, k)

$$\mathbf{W}_{j,k}^{n+1} = \mathbf{W}_{j,k}^n - \Delta t \mu_x \mu_y \mathbf{R}_{j,k}^n +$$
$$\frac{1}{2}(\Delta t)^2 \left[\left(\frac{1}{\Delta x}\right) A\mu_y \delta_x + \left(\frac{1}{\Delta y}\right) B\mu_x \delta_y\right] \mathbf{R}_{j,k}^n. \tag{3.3}$$

Setting $\mathbf{W}_{j,k}^n = \lambda^n e^{i(j\xi + k\eta)}\hat{\mathbf{W}}$ to carry out a Fourier analysis, where $\xi = k_x \Delta x$, $\eta = k_y \Delta y$ and k_x, k_y are the wave numbers in the x- and y- directions respectively, as in Morton and Paisley (1988), we find that the damping factors λ for the wave modes (ξ, η) are the eigenvalues of

$$I - 2i \cos\left(\tfrac{1}{2}\xi\right) \cos\left(\tfrac{1}{2}\eta\right) \hat{M} - 2\hat{M}^2 \tag{3.4}$$

where the matrix $\hat{M}$ is given by

$$\hat{M} = \left(\frac{\Delta t}{\Delta x}\right) \sin\left(\tfrac{1}{2}\xi\right) \cos\left(\tfrac{1}{2}\eta\right) A + \left(\frac{\Delta t}{\Delta y}\right) \cos\left(\tfrac{1}{2}\xi\right) \sin\left(\tfrac{1}{2}\eta\right) B. \tag{3.5}$$

Suppose μ is an eigenvalue of $\hat{M}$, being real because of the hyperbolicity. Then from (3.4) we have

$$|\lambda|^2 = 1 - 4\mu^2 \left[1 - \mu^2 - \cos^2\left(\tfrac{1}{2}\xi\right)\cos^2\left(\tfrac{1}{2}\eta\right)\right]. \qquad (3.6)$$

For convergence we need $|\lambda| < 1$ for all modes: and Δt should be chosen to minimise $\max\{|\lambda|,$ over all modes$\}$.

In the simplest scalar one-dimensional case, $|\lambda|^2$ reduces to $1 - 4\nu^2(1-\nu^2)\sin^4(\tfrac{1}{2}\xi)$ where $\nu = a\Delta t/\Delta x$ is the Courant number: hence $\nu < 1$ is needed for convergence and $\nu = 1/\sqrt{2}$ gives the optimal convergence rate for all modes. The situation is very different however for a typical system of equations and in two dimensions. For the linearised Euler equations in one dimension, the 3×3 matrix A has eigenvalues $u, u \pm a$ corresponding to the characteristic speeds, where a is the sound speed. Then convergence clearly requires that $(|u| + a)(\Delta t/\Delta x) < 1$: but if Δt is reduced to improve the damping of these modes, this may well worsen the damping of the modes corresponding to the lower speeds. The optimal value of Δt corresponds to equating the damping of the fastest mode with that of the slowest, which will be $|u| - a$ unless the Mach number $M = |u|/a$ satisfies $M < |1 - M|$, ie., $M < \tfrac{1}{2}$. In an obvious notation, the optimal mesh ratio is given by

$$a_f^2\left(\frac{\Delta t}{\Delta x}\right)^2\left[1 - a_f^2\left(\frac{\Delta t}{\Delta x}\right)^2\right] = a_s^2\left(\frac{\Delta t}{\Delta x}\right)^2\left[1 - a_s^2\left(\frac{\Delta t}{\Delta x}\right)^2\right],$$

i.e.,

$$\left(\frac{\Delta t}{\Delta x}\right)^2 = \frac{1}{a_f^2 + a_s^2}, \qquad \left(a_f\frac{\Delta t}{\Delta x}\right)^2 = \frac{a_f^2}{a_f^2 + a_s^2}. \qquad (3.7)$$

Hence for the Euler equations we have

$$\left[(|u| + a)\frac{\Delta t}{\Delta x}\right]^2_{opt} = \begin{cases} 1 - \frac{M^2}{M^2 + (1+M)^2} & \text{if } M \le \tfrac{1}{2} \\ \tfrac{1}{2} + \frac{M}{1+M^2} & \text{if } M \ge \tfrac{1}{2}. \end{cases} \qquad (3.8)$$

The minimum value of 0.9 for the right hand side of (3.8) occurs at $M = \tfrac{1}{2}$. It implies that the CFL number always should be taken to be at least 95 % of its maximum allowed value: such a choice will give a value of the damping coefficient of $4\nu^2(1-\nu^2)$ of 0.352, rather than the optimal value 1 attained with $\nu^2 = \tfrac{1}{2}$.

Further reductions in the efficacy of the Lax-Wendroff update occur in two dimensions. The scalar case was considered by Morton and Paisley (1988). They introduced the angle θ between the flow vector $(\nu_x, \nu_y) = (a\Delta t/\Delta x, b\Delta t\Delta y)$ and the wave vector $(\tan(\tfrac{1}{2}\xi), \tan(\tfrac{1}{2}\eta))$ in terms of which one can write

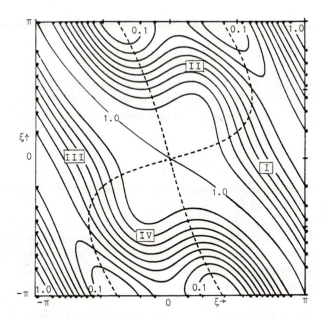

Fig. 2: Amplification factor $|\lambda(\xi,\eta)|$ for $\nu_x = 1/2$, $\nu_y = \sqrt{3}/2$: regions (I) and (III) contain those modes for which $\Delta t = \Delta t_{max}$ is optimal

$$\mu^2 = (\nu_x^2 + \nu_y^2)\left(1 - \cos^2\left(\tfrac{1}{2}\xi\right)\cos^2\left(\tfrac{1}{2}\eta\right) - \sin^2\left(\tfrac{1}{2}\xi\right)\sin^2\left(\tfrac{1}{2}\eta\right)\right)\cos^2\theta. \quad (3.9)$$

It is clear from (3.9) that no damping occurs of any of the modes for which $\theta = \frac{\pi}{2}$, which includes at one extreme the chequerboard mode for which $\xi = \eta = \pi$. Moreover, Fig. 2 shows that there is a broad region of (ξ,η) space in which there is little damping. Making the point again that there is little to be gained from reducing the CFL number below the maximum required for convergence or stability, Fig. 2 also shows the region of (ξ,η) space in which there is no advantage to such a reduction.

One can of course combine the two analyses above and deal with systems of equations in two dimensions. But the main lessons from Fourier analysis have been learned – that the CFL number must be taken near its limit and that many modes in two dimensions are little damped – and one needs to consider other effects such as non-uniformity of the mesh. For this we need an energy analysis.

3.2 *Energy analysis for non-uniform meshes*

We shall confine our attention here to a scalar equation in one dimension on the whole real line. In the simplest case, of a uniform mesh and a linear, constant speed advection equation, the Lax-Wendroff scheme reduces to

$$W_j^{n+1} = W_j^n - \nu \Delta_0 W_j^n + \tfrac{1}{2}\nu^2 \delta^2 W_j^n \tag{3.10}$$

where $\nu = a\Delta t/\Delta x$ and Δ_0 and δ^2 are the standard central difference operators. Squaring both sides, summing over j and assuming that the discrete norm $\|W\| = (\sum \Delta x W_j^2)^{\frac{1}{2}}$ is bounded, it is easily checked that one obtains

$$\|W^{n+1}\|^2 = \|W^n\|^2 - \tfrac{1}{4}\nu^2(1-\nu^2)\|\delta^2 W^n\|^2. \tag{3.11}$$

Note that this is entirely in accord with the damping factor given in the last section in terms of Fourier analysis.

Suppose we now consider the conservation law $w_t + f_x = 0$ on a non-uniform mesh and with Δt chosen locally. Then we need to work in terms of E^n, the error from the converged solution of the discrete equations, and the corresponding characteristic speed a^n which we shall assume is everywhere positive:

$$E^n := W^n - W^\infty, \qquad a^n E^n := f(W^n) - f(W^\infty). \tag{3.12}$$

To simplify notation and using $\Delta_+ W_j := W_{j+1} - W_j$, we write for the residuals

$$R_+^n = \Delta_+ f(W^n)/\Delta_+ x = \Delta_+(a^n E^n)/\Delta_+ x \tag{3.13}$$

and similarly for R_-^n. Then the update (2.4), (2.5) reduces to

$$E_j^{n+1} = E_j^n - \frac{(\Delta t)_1}{\Delta_0 x_j}\left\{\Delta_0(a_j^n E_j^n) - \frac{1}{2}\left[\frac{a_{j+}^n(\Delta t)_{2+}\Delta_+(a_j^n E_j^n)}{\Delta_+ x_j} - \frac{a_{j-}^n(\Delta t)_{2-}\Delta_-(a_j^n E_j^n)}{\Delta_- x_j}\right]\right\} \tag{3.14}$$

which just involves R_{j+}^n and R_{j-}^n. For maximum flexibility we have introduced two time-step parameters here, and also characteristic speeds calculated for the cells as in (2.5). Let us therefore define two CFL numbers, one based on nodes and one on cells,

$$\nu_1 := \frac{|a|(\Delta t)_1}{\Delta_0 x} \qquad \nu_2 := \frac{|a|_\pm (\Delta t)_{2\pm}}{\Delta_\pm x} \tag{3.15}$$

where we have omitted the subscripts j and superscripts n. We also introduce a weighted measure of the error at each time level

$$\|E^n\|_w^2 := \sum (a_j^n)^2 (E_j^n)^2, \tag{3.16}$$

so that by fixing ν_1 the weighting equals $(\nu_1 \Delta_0 x_j / (\Delta t)_1)^2$. Hence, when each side of (3.14) is squared, multiplied by the weight and summed over j, the inner products cancel as in the simple case of (3.10) to give

$$\|E^{n+1}\|_w^2 - \|E^n\|_w^2 = -\tfrac{1}{4}\nu_1^2 (1 - \nu_2^2) \sum (\Delta_+ x\, R_+^n - \Delta_- x R_-^n)^2 - $$
$$\nu_1(\nu_2 - \nu_1) \sum (\Delta_+ x\, R_+^n)^2. \tag{3.17}$$

It is easily checked that this n-level error norm is reduced if

$$\nu_2 \geq \nu_1 \quad \text{and} \quad \nu_1 \nu_2 < 1 \tag{3.18}$$

and it is often preferable to take $\nu_2 > 1$. Indeed, for oscillating residuals (corresponding to high frequency modes) the damping is optimal along the curve $\nu_1 \nu_2 = \tfrac{1}{2}$ and increases with ν_2.

Apart from showing that the Lax–Wendroff iteration will still converge on an irregular mesh when a has a constant sign, this analysis also indicates improvements to it that should be made. In particular, comparing (3.14) and (3.15) with (2.4) and (2.5), we are led to combining residuals with the simplified relative weights

$$D_\pm := (1 \mp \nu_2 \operatorname{sgn}(a_\pm)) \frac{\Delta_\pm x}{\Delta_- x + \Delta_+ x}, \tag{3.19}$$

obtained from our choice of local timesteps.

Returning to the Fourier analysis for the scalar two dimensional case but with the two CFL numbers, one can confirm that the conditions (3.18) are sufficient for convergence where in each case $\nu = (\nu_x^2 + \nu_y^2)^{1/2}$.

4. Shock updates without mesh adaption

Suppose that at the end of a shock capturing phase all the mesh lines are scanned to detect the presence of a shock, using various criteria such as those described in Morton and Paisley (1988). Suppose further that these edge crossings are grouped into sequences to form distinct non-intersecting shock structures which are fitted by cubic splines (essentially interpolatory but with some smoothing): the result will be as in Fig. 3. The question now is how to use this in subsequent updates so that the amount of smoothing can be decreased and the objective of setting all the non-shocked residuals $\mathbf{R}_c$ to zero is attained as nearly as possible.

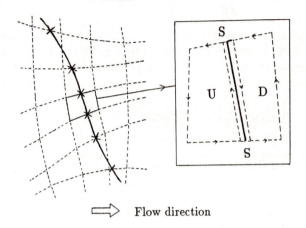

Flow direction

Fig. 3: Splitting of the residual for a shocked cell

As indicated in Fig. 3, the residual for a shocked cell when properly calculated can be divided into three parts

$$\mathbf{R}_c = \mathbf{R}_U + \mathbf{R}_S + \mathbf{R}_D, \qquad (4.1)$$

corresponding to the upstream section of the cell, the shock itself and the downstream part. When flow variables have been calculated either side of the shock points on the two edges, each of these parts can be computed using the trapezoidal rule. Then when updating a vertex of a shocked cell we have several options as to how much of (4.1) to use:-

(i)   use all of the properly calculated $\mathbf{R}_c$;
(ii)   use all the unshocked part $\mathbf{R}_U + \mathbf{R}_D$;
(iii) use only the contiguous part $\mathbf{R}_U$ or $\mathbf{R}_D$;
(iv)   ignore completely.

We can study some of the consequences of these choices by considering a one dimensional problem. Then (i) just corresponds to using the Lax–Wendroff update with its attendant proneness to oscillation so we will not consider this any further. In the energy analysis leading to the convergence result of Section 3.2, we assumed $a_j^n > 0$. If the scheme considered there, with the weights (3.18), were applied across a shock with $a_- > 0$ and

$a_+ < 0$, it is easy to see that oscillations could result: on the other hand, if the nodal value $\text{sgn}(a_j^n)$ is used for the cells either side, thus sacrificing conservation, a monotonically convergent scheme is obtained. With $\nu_2 = 1$ this is just the first order upwind scheme, so motivating the choice of (iii) or (iv) of the above options. In two dimensions the last choice has been found the most robust as well as the simplest.

Thus in the Taylor–Galerkin update after a shock has been detected, all the residuals for shocked cells are set to zero. In the subsequent shock update, at each edge-crossing of a shock, the nearest nodal values on either side are taken as the shock states to which the Rankine–Hugoniot relations are applied. The upstream values are taken as given as well as the density on the downstream side: the downstream momenta can then be updated and a shock speed calculated. This speed is used to move the shock, as in Morton and Paisley (1988): if it moves across a node, further adjustments are made to satisfy the Rankine–Hugoniot conditions at the new nodal values.

5. Numerical results

Results for two contrasting problems are presented to indicate the versatility of the algorithms described above: one is for transonic flow in a channel; the other is for hypersonic flow about a cylinder. In the shock capturing phase, it is essential to incorporate an artificial viscosity to stabilise the Lax-Wendroff iteration and to give monotone behaviour for the shock detection phase. A diffusion term is evaluated for each cell and is added to the nodes of that cell in a manner consistent with a mass-lumped treatment of the Laplacian diffusion operator using bilinear elements. The coefficient of the artificial diffusion is scaled by the nodal time-steps and the maximum eigenvalue on the characteristic cone, together with a blending function similar to that introduced by Jameson (1981), depending on the normalised second differences of both pressure and Mach number along both grid lines: it is further scaled by the residual through $\min(1, \|\mathbf{R}_c\|_{L^2})$ so that after shock fitting it is automatically switched off as convergence occurs.

Boundary conditions on a solid boundary are those employed by Morton and Paisley (1988) whereby residuals are calculated only for cells in the interior of the domain. Far field boundary conditions are determined through an examination of the Riemann invariants for the waves normal to the boundary in the locally linearised system, according to the first approximation of Engquist and Majda (1977).

In both problems the simple shock detection procedure as in Morton and Paisley (1988) is used, which involves searching for the maximum and minimum values of $\delta^2 p$ along a particular family of mesh lines. A shock point on such a line is accepted if these values occur no more than 3 mesh intervals apart and the shock position is then obtained by interpolating for $\delta^2 p = 0$. These positions are smoothed with weights depending on the shock strength using a cubic spline smoothing algorithm due to Reinsch (de Boor 1978) to give a shock curve with a dependable normal direction at each edge crossing. During the shock fitting phase of the computation, the mesh is scanned every 10 or so iterations to detect the presence of new shock points which are added to the shock list as required.

Fig. 4 shows results for Ni's (1981) problem of transonic flow along a channel with a 10% circular arc: the fitted shock is shown overlying the mesh in Fig. 4a, while isomach contours and a plot of Mach number along the channel floor are illustrated by Figs. 4b and 4c. Note that although the underlying grid was deliberately sheared in order that the shock should lie obliquely to it, the results obtained are in good agreement with those of the mesh adaption technique of Morton and Paisley (1988) and clearly show the occurrence of a Zierup singularity where the shock meets the curved channel wall. Fig. 5 shows a similar collection of results for hypersonic flow at Mach 8 past a circular cylinder on a 128×33 polar grid. Fig. 5a shows that the shock has been fitted even behind the cylinder where the shock is weak and meets the freesteam flow close to the Mach angle. The strong shear layers are well captured in the wake of the cylinder, as shown by Figs. 5b and 5c: these are not fitted and the form of the artificial viscosity is crucial in obtaining convergence there.

A comparison of convergence histories during the shock capturing phase for different choices of the two Courant numbers ν_1 and ν_2 which determine the timesteps for the first and second order updates is given in Fig. 6. For the above hypersonic problem, plots of $-\log\left(\left\|\frac{\rho^{n+1}-\rho^n}{\Delta t^n}\right\|_{L^2} \middle/ \left\|\frac{\rho^1-\rho^0}{\Delta t^0}\right\|_{L^2}\right)$ versus number of iterations are shown in Fig. 6a: these support the earlier remark that $\nu_1\nu_2$ should be taken as large as possible with ν_2 larger than ν_1. However, in this case, artificial viscosity was required to stabilise the iteration. Since this dominates the convergence rate in the neighbourhood of a shock, we also consider the same problem but with $M_\infty = 0.2$ so that the flow is everywhere subsonic and no artificial damping is required. Convergence histories for similar values of ν_1 and ν_2 are shown in Fig. 6b, where a more appropriate measure of convergence is $-\log\left(\sqrt{\sum_c \|\mathbf{R}_c^n\|^2 \middle/ \sum_c \|\mathbf{R}_c^0\|^2}\right)$. In this case, the convergence rate is dominated by ν_2, but more experimentation would be needed to draw firm conclusions about the effectiveness of the ν_1, ν_2 strategy.

6.   Conclusions

The results presented here demonstrate the effectiveness of a new shock fitting algorithm which does not employ mesh adaption, and is consequently

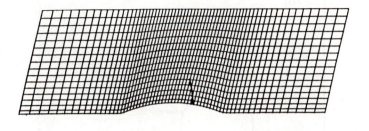

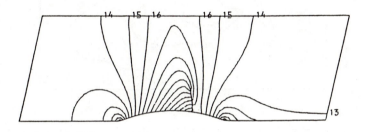

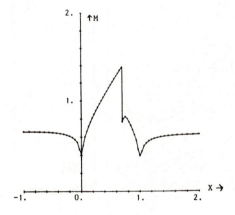

Fig. 4: Plots of Mach number for flow down a channel with $M_\infty = 0.675$.
(a) Mesh and fitted shock; (b) isomach contours, $\Delta M = 0.05$;
(c) Mach number distribution along the channel floor.

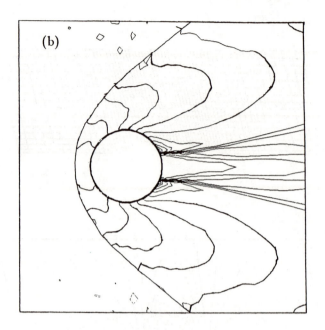

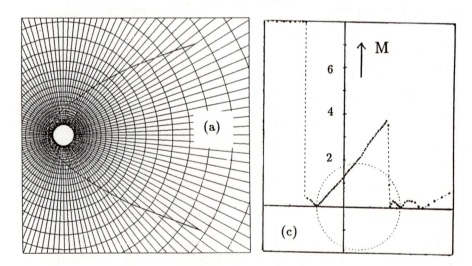

Fig. 5: Shock fitted solution for Mach 8 flow past a circular cylinder.
(a) Mesh and fitted shock;
(b) isomach contours, $\Delta M = 0.25$;
(c) Mach number distribution along the body.

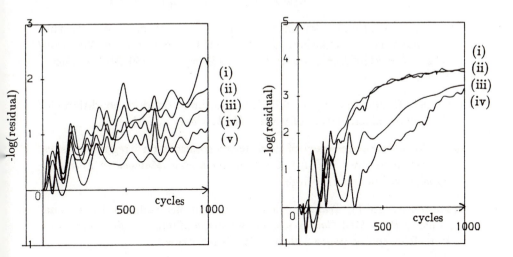

Fig. 6: Convergence histories for flow past a cylinder at:
(a) $M_\infty = 8.0$: *(i)* $\nu_1 = 0.7$, $\nu_2 = 1.4$; *(ii)* $\nu_1 = 0.7$, $\nu_2 = 0.7$; *(iii)* $\nu_1 = 0.35$, $\nu_2 = 2.0$; *(iv)* $\nu_1 = 0.35$, $\nu_2 = 0.35$.
(b) $M_\infty = 0.2$: *(i)* $\nu_1 = 0.35$, $\nu_2 = 2.0$; *(ii)* $\nu_1 = 0.7$, $\nu_2 = 1.4$; *(iii)* $\nu_1 = 0.7$, $\nu_2 = 1.0$; *(iv)* $\nu_1 = 0.7$, $\nu_2 = 0.7$; *(v)* $\nu_1 = 0.35$, $\nu_2 = 0.35$

more flexible than the earlier algorithm of Morton and Paisley (1988) based on local mesh adaption. Further algorithms are under development which make use of more local recovery techniques as in Morton and Rudgyard (1988) which should therefore be suitable for treating intersecting shocks: these have all gained much stimulus from the work of Moretti (1987).

Analysis of the update algorithm has highlighted its weaknesses as regards the spurious chequer-board modes and lower frequency modes related to it. Alternative updating algorithms are being studied with a view to overcoming these difficulties.

References

de Boor, C. (1978). *A practical guide to splines*. Springer-Verlag.

Denton, J. D. (1981). An improved time marching method for turbomachinery flow calculation. In *Proc. IMA Conference on Numerical Methods in Fluid Dynamics*, P. L. Roe, editor, pages 189–210. Academic Press.

Engquist, B. and Majda, A. (1977). Absorbing boundary conditions for the numerical simulation of waves. *Math. Comp.*, 31:629–651.

Giles, M. (1988). Accuracy of node based solutions on irregular meshes. To appear in *Proc. 11th Int. Conf. on Numerical Methods in Fluid Dynamics*, Williamsburg, Virginia.

Jameson, A. (1981). Transonic aerofoil calculations using the Euler equations. In *Proc. IMA Conference on Numerical Methods in fluid dynamics*, P. L. Roe, editor, pages 289–308. Academic Press.

Moretti, G. (1987). A technique for integrating the 2D Euler equations. *Computers and Fluids*, 15:59–75.

Morton, K. W. (1987). Finite volume and finite element methods for the steady equations of gas dynamics. In *Proc. Mathematics of Finite Elements and Applications, 1987*, J. R. Whiteman, editor, pages 353–377. Academic Press.

Morton, K. W. and Paisley, M. F. (1986). On the cell-centre and cell-vertex approaches to the steady Euler equations and the use of shock fitting. In *Proc. 10th Int. Conf. on Numerical Methods in Fluid Dynamics*, F. G. Zhuang, Y. L. Zhiu, editors, Lect. Notes in Physics 264, pages 488–493. Springer-Verlag.

Morton, K. W. and Paisley, M. F. (1988). A finite volume scheme with shock fitting for the steady Euler equations. *J. Comp. Phys.*, to appear.

Morton, K. W. and Rudgyard, M. A. (1988). Shock recovery and the cell vertex scheme for the Steady Euler equations. To appear in *Proc. 11th Int. Conf. on Numerical Methods in Fluid Dynamics*, Williamsburg, Virginia.

Ni, R.-H. (1982). A multiple grid scheme for solving the Euler equations. *AIAA J.*, 20:1565–1571.

Multigrid, defect correction and upwind schemes for the steady Navier–Stokes equations

P.W. Hemker and B. Koren

Centre for Mathematics and Computer Science,
Amsterdam, The Netherlands

1. Introduction

Several Navier–Stokes methods have been developed recently (Chakravarthy et al. 1985; Schröder and Hänel 1987; Shaw and Wesseling 1986; Thomas and Walters 1985), mainly based on existing computational methods for the Euler equations. We have followed the same approach (Koren 1988a; Koren 1988b), basing the method on the Euler code (see (Hemker and Koren 1988) for an overview). Our first objective was the efficient and accurate computation of laminar, steady, two-dimensional, compressible flows at practically relevant (i.e. high) Reynolds numbers, but (still) subsonic or low-supersonic Mach numbers. The non-isenthalpic Euler code developed earlier appeared to be a good starting point for this purpose.

The resulting method is hybrid in the sense that it can be used equally well for the steady Euler equations. An upwind finite volume technique is applied for the discretisation of the convective terms in the Navier–Stokes equations. For the diffusive terms, a central finite volume technique is applied. As a basic scheme to solve the nonlinear system of discretised equations, symmetric point Gauss-Seidel relaxation is used. Herein, one or more Newton steps are used for the collective relaxation of the four unknowns in the each finite volume. Nonlinear multigrid is applied as an acceleration technique. The process is started by nested iteration. The difficulty in inverting a higher-order accurate operator is by-passed by using defect correction as an outer iteration for the nonlinear multigrid cycling. Computational results are presented for a sub- and supersonic flat plate flow, the latter with an oblique shock wave impinging on the boundary layer. The multigrid technique appears to be efficient, and reliable results are obtained.

2. Flow model

The Navier–Stokes equations considered are

$$\frac{\partial f(q)}{\partial x} + \frac{\partial g(q)}{\partial y} - \frac{1}{Re}\{\frac{\partial r(q)}{\partial x} + \frac{\partial s(q)}{\partial y}\} = 0, \tag{2.1}$$

with $f(q)$ and $g(q)$ the convective flux vectors, Re the Reynolds number, and $r(q)$ and $s(q)$ the diffusive flux vectors. As state vector q we consider the conservative vector $q = (\rho, \rho u, \rho v, \rho e)^T$, with for the total specific energy e the perfect gas relation $e = p/(\rho(\gamma - 1)) + (u^2 + v^2)/2$. The primitive quantities used so far are: the ratio of specific heats γ, density ρ, pressure p and the velocity components u and v. The quantity γ is assumed to be constant. The convective flux vectors are defined by

$$f(q) = \begin{pmatrix} \rho u \\ \rho u^2 + p \\ \rho uv \\ \rho u(e + p/\rho) \end{pmatrix}, \quad g(q) = \begin{pmatrix} \rho v \\ \rho uv \\ \rho v^2 + p \\ \rho v(e + p/\rho) \end{pmatrix}, \tag{2.2}$$

and the diffusive flux vectors by

$$r(q) = \begin{pmatrix} 0 \\ \tau_{xx} \\ \tau_{xy} \\ \tau_{xx}u + \tau_{xy}v + \frac{\partial c^2/\partial x}{Pr(\gamma-1)} \end{pmatrix}, \quad s(q) = \begin{pmatrix} 0 \\ \tau_{xy} \\ \tau_{yy} \\ \tau_{yy}v + \tau_{xy}u + \frac{\partial c^2/\partial y}{Pr(\gamma-1)} \end{pmatrix}, \tag{2.3}$$

with Pr the Prandtl number, c the speed of sound (for a perfect gas $c = \sqrt{\gamma p/\rho}$), and with τ_{xx}, τ_{xy} and τ_{yy} the viscous stresses. Assuming the diffusion coefficients to be constant and Stokes' hypothesis to hold, the stresses are written as

$$\tau_{xx} = \frac{4}{3}\frac{\partial u}{\partial x} - \frac{2}{3}\frac{\partial v}{\partial y}, \tag{2.4}$$

$$\tau_{xy} = \frac{\partial u}{\partial y} + \frac{\partial v}{\partial x}, \tag{2.5}$$

$$\tau_{yy} = \frac{4}{3}\frac{\partial v}{\partial y} - \frac{2}{3}\frac{\partial u}{\partial x}. \tag{2.6}$$

3. Discretisation method

To still allow Euler flow ($1/Re = 0$) solutions with discontinuities , the equations are discretised in their integral form. A straightforward and simple discretisation is obtained by subdividing the integration region Ω into quadrilateral finite volumes $\Omega_{i,j}$, and requiring that the conservation laws hold for each finite volume separately:

$$\oint_{\partial\Omega_{i,j}} (f(q)n_x + g(q)n_y)ds - \frac{1}{Re} \oint_{\partial\Omega_{i,j}} (r(q)n_x + s(q)n_y)ds = 0. \quad (3.1)$$

For the evaluation of the convective flux vectors we make use of the rotational invariance of the Navier–Stokes equations. We do not do so for the diffusive flux vectors. Given our simple central discretisation of diffusive terms, use of rotational invariance for the latter is hardly advantageous. Thus, the discretised equations become

$$\oint_{\partial\Omega_{i,j}} T^{-1} f(Tq)ds - \frac{1}{Re} \oint_{\partial\Omega_{i,j}} (r(q)n_x + s(q)n_y)ds = 0, \quad (3.2)$$

with T the rotation matrix

$$T = \begin{pmatrix} 1 & 0 & 0 & 0 \\ 0 & n_x & n_y & 0 \\ 0 & -n_y & n_x & 0 \\ 0 & 0 & 0 & 1 \end{pmatrix}. \quad (3.3)$$

3.1 Evaluation of convective fluxes

For convection dominated flows, our objective, a proper evaluation of the convective flux vectors is of paramount importance. Based on previous experience, for this we prefer an upwind approach. Following the Godunov principle, along each finite volume wall we assume the convective flux vector to be constant, and to be determined by a constant left and right state only.

3.1.1 Approximation of left and right state

The approximation of the left and right state determines the accuracy of the convective discretisation. First– and higher–order accurate discretisations can be made. Considering for instance the numerical flux function

$$(f(q))_{i+\frac{1}{2},j} = f(q^l_{i+\frac{1}{2},j}, q^r_{i+\frac{1}{2},j})$$

where the superscripts l and r refer to the left and right side of volume wall $\Omega_{i+\frac{1}{2},j}$ (Fig. 3.1), first–order accuracy is obtained by taking

$$q^l_{i+\frac{1}{2},j} = q_{i,j}, \quad (3.4)$$

and

$$q^r_{i+\frac{1}{2},j} = q_{i+1,j}, \quad (3.5)$$

Higher-order accuracy can simply be obtained with the κ-schemes as introduced by van Leer (1985):

$$q^l_{i+\frac{1}{2},j} = q_{i,j} + \frac{1+\kappa}{4}(q_{i+1,j} - q_{i,j}) + \frac{1-\kappa}{4}(q_{i,j} - q_{i-1,j}), \qquad (3.6)$$

and

$$q^r_{i+\frac{1}{2},j} = q_{i+1,j} + \frac{1+\kappa}{4}(q_{i,j} - q_{i+1,j}) + \frac{1-\kappa}{4}(q_{i+1,j} - q_{i+2,j}), \qquad (3.7)$$

with $\kappa \in \mathbf{R}$ ranging from $\kappa = -1$ (fully one-sided upwind) to $\kappa = 1$ (central).

In (Koren 1988a) an optimal value for κ is found by giving an error analysis, using as model equation

$$\frac{\partial u}{\partial x} + \frac{\partial u}{\partial y} - \epsilon\left(\frac{\partial^2 u}{\partial x^2} + \frac{\partial^2 u}{\partial x \partial y} + \frac{\partial^2 u}{\partial y^2}\right) = 0. \qquad (3.8)$$

On a grid with constant mesh size h, a finite volume discretisation which uses the κ-approximation for the convective terms and which is second-order central for the diffusive terms, yields as modified equation

$$\frac{\partial u}{\partial x} + \frac{\partial u}{\partial y} - \epsilon\left(\frac{\partial^2 u}{\partial x^2} + \frac{\partial^2 u}{\partial x \partial y} + \frac{\partial^2 u}{\partial y^2}\right) +$$
$$+ h^2 \left\{ \frac{\kappa - 1/3}{4} \left[\frac{\partial^3 u}{\partial x^3} + \frac{\partial^3 u}{\partial y^3}\right] - \frac{\epsilon}{12}\left[\frac{\partial^4 u}{\partial x^4} + \frac{2\partial^4 u}{\partial x^3 \partial y} + \frac{2\partial^4 u}{\partial x \partial y^3} + \frac{\partial^4 u}{\partial y^4}\right]\right\}$$
$$= O(h^3). \qquad (3.9)$$

As optimal value for κ we take the value that gives the highest possible accuracy. From (3.9) we see that a proper diffusion- dependent κ may cancel the second-order error term, which would lead to third-order accuracy. Since convection dominated problems are our main concern, for simplicity we assume this diffusion-dependence to be negligible, which leads us to $\kappa = 1/3$.

To avoid spurious non-monotonicity, a new limiter has been constructed for the $\kappa = 1/3$ approximation (Koren 1988a). Let $q^{l(k)}_{i+\frac{1}{2},j}$ and $q^{r(k)}_{i+\frac{1}{2},j}$ be the kth component ($k = 1, 2, 3, 4$) of $q^l_{i+\frac{1}{2},j}$ and $q^r_{i+\frac{1}{2},j}$ respectively. Then a limited left and right state can be written as

$$q^{l(k)}_{i+\frac{1}{2},j} = q^{(k)}_{i,j} + \frac{1}{2}\psi(R^{(k)}_{i,j})(q^{(k)}_{i,j} - q^{(k)}_{i-1,j}), \qquad (3.10)$$

and

$$q_{i+\frac{1}{2},j}^{r(k)} = q_{i+1,j}^{(k)} + \frac{1}{2}\psi(1/R_{i+1,j}^{(k)})(q_{i+1,j}^{(k)} - q_{i+2,j}^{(k)}), \qquad (3.11)$$

with $\psi(R)$ the limiter considered, and $R_{i,j}^{(k)}$ the ratio

$$R_{i,j}^{(k)} = \frac{q_{i+1,j}^{(k)} - q_{i,j}^{(k)}}{q_{i,j}^{(k)} - q_{i-1,j}^{(k)}}. \qquad (3.12)$$

Using this notation, the limiter constructed for the $\kappa = 1/3$ approximation reads

$$\psi(R) = \frac{R + 2R^2}{2 - R + 2R^2}. \qquad (3.13)$$

3.1.2 Solution of 1D Riemann problem

Osher's scheme (Osher and Solomon 1982) has been preferred so far for the approximate solution of the standard 1D Riemann problem thus obtained. Osher's scheme has been chosen because of: (i) its continuous differentiability, and (ii) its consistent treatment of boundary conditions. (The continuous differentiability guarantees the applicability of a Newton type solution technique, which is what we make use of.) The question arises whether it is still a good choice to use Osher's scheme when diffusion also has to be modelled. Another, more widespread upwind scheme used in Navier–Stokes codes is van Leer's flux splitting scheme (van Leer 1982; Schröder and Hänel 1987; Shaw and Wesseling 1986; Thomas and Walters 1985). Reasons for its popularity are: (i) its likewise continuous differentiability, and (ii) its simplicity. The latter property is generally believed to be in contrast with Osher's scheme. (Recent work may help to reduce this difference, see e.g. (Hemker and Spekreijse 1986). In (Koren 1988a) an error analysis is given for both schemes. The analysis is confined to the steady, 2D, isentropic Euler equations for a perfect gas with $\gamma = 1$. For a subsonic flow and a first-order accurate square finite volume discretisation, the system of modified equations has been derived for both Osher's and van Leer's scheme. For both systems we considered a subsonic shear flow (the new element) along a flat plate. As a reference Lamb's approximate solution was used. Substituting Lamb's solution into the modified equation, considering the boundary layer edge at one characteristic length downstream of the leading edge, and using $Re \gg 1$, we find

$$\frac{\text{error Osher}}{\text{error van Leer}} = \left(\begin{array}{c} 1 \\ 5(1 - 2/\pi)M Re^{-1/2} \\ 1/2 \end{array} \right) \qquad (3.14)$$

where M is the outer flow Mach number. From (3.14) it appears that, when compared with Osher's, van Leer's scheme deteriorates for increasing Re.

3.2   Evaluation of diffusive fluxes

For the evaluation of the diffusive fluxes at a volume wall, it is necessary to computer ∇u , ∇v and ∇c^2 at the wall. For this we use a standard technique (Peyret and Taylor 1983). To compute for instance $(\nabla u)_{i+\frac{1}{2},j}$, we use Gauss' theorem:

$$(\nabla u)_{i+\frac{1}{2},j} = \frac{1}{A_{i+\frac{1}{2},j}} \oint_{\partial\Omega_{i+\frac{1}{2},j}} u n ds, \tag{3.15}$$

with $\mathbf{n} = (n_x, n_y)^T$; $\partial\Omega_{i+\frac{1}{2},j}$ the boundary and $A_{i+\frac{1}{2},j}$ the area of a quadrilateral dummy volume $\Omega_{i+\frac{1}{2},j}$ (Fig. 3.2) of which the vertices $\mathbf{z} = (x, y)^T$ are defined by:

$$\mathbf{z}_{i,j\pm\frac{1}{2}} = \frac{1}{2}(\mathbf{z}_{i-\frac{1}{2},j\pm\frac{1}{2}} + \mathbf{z}_{i+\frac{1}{2},j\pm\frac{1}{2}}). \tag{3.16}$$

A similar expression exists for $\mathbf{z}_{i\pm\frac{1}{2},j}$.
The line integral $\oint_{\partial\Omega_{i+\frac{1}{2},j}} u n ds$ is approximated by

$$
\begin{aligned}
\oint_{\partial\Omega_{i+\frac{1}{2},j}} u n ds \ = \ & u_{i+1,j}(\mathbf{z}_{i+1,j+\frac{1}{2}} - \mathbf{z}_{i+1,j-\frac{1}{2}}) \\
& + u_{i+\frac{1}{2},j+\frac{1}{2}}(\mathbf{z}_{i,j+\frac{1}{2}} - \mathbf{z}_{i+1,j+\frac{1}{2}}) \\
& + u_{i,j}(\mathbf{z}_{i,j-\frac{1}{2}} - \mathbf{z}_{i,j+\frac{1}{2}}) \\
& + u_{i+\frac{1}{2},j-\frac{1}{2}}(\mathbf{z}_{i+1,j-\frac{1}{2}} - \mathbf{z}_{i,j-\frac{1}{2}}),
\end{aligned} \tag{3.17}
$$

with for $u_{i+\frac{1}{2},j\pm\frac{1}{2}}$ the central expression

$$u_{i+\frac{1}{2},j\pm\frac{1}{2}} = \frac{1}{4}(u_{i,j} + u_{i,j\pm1} + u_{i+1,j} + u_{i+1,j\pm1}). \tag{3.18}$$

Similar expressions are used for the other gradients and other walls. For sufficiently smooth grids this central diffusive flux computation is second-order accurate.

4.   Solution method

To efficiently solve the system of discretised equations, symmetric point Gauss–Seidel relaxation, accelerated by nonlinear multigrid (FAS), is applied. With the scalar convection diffusion equation (3.8) as a model, local mode analysis shows that 'symmetric point Gauss–Seidel + multigrid' converges fast for the first-order discretised equation, for any value of the mesh Reynolds number h/ϵ (Koren 1988b). However, it appears to converge very slowly for the higher-order ($\kappa = 1/3$)discretised equation, for small and moderately large values of h/ϵ. It even appears to diverge for large values of h/ϵ (Koren 1988b). Clearly, the cause is the higher-order discretisation of the convection operator. No cure can be found in using some other $\kappa \in [-1, 1]$. As with the Euler equations (Hemker 1986; Koren 1988c), the difficulty in inverting the higher-order operator is by-passed by introducing iterative defect correction (IDeC) as an outer iteration for the nonlinear multigrid cycling . Let $F_h(q_h)$ denote the full, higher-order accurate operator, and $\tilde{F}_h(q_h)$ the less accurate operator that can be easily inverted. Then iterative defect correction can be written as

$$\tilde{F}_h(q_h^1) = 0,$$
$$\tilde{F}(q_h^{n+1}) = \tilde{F}_h(q_h^n) - \omega F_h(q_h^n), \quad n = 1, 2, \ldots, N. \tag{4.1}$$

where n denotes the nth iterand, and ω a damping factor. (The standard value for ω is $\omega = 1$). Special attention has been paid to the choice of the approximate operator $\tilde{F}_h(q_h)$ for the Navier–Stokes equations. The operator necessarily has only first-order accurate convection, but the amount of diffusion can be chosen freely. This freedom has been exploited by analyzing three approximate operators $\tilde{F}_h$: (i) an operator with full, second-order accurate diffusion, (ii) an operator with partial diffusion, and (iii) an operator without diffusion. The *first approximate operator* most closely resembles the higher-order operator F_h, and therefore has the best convergence properties. For sufficiently smooth problems and a second-order accurate F_h, theory (Hackbusch 1985) predicts the solution to be second-order accurate after a single IDeC-cycle for the first approximate operator. Theory does not give this guarantee for the other approximate operators. The *second approximate operator* neglects the cross derivatives in the diffusive terms, but it has full second-order diffusion, stemming from the remaining derivatives. The special feature of this operator is that, for the evaluation of the convective and diffusive fluxes in the Navier–Stokes equations, the same five-point data structure can be used. The operator combines elegance and simplicity with a rather good resemblance to the higher-order operator. The *third approximate operator* considered was already known from the Euler work. Given its successful application there, it may be expected to be suitable for very large values of the mesh Reynolds number. Local mode analyses with (3.8) as a model equation, and experiments with the Navier–Stokes equations showed the first approximate operator to have the best convergence properties indeed. The faster convergence clearly compensates for its relative complexity. The results presented in the next section have all been obtained with this operator.

Though the mesh Reynolds numbers in the computations performed were

large, we obeyed the multigrid requirement (cf. (Hackbusch 1985))
$m_r + m_p > 2m$, where m_r and m_p denote the order of accuracy of the
defect restriction and the correction prolongation respectively, and where
$2m$ denotes the order of the differential equation(s) considered. We used a
piecewise constant restriction ($m_r = 1$) and a piecewise bilinear prolongation ($m_p = 2$).

5.   Numerical results

To evaluate the computational method developed, the following flow problems have been considered: (i) a subsonic flat plate flow, and (ii) a supersonic flat plate flow with oblique shock wave-boundary layer interaction.
For the subsonic problem, the Blasius solution is used as a reference. For
the supersonic problem comparisons are made with experimental results
obtained by Hakkinen et al. (1958). For both flow problems we used:
$\gamma = 1.4$ and $Pr = 0.71$.

5.1   Subsonic flat plate flow

The geometry, the boundary conditions and the coarsest grid used for this
flow problem are given in Fig. 5.1. As far as convection is concerned,
the eastern boundary has been considered as an outflow boundary. For
diffusion the northern, southern and eastern boundary have been assumed
to be far-field boundaries with zero diffusion. For this problem we only
used grids composed of square finite volumes. As a coarsest grid in all
multigrid computations we used the 4×2 grid given in Fig. 5.1.

5.1.1   Osher's and van Leer's scheme

To compare Osher's and van Leer's scheme, we performed for both schemes
an experiment with $h-$ (mesh size) and Re variation, using the first-order
approximation only (because this will best show the differences). Results
obtained are given in Fig. 5.2. The results clearly show the superiority
of Osher's scheme, in particular under hard conditions (first-order approximation, high mesh Reynolds number). Notice that van Leer's scheme
deteriorates as predicted (compared to Osher's) for increasing Re. However, notice also that for *both* schemes the exact numerical solution (i.e. a
vanishing boundary layer) has been obtained for $Re = 10^{100}$. For van Leer's
scheme this could only be obtained by a careful treatment of the solid wall
boundary condition for the convective part (Koren 1988a).

In the further experiments we continued with Osher's scheme only.

5.1.2   Multigrid behaviour

To investigate the convergence properties of the nonlinear multigrid technique we considered the subsonic flat plate flow at $Re = 100$, using the
first-order discretised equations and Osher's approximate Riemann solver.

We are interested in the measure of grid independence of the convergence rate, the multigrid effectiveness and the influence of the order of accuracy of the prolongation. To measure the grid independence, we performed 20 FAS-cycles on a $16 \times 8-$, a $32 \times 16-$, and a 64×32 grid. For the multigrid effectiveness we performed 21 symmetric relaxation sweeps on the 64×32 grid. Further, to investigate the influence of the order of accuracy of the prolongation, we performed again 20 FAS-cycles with the 64×32 grid as finest grid, but now with the piecewise constant correction prolongation ($m_p = 1$, so violating the rule $m_r + m_p > 2m$). The results are given in Fig. 5.3. They clearly show that, for the flow considered, the multigrid method is nearly grid-independent and highly effective. The effect of the order of accuracy of the prolongation appears to be negligible.

5.1.3 Convergence to first- and second-order accuracy

Theory predicts that a single FAS-cycle may be sufficient for obtaining first-order accuracy (Hemker 1986). Further, as mentioned before, for smooth problems theory predicts a single IDeC-cycle to be sufficient for obtaining second-order accuracy (Hackbusch 1985). To investigate the convergence properties with respect to these two predictions we computed again solutions on the $16 \times 8-, 32 \times 16-$, and 64×32 grids for $Re = 100$. We performed the computations for successively the first-order and the (non-limited) $\kappa = 1/3$ approximation. Solutions obtained after 1 FAS-cycle and 1 IDeC-cycle (with inside the latter only a single FAS-cycle) are given in Fig. 5.4a and 5.4b respectively. Assuming the Blasius solution to be the exact solution, it can be verified that the results obtained (more or less) satisfy the theoretical predictions. In order to compare, for both discretisations the fully converged solutions (square markers) have been given. Additionally, as a contrast, in Fig. 5.4a, the single 64×32 grid solutions as obtained after 1,2,3 and 4 symmetric relaxation sweeps have been given (for the first-order discretisation only). The latter solutions clearly show once more the effectiveness of the multigrid technique.

5.2 Supersonic flat plate flow with oblique shock wave–boundary layer interaction

As reference test case from Hakkinen et al. (1958) we considered the experiment performed at $Re = 2.96 \times 10^5$. At first we tried to make a satisfactory grid. Since the present code has the possibility to compute Euler flows, it is easy to optimise the grid for convection only. For the present test case this led via the 80×32 grid shown in Fig. 5.5a to the 80×32 grid in Fig. 5.5b. The corresponding inviscid surface pressure distributions as obtained with the first-order, the non-limited $\kappa = 1/3$ and the limited $\kappa = 1/3$ approximation, are given in Fig. 5.6. The poor solution quality on the rectangular grid is clear. (For boundary conditions etc., we refer to

(Koren 1988b).)

Together with the measured data, the computed viscous surface pressure distributions are given in Fig. 5.7. (For convergence histories we refer to (Koren 1988b)). First we consider the results obtained on the rectangular grid. Given the bad inviscid solutions, obtained on the regular grid, it should be noticed that the good resemblance of the experimental and the second-order accurate viscous surface pressure distribution is absolutely fake. Since for this standard test case most authors use rectangular grids, and since most codes smear out discontinuities which are not aligned with the grid, a lot of good resemblance ever found for this test case might in fact be deceptive. Considering the results obtained on the oblique grid and comparing at first the computed surface pressure distributions, we see that diffusion has done its job in qualitatively different ways. In downstream direction, the second-order pressure distribution in the interaction region shows successively: a compression, a plateau and another compression. The computed second-order accurate surface pressure distribution is characteristic for a shock wave–boundary layer interaction with separation bubble (i.e. with separation and reattachment), whereas the first-order distribution typically is the distribution belonging to a non-separating flow. Given the occurrence of a separation bubble in the experimental results indeed, the first-order solution (on this 80 × 32 grid) has to be rejected. Comparing the second-order and measured surface pressure distribution, it appears that the latter is more strongly diffused. An explanation for this quantitative difference is lacking. Due to all kinds of uncertain influences a detailed quantitative comparison is probably impossible. Uncertain factors in the experiment are for instance: cross flow influences (3D effects), non-observed though influential turbulence, some slight heat transfer through the wall etc. Uncertain influences in the computation are for instance: a possibly too crude boundary condition treatment somewhere, the neglect of temperature dependence in the diffusion coefficients, and so on.

6.   Conclusions

In this paper we showed that the geometric multigrid method (combined with iterated defect correction) is a feasible method for the efficient solution of the steady full Navier–Stokes equations.

An important practical result of the present paper is the illustrated importance of carefully checking the reliability of a computed Navier–Stokes solution. In particular, the reliability should be checked with respect to the numerical errors introduced by the discretisation of the convective part. *This seems a trivial remark, but it appears that one is not sufficiently aware of this problem in practice.* The present approach allows an easy check of false diffusion: the same code can be used for both viscous and inviscid flow computations.

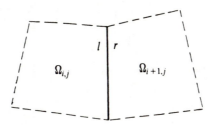

Fig. 3.1. Volume wall $\partial\Omega_{i+\frac{1}{2},j}$

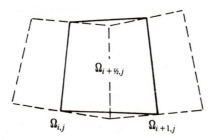

Fig. 3.2. Shifted volume $\Omega_{i+\frac{1}{2},j}$

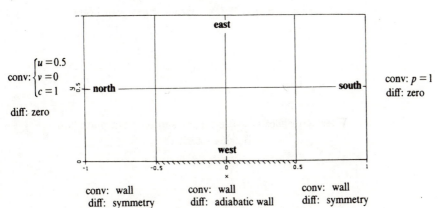

Fig. 5.1. Geometry, boundary conditions and coarsest grid subsonic flat plate flow (conv: convection, diff: diffusion)

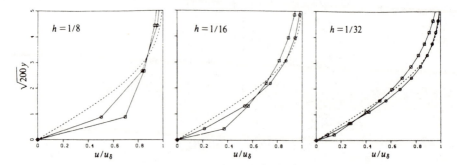

a. h−variation at $Re = 100$ ($\bigcirc$: Osher, $\square$: van Leer).

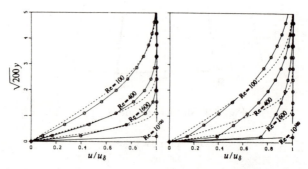

b. Re−variation at $h = 1/32$ (left: Osher, right: van Leer).

Fig. 5.2. Velocity profiles at $x = 0$ for the subsonic flat plate flow.
(——: Blasius solution)

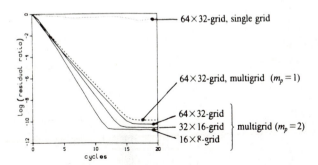

Fig. 5.3. Multigrid behaviour

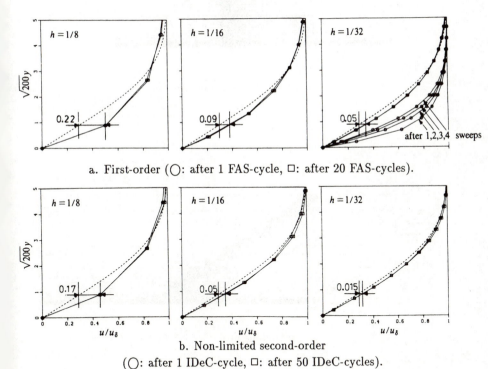

a. First-order (◯: after 1 FAS-cycle, ☐: after 20 FAS-cycles).

b. Non-limited second-order
(◯: after 1 IDeC-cycle, ☐: after 50 IDeC-cycles).

Fig. 5.4. Velocity profiles at $x = 0$ for the subsonic flate flow.
(——: Blasius solution)

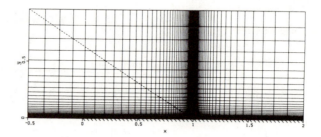

a. Rectangular grid (——: shock wave).

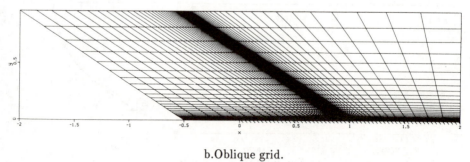

b.Oblique grid.

Fig. 5.5. Finest grids supersonic flat plate flow

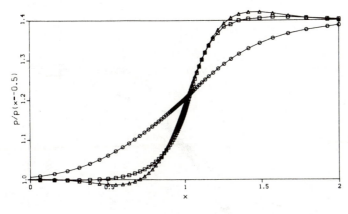

a. On rectangular grid.

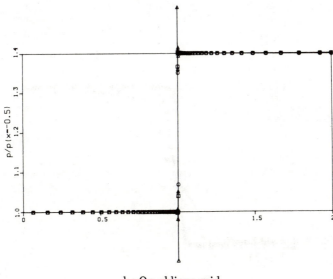

b. On oblique grid.

Fig. 5.6. Inviscid surface pressure ditributions supersonic flat plate flow
(○: first-order, △: non-limited higher-order, □: limited higher-order)

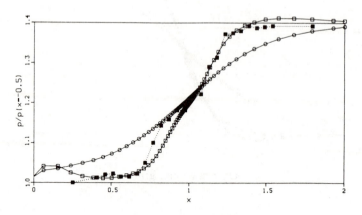

a. On rectangular grid.

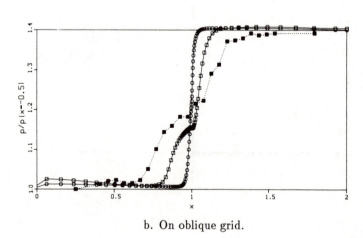

b. On oblique grid.

Fig. 5.7. Viscous surface pressure distributions supersonic flat plate flow
(○: first-order, □: limited second-order, ■ : measured)

Acknowledgement

This work was supported, in part, by the European Space Agency (ESA), via Avions Marcel Dassault–Bréguet Aviation (AMD-BA).

References

Chakravarthy, S., Szema, K., Goldberg, U., Gorski, J., and Osher, S. (1985). Application of a new class of high accuracy TVD schemes to the Navier–Stokes equations. AIAA paper 85–0165.

Hackbusch, W. (1985). *Multi-Grid Methods and Applications*. Springer, Berlin.

Hakkinen, R., Greber, I., Thrilling, L., and Abarbanel, S. (1958). The interaction of an oblique shock wave with a laminar boundary layer. NASA-memorandum 2-18-59 W.

Hemker, P. (1986). Defect correction and higher order schemes for the multigrid solution of the steady Euler equations. In *Proceedings of the Second European Conference on Multigrid Methods, Cologne 1985*. Springer, Berlin.

Hemker, P. and Koren, B. (1988). Defect correction and nonlinear multigrid for the steady Euler equations. In *Lecture Series on Computational Fluid Dynamics*, Von Karman Institute for Fluid Dynamics, Rhode–Saint–Genèse, Belgium.

Hemker, P. and Spekreijse, S. (1986). Multiple grid and Osher's scheme for the efficient solution of the steady Euler equations. *Appl. Num. Math.*, 2:475–493.

Koren, B. (1988a). Upwind schemes for the Navier–Stokes equations. In *Proceedings of the Second International Conference on Hyperbolic Problems, Aachen 1988*. Vieweg, Braunschweig.

Koren, B. (1988b). Multigrid and defect correction for the steady Navier–Stokes equations. In *Proceedings of the Fourth GAMM Seminar on Robust Multigrid Methods, Kiel 1988*. Vieweg, Braunschweig.

Koren, B. (1988c). Defect correction and multigrid for an efficient and accurate computation of airfoil flows. *Journal of Computational Physics*. (to appear in 1988).

van Leer, B. (1982). Flux–vector splitting for the Euler equations. In *Proceedings of the 8th International Conference on Numerical Methods in Fluid Dynamics, Aachen 1982*. Springer, Berlin.

van Leer, B. (1985). Upwind–difference methods for aerodynamic problems governed by the Euler equations. In *Proceedings of the 15th AMS–SIAM Summer Seminar on Applied Mathematics*, Scripps Institution of Oceanography, 1983. AMS, Providence, Rhode Island.

Osher, S. and Solomon, F. (1982). Upwind–difference schemes for hyperbolic systems of conservation laws. *Mathematics of Computation*, 38:339–374.

Peyret, R. and Taylor, T. (1983). *Computational Methods for Fluid Flow*. Springer, Berlin.

Schröder, W. and Hänel, T. (1987). An unfactored implicit scheme with multigrid acceleration for the solution of the Navier–Stokes equations. *Computers and Fluids*, 15:313–336.

Shaw, G. and Wesseling, P. (1986). Multigrid solution of the compressible Navier–Stokes equations on a vector computer. In *Proceedings of the 10th International Conference on Numerical Methods in Fluid Dynamics, Beijing 1986*. Springer, Berlin.

Thomas, J. and Walters, R. (1985). Upwind relaxation algorithms for the Navier–Stokes equations. AIAA paper 86–1501.

Recent developments of the Taylor–Galerkin method for the numerical solution of hyperbolic problems

J. Donea

V. Selmin

Joint Research Centre, Ispra, Italy

Aeritalia, Torino, Italy

L. Quartapelle

Istituto di Fisica del Politecnico di Milano, Italy

1. Introduction

A great deal of effort has been expended in recent years on the development of finite element methods for the solution of fluid dynamical problems. In the particular case of unsteady inviscid flows, an early study by Morton and Parrott (1980) analysed the issue of assuring a proper coupling between the time discretisation of hyperbolic equations and the spatial approximation afforded by a Galerkin method. As clearly shown in that paper, for each specific time integration algorithm, a different modification to the weighting functions is required to take full advantage of a finite-element-based spatial discretisation. Motivated by these findings, the Taylor–Galerkin (TG) method was introduced to achieve time-accurate finite element solutions to transient hyperbolic problems (Donea 1983).

Basically, in the TG method the governing partial differential equation is first discretised in time by means of a Taylor series expansion. The successive time derivatives of the unknown are then expressed in terms of spatial derivatives, similarly to the Lax–Wendroff difference method. The time-discretised equation thus obtained is subsequently discretised in space by means of the standard Galerkin finite element method. In this way, parameter-free, second- and higher-order accurate schemes are obtained which possess very low dissipation errors and excellent phase-speed characteristics (Donea 1984). The conceptual simplicity of the method makes its application very easy and the authors have contributed to the development of various TG schemes for solving advection-diffusion problems (Donea et al. 1984) and systems of nonlinear hyperbolic equations in one dimension (Selmin et al. 1984; Selmin et al. 1985) and two dimensions (Selmin et al. 1986). A complete analysis of the numerical properties of the Taylor–Galerkin schemes has been provided by Donea, Quartapelle and Selmin (1987).

Since its appearance, several investigators have studied the relationship between the Taylor–Galerkin method and other finite element formulations. For example, the equivalence of some TG schemes to Petrov–Galerkin methods of characteristic type has been discussed by Morton (1983; 1985). Furthermore, the second- and third-order TG schemes of Lax–Wendroff type have been discussed in connection with more general weak formulations by Baker and Kim (1986) and Morton (1987) and compared with least-squares approximations by Nguyen and Reynen (1984) and by Carey and Jiang (1988). At the same time, the simplest second-order accurate TG scheme, which merely amounts to a weak variational restatement of the Lax–Wendroff finite difference method, has been employed for the calculation of time-dependent compressible flows in one dimension (Thornton and Ramakrishnan 1984) and for the solution of two-dimensional hyperbolic equations (Bey et al. 1985; Löhner et al. 1984).

Subsequently, the method has been combined with other computational strategies such as adaptive mesh refinement (Morgan et al. 1987), adaptively moving mesh (Oden et al. 1986; Oden et al. 1987) and domain decomposition (Nakahashi 1986). Further applications include the solution of the Euler equations in three dimensions (Billey et al. 1987), the calculation of chemically reacting compressible flows (Chung and Sohn 1987) and the solution of the incompressible Navier–Stokes equations by means of time splitting techniques (Koschel et al. 1986; Laval 1988). The Taylor–Galerkin method appears now to be accepted as a standard technique in the finite element community, particularly for the accurate computation of transient solutions, with the only exception of some *a priori* criticism by Hughes and Mallet (1986) .

While the implementation of the second-order Taylor–Galerkin method is immediate for the case of linear equations, its use becomes computationally expensive when systems of nonlinear equations have to be considered. Moreover, as with the Lax–Wendroff difference methods, the TG method suffers from nonlinear instabilities which manifest themselves by the presence of bounded spatial oscillations in shock regions.

In this contribution, we shall be concerned with techniques aimed at overcoming, as far as possible, the difficulties associated with a straightforward application of the original Taylor–Galerkin schemes to the solution of the nonlinear hyperbolic problems, such as the Euler equations of gas dynamics. In Section 2 we describe the second-order accurate TG scheme together with its two-step implementation. The latter represents the finite element equivalent of the two-step Lax–Wendroff method originally introduced in the finite difference literature by Richtmyer (1967) and already employed in a finite element context (Angrand et al. 1983; Oden et al. 1986). In Section 3 we describe a method for introducing a locally modulated dissipation in order to construct a nonoscillatory shock-capturing scheme of the Taylor–Galerkin type. Numerical results for one- and two-dimensional problems are presented in Section 4 to illustrate the performance of the methods discussed in the paper.

2.   Second-order Taylor–Galerkin method

Consider a system of nonlinear hyperbolic equations written in the conservative form

$$V_t + \nabla \cdot \mathbf{F}(V) = 0 \tag{2.1}$$

where V is the vector of the unknowns and $\mathbf{F}(V)$ is the corresponding flux vector.

2.1   Basic scheme

A second-order accurate Taylor–Galerkin scheme is obtained by discretising the governing equation in time as in the classical Lax–Wendroff method. For system 2.1, a Taylor expansion in the time step Δt yields

$$
\begin{aligned}
V^{n+1} &= V^n + \Delta t\, V_t^n + \tfrac{1}{2}(\Delta t)^2\, V_{tt}^n + O\left[(\Delta t)^3\right] \\
&= V^n - \Delta t\, \nabla \cdot \mathbf{F}^n - \tfrac{1}{2}(\Delta t)^2\, \nabla \cdot [\mathbf{F}_t^n] \\
&= V^n - \Delta t\, \nabla \cdot \mathbf{F}^n + \tfrac{1}{2}(\Delta t)^2\, \nabla \cdot [\mathbf{A}^n \nabla \cdot \mathbf{F}^n].
\end{aligned}
\tag{2.2}
$$

Here, $\mathbf{A}(V) = \partial \mathbf{F}(V)/\partial V$ is the Jacobian matrix, $\mathbf{F}^n \equiv \mathbf{F}(V^n)$ and $\mathbf{A}^n \equiv \mathbf{A}(V^n)$. To introduce a spatial approximation by means of the finite element method, (2.2) is first written in weak form. Denoting by W an appropriate set of weighting functions and integrating by parts the spatial derivative terms in (2.2), the following weighted residual formulation is obtained

$$\int_\Omega W \left(\frac{V^{n+1} - V^n}{\Delta t} \right) d\Omega = \int_\Omega \nabla W \cdot \mathbf{F}^{n+\frac{1}{2}}\, d\Omega - \int_\Gamma W\, \mathbf{n} \cdot \mathbf{F}^{n+\frac{1}{2}}\, d\Gamma. \tag{2.3}$$

where

$$\mathbf{F}^{n+\frac{1}{2}} \equiv \mathbf{F}^n + \tfrac{1}{2}\Delta t\, \mathbf{F}_t^n = \mathbf{F}^n - \tfrac{1}{2}\Delta t\, \mathbf{A}^n \nabla \cdot \mathbf{F}^n. \tag{2.4}$$

In (2.3), $\mathbf{n}$ denotes the outward unit normal to the boundary Γ of the integration domain Ω. The spatial discretisation of (2.3) is then performed according to the standard Galerkin finite element method. The unknown vector V and the test functions W are locally approximated in terms of nodal values through shape functions $\phi_i(\mathbf{x})$. The global vector V^{n+1} of the nodal unknowns at the advanced time level is then obtained from the linear system of equations

$$M \left(V^{n+1} - V^n \right) = \Delta t\, R^n \tag{2.5}$$

where M represents the consistent mass matrix, with coefficients

$$M_{ij} \equiv \int_{\Omega} \phi_i \phi_j \, d\Omega \tag{2.6}$$

and R^n denotes the nodal load vector, with elementary components given by

$$R_i^n \equiv \int_{\Omega} \nabla \phi_i \cdot \mathbf{F}^{n+\frac{1}{2}} \, d\Omega - \int_{\Gamma} \phi_i \, \mathbf{n} \cdot \mathbf{F}^{n+\frac{1}{2}} \, d\Gamma. \tag{2.7}$$

It is important to note that the mass matrix M must be kept in its consistent form to preserve the good phase-speed characteristics associated with the finite element approximation. Therefore, the considered scheme is actually implicit even though it is derived from an explicit time integration algorithm. However, owing to the symmetric and diagonally dominant character of the consistent mass matrix, its inversion can be conveniently approximated by Jacobi iteration, thus avoiding the need for solving a system of linear algebraic equations at each time level.

2.2   Two-step implementation

A major drawback of the above second-order accurate method is that the evaluation of the load vector R^n in (2.7) is quite cumbersome and time consuming for systems of equations due to the presence of the Jacobian matrix $\mathbf{A}^n$ in the flux vector $\mathbf{F}^{n+\frac{1}{2}}$. It therefore appears to be advantageous from the viewpoint of computational efficiency to implement the scheme using a two-step procedure as originally proposed by Richtmyer and employed for the first time in a finite element context by Angrand (1983). Considering the case of quadrilateral elements with bilinear local approximations in two dimensions, the two-step implementation of the second-order Taylor–Galerkin scheme is obtained as follows:

First Step: For each element e, calculate an element-wise constant vector $\overline{V}$ from

$$A_e \overline{V}_e = \sum_{k=1}^{4} \int_{\Omega_e} \phi_k \, d\Omega \, V_k^n - \tfrac{1}{2}\Delta t \int_{\Omega_e} \nabla \cdot \mathbf{F}^n \, d\Omega \tag{2.8}$$

where A_e denotes the area of element e.

Second Step: For each node, calculate V^{n+1} by solving the following system of equations

$$\sum_{j=1}^{N} M_{ij} \left(V_j^{n+1} - V_j^n \right) / \Delta t = \int_{\Omega} \nabla \phi_i \cdot \overline{\mathbf{F}} \, d\Omega - \int_{\Gamma} \phi_i \, \mathbf{n} \cdot \overline{\mathbf{F}} \, d\Gamma \tag{2.9}$$

Here, N denotes the total number of nodal points and $\overline{\mathbf{F}} \equiv \mathbf{F}(\overline{V})$ is evaluated using the values of $\overline{V}$ resulting from the first step of the calculation.

3. Taylor–Galerkin schemes with modulated dissipation

In this section, we first show how a nonoscillatory first-order accurate scheme can be constructed by a simple modification of the second-order TG scheme. The low-order scheme has the same *phase* accuracy as the higher-order method, and is implemented using a two-stage procedure which separates the effects of convective transport and dissipation. In this way, a local modulation of the dissipation can be introduced without altering the phase response of the scheme. Therefore, the resulting scheme succeeds in combining the high resolution afforded by the second-order Taylor–Galerkin approximation in the smooth part of the solution with the ability of a first-order method to damp out oscillations near shocks or other types of discontinuity. Finally, it is shown that various classical limiting procedures, such as the artificial viscosity method and the Total Variation Diminishing (TVD) method, are easily adapted for use in the present finite element context.

3.1 First-order scheme

Consider the second-order TG scheme (2.5)

$$MV^{n+1} = MV^n + \Delta t\, R^n \tag{3.1}$$

and replace in the left-hand side the consistent mass matrix M by the diagonally lumped mass matrix L defined by $L_{ii} = \sum_j M_{ij}$ and $L_{ij} = 0$ for $i \neq j$. This yields the following first-order accurate scheme

$$LV^{n+1} = MV^n + \Delta t\, R^n, \tag{3.2}$$

which can be rewritten in the form

$$LV^{n+1} = LV^n + (M - L)V^n + \Delta t\, R^n. \tag{3.3}$$

It is readily verified that the second term in the right-hand side of (3.3) represents an added dissipation, so that the first-order scheme can be interpreted as a second-order approximation to a parabolic equation of the form

$$V_t + \nabla \cdot \mathbf{F}(V) = \frac{h^2}{6\Delta t} D^2 V, \tag{3.4}$$

where D^2 represents a discretised version of the Laplacian operator and h is a characteristic dimension of the element. In the case of a one-dimensional linear advection equation for a scalar variable u, scheme (3.3) yields the following equation for the vector of nodal values $U \equiv (U_j, j = 1, 2, \ldots)$

$$U^{n+1} = U^n - \nu \Delta_0 U^n + \left(\tfrac{1}{2}\nu^2 + \beta\right) \delta^2 U^n, \tag{3.5}$$

where ν is the Courant number and $\beta = 1/6$. Scheme (3.5) is clearly of the Lax–Wendroff type with an added dissipation term. It is stable for $|\nu| \leq \sqrt{2/3}$ and over this whole range it is monotone (Harten, Hyman and Lax 1976).

3.2 *Two-stage implementation*

For the simulation of truly transient problems, it is convenient to implement the first-order scheme (3.2) according to the following two-stage procedure which separates the effect of convective transport from that associated with the added dissipation

$$MV^* = MV^n + \Delta t \, R^n, \tag{3.6}$$
$$LV^{n+1} = MV^* = LV^* + (M - L)V^*. \tag{3.7}$$

The first stage corresponds to the second-order TG method and is characterized by the same (complex) amplification factor, whereas the second only introduces a real multiplicative coefficient into the amplification factor of the preceding stage. The advantage of such a procedure over the original single-step scheme (3.2) or (3.3) is that it preserves the second-order phase accuracy even upon introduction of a modulation into the dissipative term.

3.3 *Modulated dissipation*

At a given node i, the dissipative term of the first-order scheme is expressed in the form

$$\left(D^2 V\right)_i = \sum_k M_{ik} V_k - L_{ii} V_i \tag{3.8}$$

or, since $L_{ii} = \sum_k M_{ik}$,

$$\left(D^2 V\right)_i = \sum_k M_{ik} \left(V_k - V_i\right) \tag{3.9}$$

Now, the dissipation can be modulated by modifying (3.9) as follows

$$\left(D^2 V\right)_i = \sum_k d_{ik} \, M_{ik} \left(V_k - V_i\right) \tag{3.10}$$

where d_{ik}, with $0 \leq d_{ik} \leq 1$, is the modulating coefficient. By virtue of the present construction, the dissipative operator consists of segment contributions (element sides for triangular meshes) and the one-dimensional character of the segments allows for an easy adaptation of the procedures devised in one dimension for limiting the dissipative effect, particularly for the case of the TVD method. In the following subsections, we give a short exposition of two limiting procedures, namely the artificial viscosity method and the characteristic TVD method, which have been implemented in the present finite element formulation.

3.4 Artificial viscosity method

In the method of artificial viscosity the adjustable parameter d is expressed in terms of a sensor which recognizes discontinuities in the flow. An effective sensor of the presence of shocks can be constructed by considering the second derivative of the pressure (Jameson 1983). To this end, for a given segment i–k one introduces the quantities

$$d_i = \left| \frac{p_k - 2p_i + p_{i-}}{p_k + 2p_i + p_{i-}} \right| \quad \text{and} \quad d_k = \left| \frac{p_{k+} - 2p_k + p_i}{p_{k+} + 2p_k + p_i} \right|, \tag{3.11}$$

in which

$$p_{i-} = p_k - 2\,(\mathbf{x}_k - \mathbf{x}_i)\cdot(\nabla p)_i \quad \text{and} \quad p_{k+} = p_i + 2\,(\mathbf{x}_k - \mathbf{x}_i)\cdot(\nabla p)_k. \tag{3.12}$$

Here, $(\nabla p)_i$ denotes the pressure gradient at node i. The coefficient of artificial viscosity for segment i–k is then evaluated from

$$d_{ik} = \min(\chi \max(d_i, d_k), 1) \tag{3.13}$$

where χ is an adjustable parameter. In this way, d_{ik} is maximum on both sides of a shock, but is zero inside it.

The main drawback of the artificial viscosity method, i.e., the presence of a free parameter in (3.13), is avoided in the symmetric characteristic TVD formulation which will now be described.

3.5 Characteristic TVD method

As shown by Harten (1982a; 1982b), Yee (1986) and others, one way to obtain shock-capturing nonoscillatory schemes without adjustable parameters is to use the following strategy: i) Find the locations where the second-order scheme produces oscillations; ii) Insert there the maximum dissipation to render the scheme monotone; iii) Reduce or compensate the dissipation in the other parts of the flow.

In one space dimension, the above strategy is implemented using the variation of a sensor over three contiguous elements. In 2D we work with segments and, upon introduction of the extrapolated points $i-$ and $k+$, construct the following variations

$$\Delta_{i-}q = q_i - q_{i-}, \qquad \Delta_{ik}q = q_k - q_i, \qquad \Delta_{k+}q = q_{k+} - q_k \tag{3.14}$$

where q_{i-} and q_{k+} are determined by expressions similar to those in (3.12). In tackling nonlinear hyperbolic systems, such as the Euler equations, it turns out to be effective to implement the TVD method using some characteristic variables. Since the governing system is hyperbolic, any linear combination

$$C = \mathbf{c} \cdot \mathbf{A} \tag{3.15}$$

of the Jacobian matrices $\mathbf{A} = (A_x, A_y)$ possesses real eigenvalues λ^l ($l = 1, \ldots, 4$), and the associated right eigenvectors form the matrix T which diagonalises C, i.e.,

$$T^{-1}CT = \Lambda, \qquad \Lambda_{lm} = \lambda^l \delta_{lm}, \tag{3.16}$$

Here, we select the unit vector $\mathbf{c}$ as follows:

$$\mathbf{c} = (\mathbf{x}_k - \mathbf{x}_i)/|\mathbf{x}_k - \mathbf{x}_i|.$$

Let V_{ik} be the average of V_i and V_k proposed by Roe (1981), and denote by λ_{ik}^l and T_{ik} the quantities λ^l and T evaluated at V_{ik}. We then define α_{ik}^l as the component of $\Delta_{ik}V$ in the l-th characteristic direction, i.e.,

$$\Delta_{ik}V = T_{ik}\alpha_{ik}, \qquad \alpha_{ik} = T_{ik}^{-1}\Delta_{ik}V. \tag{3.17}$$

With the above notation, the contribution of segment i–k to the dissipation term of node i can be written in the following form

$$d_{ik}M_{ik}(V_k - V_i) = T_{ik}\Psi_{ik}, \tag{3.18}$$

where

$$\Psi_{ik}^l = M_{ik}\left[\alpha_{ik}^l - Q_{ik}^l\right] \tag{3.19}$$

and Q_{ik}^l is a limiting function, examples of which are:

$$Q_{ik}^l = \min \mathrm{mod}\left(2\alpha_{i-}^l, 2\alpha_{ik}^l, 2\alpha_{k+}^l, \tfrac{1}{2}(\alpha_{i-}^l + \alpha_{k+}^l)\right) \tag{3.20}$$

and

$$Q_{ik}^l = \min \mathrm{mod}\left(6\alpha_{i-}^l, \alpha_{ik}^l, 6\alpha_{k+}^l\right). \tag{3.21}$$

The min mod function of a list of arguments is equal to the smallest number in absolute value if all arguments are of the same sign and is equal to zero if any two arguments are of opposite sign.

4.   Numerical examples

In the examples which follow, transient problems have been solved using the two-stage scheme (3.6)–(3.7), whereas steady-state solutions have been calculated using the single-step scheme (3.2).

4.1   Shock tube problem

To compare the effect of the two limiting procedures described above, we present numerical results for the classical shock-tube problem considered by Sod (1978) (Fig. 1). We note that nonoscillatory solutions are obtained using the artificial viscosity method (Fig. 1a), the quality of the solution depending on the value chosen for the parameter χ (here $\chi = 3$). In the case of the TVD method, the solution depends on the choice of the limiting function (Figs. 1b, 1c). The limiter (3.20) appears to be more dissipative than the limiter (3.21).

It is worth mentioning that using the single-step scheme (3.2) instead of the two-stage scheme (3.6)–(3.7) would yield a far less accurate representation of the rarefaction wave and of the contact discontinuity.

4.2   Wedge problem

The geometry of the problem and the employed finite element mesh are shown in Fig. 2. The initial condition is a uniform supersonic flow with a Mach number of 1.6. Transient solutions have been computed using the artificial viscosity method with $\chi = 2$ (Fig. 2a) and the TVD method with the limiter (3.21) (Fig. 2b). The contours of the Mach number and of the entropy at a selected time are shown in the figure. The TVD method yields a rather sharp representation of the discontinuities, but it appears to be slightly more dissipative than the artificial viscosity method in the regular part of the flow.

4.3   Steady transonic flow past an airfoil

To further illustrate the tendency of the TVD method when combined with the present first-order scheme to introduce an excessive dissipation, we present in Fig. 3 results for the steady transonic flow past a NACA-0012 airfoil ($M_\infty = 0.85, \alpha = 0$). Here are compared profiles of the pressure coefficient obtained with the artificial viscosity method (Fig. 3a, $\chi = 2$) and the TVD method with limiter (3.21), using full dissipation (Fig. 3b) and a fourfold reduced dissipation (Fig. 3c). The last method is found to deliver a solution very close to the one calculated with the artificial viscosity method and free of numerical oscillations.

5.   Conclusion

In the present paper the Taylor–Galerkin method has been adapted to compute transient solutions to multidimensional problems governed by nonlinear hyperbolic equations. In particular, a dissipation operator has been introduced in a mass-matrix-based form which enables the generalization to unstructured finite element meshes of concepts, such as the TVD method, formulated for one-dimensional problems. At this stage the results obtained for steady and unsteady two-dimensional calculations are most encouraging and work is in progress to extend the present formulation to deal with three-dimensional problems.

Donea, Selmin, and Quartapelle

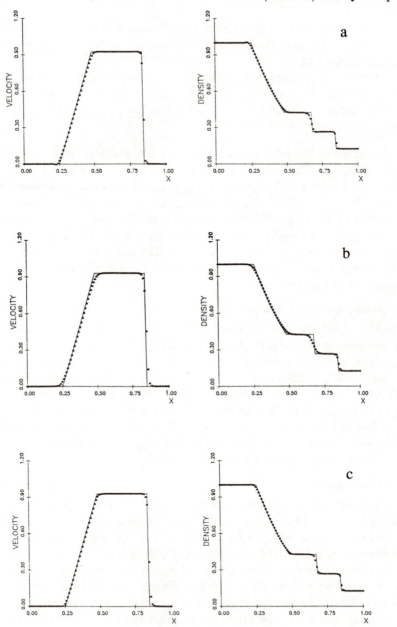

Fig. 1: Sod's shock tube problem; (a) artificial viscosity method ($\chi = 3.0$); (b) characteristic TVD method with the first limiter; (c) *idem* with the second limiter

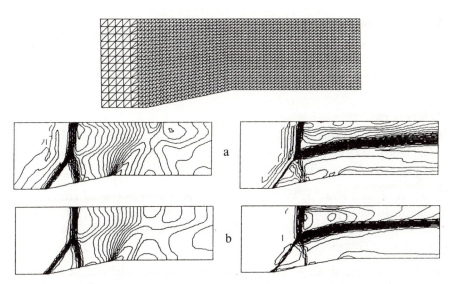

Fig. 2: Wedge problem: finite element discretisation and contours of the Mach number (left) and of the entropy (right); (a) artificial viscosity method ($\chi = 2.0$); (b) characteristic TVD method with the second limiter

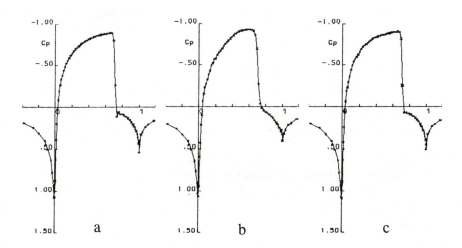

Fig. 3: Steady transonic flow past a NACA-0012 airfoil at $M_\infty = 0.85$: pressure coefficient profiles; (a) artificial viscosity method ($\chi = 2.0$); (b) characteristic TVD method with the second limiter; (c) *idem* with dissipative effect reduced by a factor of four

Acknowledgements
Part of the numerical calculations reported in this work were performed by the second author during his stay at the INRIA laboratory at Sophia Antipolis. He is grateful to Dr. Alain Dervieux for many stimulating discussions.

References

Angrand, F., Dervieux, A., Boulard, V., Periaux, J. and Vijayasundaram, G. (1983). Transonic Euler simulations by means of finite element explicit schemes. AIAA paper 83-1924.

Baker, A. J. and Kim, J. W. (1986). Analysis of a Taylor weak-statement algorithm for hyperbolic conservation laws. *Technical Report CFDL/86-1*, Computer Fluid Dynamics Laboratory, Unversity of Tennessee, Knoxville, Tennessee.

Bey, K. S., Thornton, E. A., Dechaumpai, P. and Ramakrishnan, R. (1985). A new finite element approach for prediction of aerothermal loads. AIAA Paper 85-1533-CP, *AIAA 7th Computational Fluid Dynamics Conference*, Cincinnati, Ohio, 16–18 July 1985.

Billey, V., Dervieux, A., Fezoui, L., Periaux, J., Selmin, V. and Stoufflet, B. (1987). Recent improvements in Galerkin and upwind Euler solvers and application to 3-D transonic flow in aircraft design. *Eighth International Conference on Computing Methods in Applied Sciences and Engineering*, Versailles, 14–18 December 1987.

Carey, G. F. and Jiang, B. N. (1988). Least-squares finite elements for first-order hyperbolic systems. *Int. J. Numer. Meths. Eng.* 26 : 81–93.

Chung, T. J., Kim, Y. M. and Sohn, J. L. (1987). Finite element analysis in combustion phenomena. *Int. J. Numer. Meths. Fluids*, 7 : 989–1012.

Donea, J. (1983). A Taylor–Galerkin method for convective transport problems. *Numerical Methods in Laminar and Turbulent Flow*, Eds. C. Taylor, J. A. Johnson and W. R. Smith, *Proceedings of the Third International Conference*, Seattle, 8–11 August 1983, 941–948. Pineridge Press, Swansea, U. K..

Donea, J. (1984). A Taylor–Galerkin method for convective transport problems. *Int. J. Numer. Meths. Eng.* 20 : 101–120.

Donea, J., Giuliani, S., Laval, H. and Quartapelle, L. (1984). Time-accurate solution of advection-diffusion problems. *Comp. Meths. Appl. Mech. Eng.* 45 : 123–146.

Donea, J., Quartapelle, L. and Selmin, V. (1987). An analysis of time discretization in the finite element solution of hyperbolic problems. *J. Comput. Phys.* 70:463–499.

Harten, A. (1982a). A high-resolution scheme for the computation of weak solutions of hyperbolic conservation laws. *NYU Report*, New York University, March 1982.

Harten, A. (1982b) On a class of high resolution total-variation-stable finite-difference schemes. *NYU Report*, New York University, October 1982.

Harten, A., Hyman, J. M. and Lax, P. D. (1976). On finite-difference approximations and entropy conditions for shocks. *Comm. Pure Appl. Math. XXIX*, 297–322.

Hughes, T. J. R. and Mallet, M. (1986). A new finite element formulation for computational fluid dynamics: III. The generalized streamline operator for multidimensional advection-diffusion systems. *Comp. Meths. Appl. Mech. Eng.* 58:305–328.

Jameson, A. (1983). Numerical solution of the Euler equations for compressible inviscid fluids. *Report MAE 1643*, Princeton University.

Koschel, W., Lotzerich, M. and Vornberger, A. (1986). Explicit method for solving Navier Stokes equations using a finite element formulation. *Notes in Numerical Fluid Mechanics*, Vol. 14, 148–160, Ed. E. H. Hirschel. Vieweg Publishing, Wiesbaden, Germany.

Laval, H. (1988). Taylor–Galerkin solution of the time-dependent Navier–Stokes equations. *International Conference on Computational Methods in Flow Analysis*, Okayama, Japan, 5–8 September 1988.

Löhner, R., Morgan, K. and Zienkiewicz, O. C. (1984). The solution of non-linear hyperbolic equation systems by the finite element method. *Int. J. Numer. Meths. Fluids* 4:1043–1063.

Morgan, K., Peraire, J., Thareja, R. R. and Stewart, J. R. (1987). An adaptive finite element scheme for the Euler and Navier–Stokes equations. AIAA Paper 87-1172-CP, *AIAA 8th Computational Fluid Dynamics Conference*, Honolulu, Hawaii, 9–11 June 1987.

Morton, K. W. (1983). Introduction to computational fluid dynamics. *von Karman Institute for Fluid Dynamics, Lecture Series 1983-01*, Brussels.

Morton, K. W. (1985). Generalized Galerkin methods for hyperbolic problems, *Comp. Meths. Appl. Mech. Eng.* 52.

Morton, K. W. (1987). Finite volume and finite element methods for the steady Euler equations of gas dynamics. *MAFELAP Conference*, Uxbridge, U. K., 28 April–1 May, 1987.

Morton, K. W. and Parrott, A. K. (1980). Generalized Galerkin methods for first-order hyperbolic equations. *J. Comput. Phys.* 36 : 249–270.

Nakahashi, K. (1986). FDM-FEM zonal approach for computations of compressible viscous flows. *Lecture Notes in Physics 264*, 494–498, Eds. F. G. Zhuang and Y. L. Zhu, *Proceedings of the Tenth International Conference on Numerical Methods in Fluid Dynamics*, Beijing, China, 23–27 June 1986.

Nguyen, H. and Reynen, J. (1984). A space-time least-square finite element scheme for advection–diffusion equations. *Comp. Meths. Appl. Mech. Eng.* 42 : 331–342.

Oden, J. T., Strouboulis, T. and Devloo, Ph. (1986). Adaptive finite element methods for the analysis of inviscid compressible flow: I. Fast refinement/unrefinement and moving mesh methods for unstructured meshes. *Comp. Meths. Appl. Mech. Eng.* 59 : 327–362.

Oden, J. T., Strouboulis, T. and Devloo, Ph. (1987). Adaptive finite element methods for high-speed compressible flows. *Int. J. Numer. Meths. Fluids* 7 : 1211–1228.

Richtmyer, R. D. and Morton, K. W. (1967). *Difference Methods for Initial-Value Problems*. Wiley, New York.

Roe, P. L. (1981). Approximate Riemann solvers, parameter vectors, and difference schemes. *J. Comput. Phys.* 43 : 357–372.

Selmin, V., Donea, J. and Quartapelle, L. (1984). Taylor–Galerkin method for nonlinear hyperbolic equations. *Numerical Methods for Transient and Coupled Problems*, Eds. P. Bettess, E. Hinton, R. W. Lewis and B. A. Schrefler, *Proceedings of the International Conference*, Venice, 9–13 July 1984, 816–827. Pineridge Press, Swansea, U. K..

Selmin, V., Donea, J. and Quartapelle, L. (1985). Finite element methods for nonlinear advection. *Comp. Meths. Appl. Mech. Eng.* 52 : 817–845.

Selmin, V. and Quartapelle, L. (1986). Finite element solution to the Euler equations. *Lecture Notes in Physics 264*, 559–565, Eds. F. G. Zhuang and Y. L. Zhu, *Proceedings of the Tenth International Conference on Numerical Methods in Fluid Dynamics*, Beijing, China, 23–27 June 1986.

Sod, G. A. (1978). A survey of several finite difference methods for systems of nonlinear hyperbolic conservation laws. *J. Comput. Phys.* 27:1–31.

Thornton, E. A. and Ramakrishnan, R. (1984). One-dimensional time-dependent compressible flow solutions. *Fifth International Symposium on Finite Elements and Flow Problems*, Eds. G. F. Carey and J. T. Oden, Austin, Texas, 1984.

Yee, H. C. (1986). Numerical experiments with a symmetric high-resolution shock-capturing scheme. *NASA TM88325*.

Numerical grid generation in 3-D Euler–flow simulation

J. W. Boerstoel

National Aerospace Laboratory NLR,
The Netherlands

1. Introduction

In 1980–1985, it became clear that, when applying CFD (Computational Fluid Dynamics) technology in aerodynamics, the construction of computation grids in the flow domain around complex 3D aerodynamic configurations was a major obstacle, see e.g. the survey report AGARD-AR-209 (1984). Since then, much work has been done at many places towards a solution of the grid generation problem. A vast collection of recent results may be found in the proceedings of two conferences on numerical grid generation in 1982 and in 1986, see Thompson (1982), and Häuser and Taylor (1986). Grid generation is nowadays a major topic at many CFD conferences, see e.g. the proceedings of the AIAA 8th Computational Fluid Dynamics Conference (1987).

The purpose of this paper is to sketch technical problems with grid generation, and to give a brief overview of proposed solutions. A simple example will illustrate that the core of the grid generation problem is, in fact, a topology problem. Further, the usefulness of various numerical grid generation techniques for aerodynamic work will be illustrated.

When judging the usefulness of grid generation techniques, it is necessary to specify from what viewpoint such a judgement is made. Here, this viewpoint is that of a designer of a system of computer programs for the numerical simulation of Euler (and Navier–Stokes) flows around complex 3D aerodynamic configurations. Moreover, this system must be operational in both industrial aircraft design environments and in research environments. Further, the system should be operational on various computer workstation network systems of sufficient processing power. These considerations are a starting point for this paper.

The paper consists of five sections, as follows.

- Section 1 – Introduction.

- Section 2 – Flow simulation system overview.

- Section 3 – Design of grid generators. Literature review.

- Section 4 – Applications of grid generation. Experiences.

- Section 5 – Conclusions.

2.   Flow simulation system overview

For an appreciation of how the construction of blocked grids is embedded in the total numerical flow simulation, it is useful to take a look at such a simulation process from an informatics point of view. An overview of the NLR flow simulation system is presented in this section. The structure of this system is more or less representative of what such systems look like, so that a few general conclusions can be drawn from it.

Mathematically, a flow simulation boils down to solving numerically an initial boundary value problem for the Euler (or Navier–Stokes) conservation equations on a flow domain bounded by the aerodynamic configuration surface. Initial and boundary conditions are numerically specified to give a well-posed problem. The simulation may then be considered to consist of the execution of the three mathematical tasks of Fig. 2.1:

- construction of a computation grid in the flow domain,

- discretisation of the conservation equations, and of the initial and boundary conditions, and

- numerical solution of the resulting discrete equation system.

The mathematical model is mapped into a flow simulation system like that of Fig. 2.2. This system is a collection of computer codes and data interfaces that is operational on a computer network with the indicated components.   The major function of the system is to produce, for a given aerodynamic configuration, results of flow simulations of sufficient accuracy in acceptable turnaround times.

The total simulation task may be divided into subtasks as shown in Fig. 2.2. Data interfaces (files, common blocks, plots and tables) are used to exchange information between the various subtasks.   There are eight subtasks,

1. topological block subdivision of the flow domain,

2. geometrical block subdivision of the flow domain,

3. grid generation on blocked flow domains,

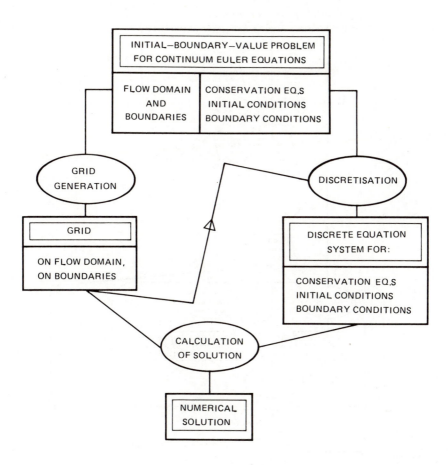

Fig. 2.1: Euler–flow simulation as the execution of mathematical tasks

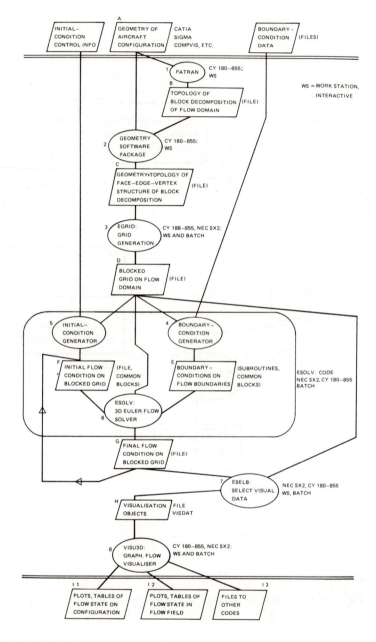

Fig. 2.2: Task decomposition and subtask data interfaces in NLR Euler–flow simulation system

4. boundary condition generation (e.g. solid wall conditions, free stream conditions, propeller–disk conditions, block interface conditions, boundary layer effect conditions, exhaust exit conditions, tunnel entrance and tunnel exit conditions, etc.),

5. intitial condition generator (uniform flow conditions, previous calculation result conditions),

6. Euler–flow or Navier–Stokes flow simulation,

7. selection of visualisation data from grid generation and from flow simulation, and

8. graphical interactive flow visualisation on a workstation or on a terminal.

There are nine major data interfaces between the subtasks, namely

A. the data of the aerodynamic geometry of the aircraft configuration, obtained from a solid model or CAD software package like CATIA, SIGMA, COMPVIS, etc. (file),

B. topology data of a block decomposition of the flow domain (file),

C. geometry data of the block decomposition of the flow domain (file),

D. the blocked grid on the flow domain (file),

E. the boundary condition data on the blocked grid (produced in the flow solver from given boundary data) (common block),

F. the initial flow condition (produced in the flow solver on the blocked grid) (common block),

G. the final flow condition produced by the flow solver (file),

H. those parts of the grid and the flow data that have to be visualised with the visualisation code (file VISDAT), and

I. plots, tables, and files of desired parts of the final flow produced interactively on a workstation or on a terminal.

System layouts like that of Fig. 2.2 show that informatics considerations play a dominant role when analysing the question which numerical grid generation techniques are useful for a flow simulation system designer.

- Grid generation is only a subtask in a flow simulation task. Hence, it is necessary to take into account the use of grids in the rest of the simulation process. For example, a choice between the use of structured or unstructured grids requires at least some analysis of how such grids perform in the flow solver and the flow visualizer.

- Performance characteristics and limits of current and future computer workstation network systems must be taken into account. For example, the maximum memory capacity and the maximum processing speed of (super)computers must be carefully taken into account, visualisation of 3D grids requires much care, and the maximum effective transfer speeds of data over networks is a third critical element.

These considerations pose requirements and constraints on the grid generator task and its results (grids), and limit the range of useful numerical grid generation techniques proposed in the literature to a small range.

3.   Design of grid generator, selection of numerical grid generation techniques

3.1   Grid generator design problem

The designer of a grid generation procedure has to answer many questions. How is a grid generator code to be built that allows a CFD engineer to design a grid around a given 3D aerodynamic configuration, that is operational on computer networks of sufficient processing power, and make efficient use of the networks? What kinds of 3D configurations are given? What performance criteria are of interest (grid quality, turnaround time)? A part of this problem area concerns the question of which numerical grid generation techniques can be usefully mapped into grid generator codes.

3.2   Requirements

Grid generators for Euler– (and Navier–Stokes) flow simulation around possibly complex 3D aerodynamic configurations should satisfy a number of major functional and performance requirements to be useful in industrial and research environments.

On the input side, the grid generator should be loosely coupled to any desired CAD solid model or geometry software package (CATIA, CADAM, SIGMA, PATRAN, etc.) via a simple data interface (file). This data interface should transfer numerical geometry definitions of aerodynamic configurations. The grid generator should be able to accept the geometry of any complex aerodynamic configuration as input.

On the output side, the grid generator should be loosely coupled to flow solvers and to visualizers by also a very simple data interface (file), to transfer grids.

Desired turnaround times for the execution of a grid generation task are of the order of a few days for the design of a new grid around a complex 3D configuration, and of the order of 2 hours for a local improvement of a given grid. This requirement induces a need for fast numerical grid generation algorithms. The grid generator should provide the grid designer with means to tune the mesh sizes in the grid to what he expects to be needed for sufficient accurate flow simulations. Mesh size tuning algorithms should be transparent to the grid designer, and be local in nature (which means that local grid modifications should affect grids in limited parts only).

Grid generators should be operational on any computer workstation network system with a processing power exceeding a certain minimum. This portablility requirement means that they should be preferably coded in ANSI Fortran 77.

Further, numerical grid generation techniques should have good vector performance in ANSI Fortran 77. This means that data structures should be well-ordered, and algorithms should be vectorizable. Finally, parallel processing should be possible in future.

3.3 Starting points in the conceptual design

Numerical grid generation techniques in a grid generator are based on a few design concepts that are often generally accepted and implemented. A few of them are discussed in this section.

Infinite flow domains are made finite by specifying a bounding surface in the flow, that subdivides the flow domain into a finite domain to be covered by a grid, and an infinite domain. The flow in the infinite domain is usually simulated by a truncated asymptotic expansion, which gives a boundary condition on the bounding surface (e.g. a uniform flow). Hence, in grid generation, *flow domains are finite*.

There are at present two alternative major concepts in numerical grid generation, *blocked grids* and *unstructured grids* (c.f. e.g. Lee et al. (1980), and Baker (1987)). An appreciation of what key problem is involved here, is obtained by first analysing two simple examples.

Ideally, one whould like to map any finite flow domain onto one *computational cube* by a sufficiently smooth transformation because, under such a transformation, a uniform decomposition of the cube into small cubes will map into a smooth decomposition of the flow domain into hexahedra cells having simple data structures: the cell data may be stored in 3D arrays. Such data structures are necessary to achieve efficiency in (vector) calculation processes. Moreover, these simple data structures and the smoothness of the grid allow the application of efficient standard CFD techniques.

However, the finite flow domains around complex configurations can in general not be mapped onto one computational cube because of difference in the *topological structure* of the flow domain and the cube. This difference concerns the connectivity relations between vertices, edges, and faces of these two volumetric configurations, c.f. Fig. 3.1. The finite flow domain of Figs. 3.1c, d has a topology structure (connectivity relation between vertices, edges and faces) that is very different from that of a cube (Figs. 3.1a, b). In such cases, it is in general necessary to give up the regular 3D data structure, or to give up the use of hexahedronal cells.

Incompatibility of the topology of the vertex–edge–face structure of finite flow domains with the topology of a cube is considered here the key problem.

In unstructured grid (finite element) approaches, this incompatibility problem is solved by replacing hexahedra cells by tetrahedra cells, and allowing irregularities in the data structures. Moreover, grid smoothness is usually given up, because this concept cannot easily be given a useful practical numerical form. As a consequence, numerical procedures in flow solvers should perform well on nonsmooth grids with irregular data structures. In blocked grid approaches, the solution of the incompatibility problem is sought in a decomposition of the finite flow domain into a limited number (1–0(100)) of "large" hexahedra called *blocks*. Subsequently, within each block, a 3D array of hexahedra cells is constructed. The collection of all cells in the finite flow domain may thus be grouped into a 4D array, with one index for block identification, and the remaining three indices for a 3D data structure in each block. Moreover, it is easy to create in each block cell distributions that are smooth in numerically well-defined senses. The concept of grid smoothness over block faces requires special attention.

A *choice* between these two approaches may be based on the following considerations.

a. Blocked grids allow the use of regular data structures and of concepts of grid smoothness to be built in the CFD algorithms of a block. This leads to more efficient algorithms than those for unstructured grids.

b. Unstructed grids are often claimed to be able to cover topologically complex flow domains. Nowadays, blocked grids can do this, too, so that this issue is no longer a basis for preferring the one above the other.

c. Blocked grids will facilitate the use of different flow models in different blocks.

d. In different blocks, different mesh refinement strategies (e.g. mesh size halving, see section 4) may be easily incorporated.

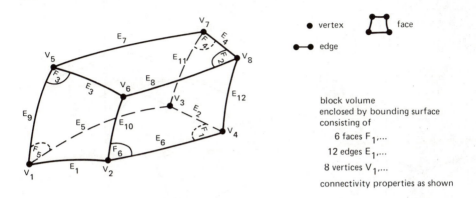

Fig. 3.1a: Topology of block (hexahedron)

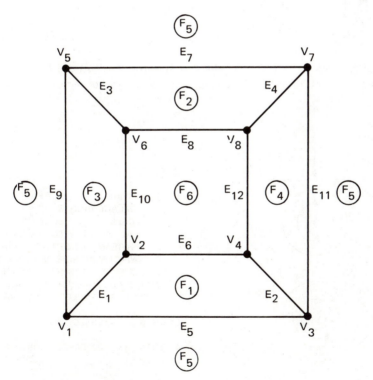

Fig. 3.1b: Labelled planar graph of block, c.f. Fig. 3.1a

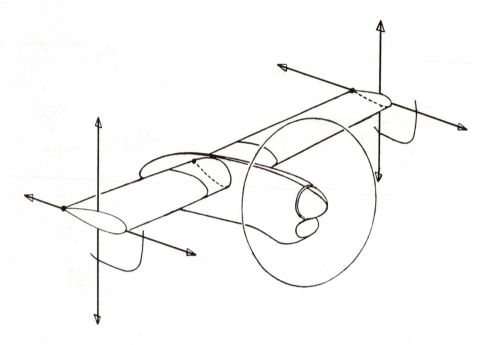

Fig. 3.1c: Flow domain around wing/nacelle/propeller, enclosed by a surface, consisting of two tunnel–side–wall segments, which are connected far from the configuration

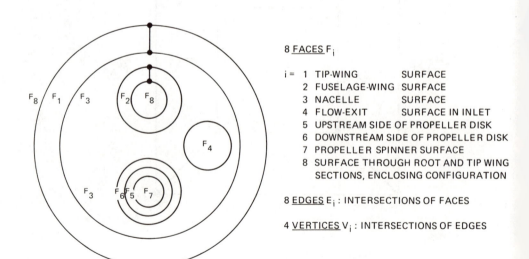

8 FACES F$_i$

i = 1 TIP-WING   SURFACE
    2 FUSELAGE-WING SURFACE
    3 NACELLE   SURFACE
    4 FLOW-EXIT   SURFACE IN INLET
    5 UPSTREAM SIDE OF PROPELLER DISK
    6 DOWNSTREAM SIDE OF PROPELLER DISK
    7 PROPELLER SPINNER SURFACE
    8 SURFACE THROUGH ROOT AND TIP WING
       SECTIONS, ENCLOSING CONFIGURATION

8 EDGES E$_i$: INTERSECTIONS OF FACES

4 VERTICES V$_i$: INTERSECTIONS OF EDGES

Fig. 3.1d: Labelled planar graph of finite flow domain of Fig. 3.1c

e. The regular data structures of the blocked grid approach will offer better vectorization possibilities in grid generator and flow solver codes.

f. It may be expected that the blocked grid approach will naturally lead to parallel executions of calculations per block on different vectors processors.

g. Blocked grids will require special block–face–boundary algorithms in grid generators and flow solvers, to couple grids and flows over block faces. In the unstructured grid approach, such special algorithms are not required.

h. Most of the aeronautical CFD literature is based on the use of smooth grids with a regular data structure.

Based on these considerations, at NLR the blocked grid approach was selected for implementation in a grid generator (Boerstoel 1986).

In order to allow easy implementation of numerical boundary conditions at block interfaces, blocks are *patched to each other face-to-face*, and grids are made block- and configuration-boundary conforming at NLR. Hence overlapping blocks have not been used up till now.

3.4 Block decomposition and grid generation

Given the requirements and choices made above, we have next to consider the selection of a constructive technique for grid generation.

The grid generation task will be subdivided into two steps. First, decompose the flow domain into a limited number of large blocks, and define the topology and geometry of this decomposition. Next, construct in each block a grid, with over block faces desired continuity properties for the grid. This subdivision is useful, because the block decomposition task requires considerably smaller data structures than those of complete grids. Moreover, this block decomposition task can be decoupled to a considerable degree from the rest of the grid generation task: block decomposition involves only manipulation of topological and geometrical aspects of a blocked flow domain, but not yet grid constructions.

3.5 Block decomposition and block geometry definition

Block decomposition is the subdivision of a given finite flow domain into a limited number (1 to about 100) of blocks. Input is the geometry of a flow domain boundary. This includes the geometry of the aerodynamic configuration. Output is the geometricial shape of the collection of vertices, edges, and faces of blocks.

The topological aspects of block decompositions are an area of active research. Lee et al. (1980) and Lee (1982) and Coleman (1982) report about connection rules of grids over block faces, and about grid singularities. Benek, Buning and Steger (1985) propose techniques with overlapping blocks. Weatherill and Forsey (1984) propose non-overlapping blocks, and introduce the concept of a topology file. This is a list of all block faces giving, for each face, a face number, the type of connection of the grid points over the face, adjacent block numbers, and orientation of the two sides of the face in the adjacent blocks.

Each block is a hexahedra volume of the flow domain (Fig. 3.1a), with the vertex–edge–face connectivity structure of that of a cube. Blocks are always assumed to have the connectivity structure of a cube, but they can be geometrically singular. (See Fig. 3.2 for a list of all kinds of possible geometrical degeneracies of a block.) Degenerated blocks may be useful when the usual connectivity relations between vertices, edges, and faces are geometrically a problem.

To limit the complexity of possible block–face–edge–vertex structures, it may be assumed that blocks are geometrically packed face-to-face. This assumption will be discussed further in section 4.

In the grid generator and the flow solver, simple techniques are needed to map local coordinate systems in blocks onto each other via common block faces. This may be done as follows. The topology of a block decomposition is described by storing labels of blocks, faces, edges and vertices in arrays, as follows.

Entity	Label	Labels stored in array	Remark
block	B_l	$B = \{B_l \mid l = 1, N_B\}$	$N_B \geq 1$,
face	F_m	$F = \{F_m \mid m = 1, N_F\}$	$N_F \geq 6$,
edge	E_p	$E = \{E_p \mid p = 1, N_E\}$	$N_E \geq 12$,
vertex	V_q	$V = \{V_q \mid q = 1, N_V\}$	$N_V \geq 8$.

Each block label B_l is made equivalent to a 6-element array containing the six face labels F_{m_i} of that block:

$$B_l = \{F_{m_1}, F_{m_2}, F_{m_3}, F_{m_4}, F_{m_5}, F_{m_6}\}, \quad m_i \in [1, N_F].$$

Similarly, each face label F_m is made equivalent to a 4-element array containing the four edge labels E_{p_i} of that face:

$$F_m = \{E_{p_1}, E_{p_2}, E_{p_3}, E_{p_4}\}, \quad p_i \in [1, N_E].$$

Each edge label E_p is made equivalent to a 2-element array containing the two vertex labels of that block:

$$E_p = \{V_{q_1}, V_{q_2}\}, \quad q_i \in [1, N_V].$$

In computer codes, the labels in the arrays are identified with the values of their integer subscripts,

$$B_l \equiv l, \quad F_m \equiv m, \quad E_p \equiv p, \quad V_q \equiv q.$$

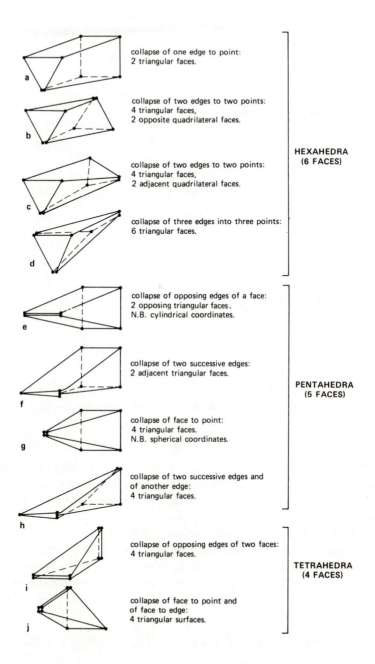

collapse of one edge to point:
2 triangular faces.

collapse of two edges to two points:
4 triangular faces,
2 opposite quadrilateral faces.

collapse of two edges to two points:
4 triangular faces,
2 adjacent quadrilateral faces.

collapse of three edges into three points:
6 triangular faces.

HEXAHEDRA
(6 FACES)

collapse of opposing edges of a face:
2 opposing triangular faces.
N.B. cylindrical coordinates.

collapse of two successive edges:
2 adjacent triangular faces.

collapse of face to point:
4 triangular faces.
N.B. spherical coordinates.

collapse of two successive edges and
of another edge:
4 triangular faces.

PENTAHEDRA
(5 FACES)

collapse of opposing edges of two faces:
4 triangular faces.

collapse of face to point and
of face to edge:
4 triangular surfaces.

TETRAHEDRA
(4 FACES)

Fig. 3.2: Geometrically degenerated blocks

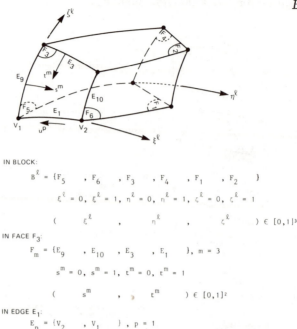

IN BLOCK:

$$B^{\ell} = \{ F_5 \quad , \; F_6 \quad , \; F_3 \quad , \; F_4 \quad , \; F_1 \quad , \; F_2 \quad \}$$

$$\xi^{\ell} = 0, \; \xi^{\ell} = 1, \; \eta^{\ell} = 0, \; \eta^{\ell} = 1, \; \zeta^{\ell} = 0, \; \zeta^{\ell} = 1$$

$$(\quad \xi^{\ell} \quad , \quad \eta^{\ell} \quad , \quad \zeta^{\ell} \quad) \in [0,1]^3$$

IN FACE F_3:

$$F_m = \{ E_9 \quad , \; E_{10} \quad , \; E_3 \quad , \; E_1 \quad \}, \; m = 3$$

$$s^m = 0, \; s^m = 1, \; t^m = 0, \; t^m = 1$$

$$(\quad s^m \quad , \quad , \quad t^m \quad) \in [0,1]^2$$

IN EDGE E_1:

$$E_p = \{ V_2 \quad , \; V_1 \quad \}, \; p = 1$$

$$u^p = 0, \; u^p = 1$$

$$(\quad u \quad) \in [0,1]$$

Fig. 3.3: Definition of boundary-conforming local coordinate systems in blocks, faces and edges, by pairing and ordering labels in arrays specifying connectivity relations

The array structures B, F, E and V are then reduced to two-dimensional integer index arrays. These integer index arrays are used in computer codes at NLR to relate local coordinate systems in blocks, faces, and edges to each other as follows.

In each block B_l, the ordering of the faces F_{m_i} in array B_l are made corresponding to the ordering and ranges of local curvilinear face-conforming coordinates (ξ^l, η^l, ζ^l) in that block. Observe that these coordinate system properties are block-topology dependent. The correspondence rule is defined by the scheme of the first four lines in Fig. 3.3. This scheme means that, in block B_l,

$$\xi^l = 0 \text{ on face } F_{m_1}, \xi^l = 1 \text{ on face } F_{m_2}, \eta^l = 0 \text{ on face } F_{m_3}, \text{ etc.}$$

and that ξ^l runs from 0 to 1 when going in from face F_{m_1} to face F_{m_2}, etc.

In each face F_m, the ordering and ranges of local curvilinear edge-conforming coordinates (s^m, t^m) in that face are made corresponding to the ordering of edge labels in array F_m, according to the scheme of the middle four lines of Fig. 3.3. In each edge E_p, the range of a curvilinear coordinate u^p in that edge is made corresponding to the ordering of the two vertex labels in array E_p, according to the scheme of the last four lines in Fig. 3.3.

It may be seen that, for each face F_m belonging to two different blocks B_l, the correspondence between the different block coordinates (ξ^l, η^l, ζ^l) and the face coordinates (s^m, t^m) may be different. Similarly, for each edge E_p belonging to two or more faces, the correspondences between the different face coordinates (s^m, t^m) and the edge coordinate u^p may be different. However, the correspondences between different coordinate systems at a face, and edge, or a vertex may in a unique way be extracted out of the contents of the topological data arrays B, F, E and V.

During block decomposition, a designer of a blocked grid for a particular flow calculation has to take into account three different kinds of requirements and limitations.

- The *topological* requirements concern the connectivity relations between the vertices, edges, faces and blocks of face-patched block decompositions, and the choice of local coordinate system orientations in edges, faces and blocks for indexing purposes, and for geometry descriptions of edges and faces.

- The *geometrical* requirements concern the specification of the position of vertices, and of the geometry of edges as line segments and of faces as surface segments. This includes the specification of block faces on the aerodynamic configuration and on the boundary of the computation domain. Further, the use of geometrically degenerated blocks must be considered, to solve problems with connectivity relations.

- From the point of view of *approximation accuracy* of flow calculation results, there are requirements with respect to desired densities of grid points in blocks, and grid quality (including grid smoothness). Also of interest is the application of various boundary conditions (in the flow solver, the kind of boundary condition can be specified per block face), and the use of different flow models (Euler, Navier–Stokes) in different blocks.

3.6   Grid generation

It is clear from the literature that, for block-structured grids, elliptic and algebraic grid generation techniques are conceptually attractive (Boerstoel 1986). Algebraic techniques may be used to generate initial grid point distributions. The elliptic techniques may be used to fine tune grid point distributions.

Elliptic grid generation is based on the numerical solution of Dirichlet or Neumann boundary value problems for three elliptic partial differential equations, whose solution defines a one-to-one mapping from the unit cube onto a block. A uniform grid in the unit cube should give, under this mapping, a desired grid in the block. A popular version is the technique proposed by Winslow and Thompson (1982).

Algebraic grid generation is based on the application of transfinite interpolation formulas in three-dimensional blocks. Conceptually, a uniform grid in the three independent variables in the interpolation formulas should produce a desired grid in the block. Transfinite interpolation formulas can be given the property to interpolate exactly in given continuous data of the block-face geometry. An introduction to transfinite interpolation theory is presented by Gordon and Thiel (1982). Gordon (1969) has given a detailed technical, excellent survey of the topic. The technique has been applied with at least reasonable success by e.g. Eiseman (1982), Eriksson and Rizzi (1983), and Smith, Kudlinski, and Everton (1984). At NLR, the technique is also applied (Boerstoel 1986).

The elliptic differential equations used at NLR are somewhat simpler than those commonly used. They are (Boerstoel 1986)

$$Lx = 0, \quad Ly = 0, \quad Lz = 0,$$

$$Lx = \left(w_1^2 \|P_\xi\| x_\xi\right)_\xi + \left(w_2^2 \|P_\eta\| x_\eta\right)_\eta + \left(w_3^2 \|P_\zeta\| x_\zeta\right)_\zeta,$$

$$\|P_\xi\| = (x_\xi^2 + y_\xi^2 + z_\xi^2)^{\frac{1}{2}}, \quad \text{etc.,}$$

where the w_i are user-specified positive smooth weight functions of (ξ, η, ζ) that should be chosen approximately inversely proportional to desired grid point distances (mesh sizes). These weight functions thus provide direct smoothness and mesh size control in a conceptually simple way.

Elliptic grid generation procedures have drawbacks to be taken into account. It must be expected that mesh size control will usually give rise to "air cushion" effects: if the mesh must be retuned, local changes in control functions will in general produce grid changes in at least an entire block, so that the design of a good grid is sometimes tricky. Further, the ellipticity of the differential equations (often) guarantees the absence of grid folding in continuous solutions. However, the discrete solutions of discrete elliptic difference equations are then not always free of folding, so that grid designers sometimes have to make their grids fold free using additional techniques (fixes).

A grid generator for complex flow domains may thus be based on the following construction principles..

- Make flow domains finite, if necessary.

- Apply block decomposition (topologically, geometrically). Patch blocks. Make blocks configuration-boundary conforming.

- Generate initial grid point distributions by algebraic techniques.

- Fine tune grid point distributions by simple elliptic techniques.

- Apply transfinite interpolation and elliptic grid point distribution techniques first in the edges, then in block faces, and finally in block interiors of a block decomposition of a flow domain, to obtain a well-decomposed grid construction procedure.

For further details, see Boerstoel (1986).

4. Examples and experiences

In this section, the example of the construction of a blocked grid around a wing–nacelle–propeller configuration in a wind tunnel is used to illustrate a number of technical issues, and to report experiences.

The geometry of the wing–nacelle–propeller configuration is presented in Fig. 4.1. The propeller (not shown here) is modelled in the flow solver as an actuator disk, with a flat annular ring surface around the spinner as geometry. The tunnel side walls (not shown here) at the two tips of the unswept and untapered wing are parallel to the vertical symmetry plane of the configuration, and extend to infinity. The nacelle has underneath the wing an outlet, at about half-chord position.

The flow domain around the configuration and between the tunnel walls was decomposed into 96 blocks. Each vertex, edge, face and block in the block structure was numbered (labelled). An experienced engineer can construct block decompositions of a flow domain around a complex configuration like that discussed here in a few days (10–40 hours). The connectivity relations were stored in the tables (arrays) as explained in Fig. 3.3. These tables were automatically produced with auxiliary Fortran 77 software.

For the definition of the topology of a block decomposition, accurate geometry is not required, so that for this particular purpose PATRAN could be used. PATRAN is one of the few solid model software packages with deformed cubes as building elements. Hence, decomposed flow domains automatically consist of deformed cubes, a useful property when complex spatial decompositions are required.

Having defined the topology of a block decomposition, the geometry of a block decomposition was accurately defined. This means that the geometrical shape and position of each block face was numerically defined as a

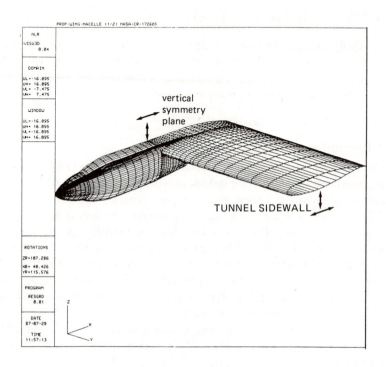

204

Boerstoel

Fig. 4.1: Wing–nacelle geometry

smooth surface by a slope-continuous function for the surface coordinates, with two parameters as independent variables.

After completion of the topology and geometry definition, the grid was designed, first in each edge, then in each face, and finally in each block. This process was repeated in order to iterate out to weight functions w_i giving a desired grid.

The distribution of grid points on the vertical symmetry plane in the flow and on the configuration surface (including the propeller disk) is shown in Figs. 4.2a–d. See further Figs. 4.3–4.4. The following is illustrated.

- Grid lines over block faces are continuous, but in general not slope-continuous. Inside blocks, faces, and edges, grids are smooth, however. Block edges in the grid may be traced as lines over which other grid lines are slope-discontinuous.

- Degenerated block faces (triangular geometry) are visible on the spinner. In the degenerated blocks around the spinner and at the propeller disk, the grid has the topology of a cylindrical coordinate system, with as axis the grid line from the spinner leading edge in upstream direction (Fig. 4.2c).

- Because of the face-to-face packing of blocks adopted here, block edges are continuous lines through the flow domain interior. In grid plots, they may be traced as grid lines where other grid lines are slope-discontinuous. Fig. 4.2 shows a number of such edges propagating through the entire flow interior.

The figures illustrate various rules applied during the design of the grid. These rules were found useful when using the grids in Euler–flow calculations. The rules concern the concept of grid quality. Grid quality may be measured by stretch factors, skewness angles, and slenderness ratios.

- A stretch factor is defined as the ratio of the lengths of two successive meshes in the grid along one grid line. A skewness angle is defined as the blunt angle between two grid lines, minus 90°. A slenderness ratio is the ratio of the lengths of two meshes of a grid cell in two different directions.

- In the interior of each block, each face, and each edge (where the grid is formally smooth, because the grid point distributions are a solution of second-order partial difference equations), stretch factors should be preferably in the range 0.9–1.1, with 1 as the ideal factor. The grid in the figures have sometimes somewhat larger stretch factors. Stretch factors are controlled by block dimensions, by the number

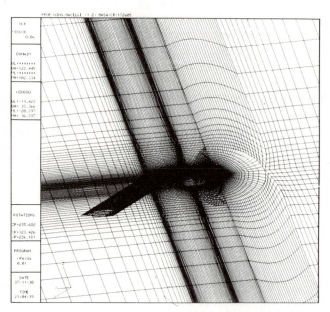

Fig. 4.2a: Grid and block faces in vertical symmetry plane and on configuration

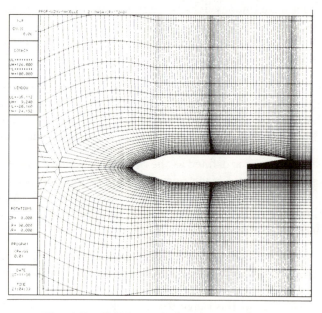

Fig. 4.2b: Grid in vertical symmetry plane

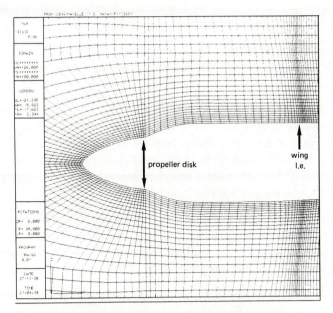

Fig. 4.2c: Grid in vertical symmetry plane

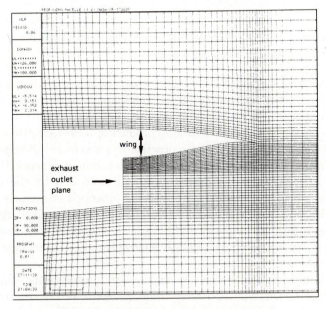

Fig. 4.2d: Grid in vertical symmetry plane

of grid points in a block coordinate direction, and by the weight functions in the elliptic grid generation procedure.

- Over block faces, stretch factors should be preferably in the range 0.5–2.0, because then the Euler flow solver can handle such factors easily.

- In the interior of each block and each face, skewness angles and slenderness ratios are allowed to deviate from their nominal values of 0 degrees and 1.0. The Euler–flow solver can handle skewness angles as large as $50° - 60°$ (Fig. 4.2b), and slenderness ratios of 100 (Figs. 4.2a,d), if skewness angles and slenderness ratios do not vary too fast in the interior of each block and each face.

- At the leading edge of the wing, the grid has an H–type topology, because blocks leading to a C–type topology cannot so easily be combined with blocks leading to good grid resolution around the spinner, nacelle and propeller disk.

- In Euler–flow solver runs, it was observed that a good grid quality near the wing leading edge is needed to obtain a convergent calculation process and accurate results. A reasonable grid is shown in Figs. 4.3–4.4.

- Similarly, at the wing trailing edge, the grid should have the same good quality, see Figs. 4.2d and 4.4.

This section is completed by discussing possible improvements in the presented concepts and procedures. The elliptic mesh tuning technique allows, in principle, accurate grid design by defining iteratively the weight functions in the difference equations producing a desired grid. As mentioned at the end of section 3.6, in practice this was found to be sometimes at some places in a grid a tricky and time consuming process due to "air cushion" effects, when the desired mapping described by the difference equations, boundary conditions and control functions was nearly singular. This is the case, for example, at trailing edges of wings in 0–type grids. Therefore, it may be useful to build the mesh tuning in the linear transfinite interpolation procedures, and to remove the elliptic techique from the grid generation procedure.

Grid folding could become a second reason to remove the elliptic technique. Discrete Dirichlet boundary value problems for the discrete elliptic difference equation systems do not always have a fold-free grid as solution. It is hoped that linear interpolation procedures can be made more robust in this respect.

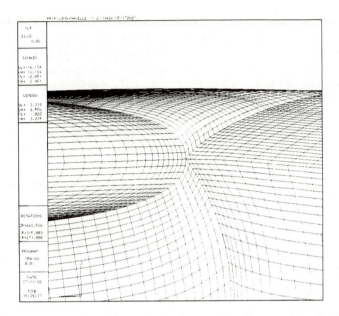

Fig. 4.3: Oblique view of grid on wing and nacelle near intersection

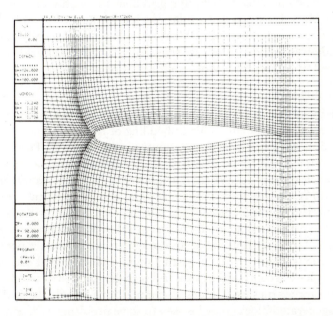

Fig. 4.4: Grid in flow around wing at approximately the propeller slipstream
boundary surface

The face-to-face packing of blocks can be generalized somewhat, by allowing new blocks that are the union of existing blocks. The topology of each new block should again have a vertex–edge–face structure of a cube. Such "unstructured block packings" will complicate the data structures for the interfacing of blocks, but will solve the following problems encountered in practice.

- It was found that, in the flow solver, degenerated blocks may lead to reduced convergence speeds and/or accuracy loss, due to dense packings of grid lines. It will often be possible to eliminate degenerated blocks by combining them with other blocks to a non-degenerated block, so that dense packings of grid lines can be avoided.

- Blocks were found to have in practice a rather small number of grid points in the three block coordinate directions, so that the flow solver and grid generator codes have a somewhat low vector speed due to short vectors. Vector lengths and speeds will increase if unions of blocks are grouped together to a new block.

It is expected that improved mesh size control (and thus accuracy control) should be made possible by allowing grid lines to be only partially continuous over block faces, in the regular fashion shown in Fig. 4.5. Further, one may allow in the interior of blocks subblocks with locally finer grids, if this is desirable for accuracy reasons. Such locally finer grids require in the flow solver appropriate prolongation and restriction operators.

5. Conclusions

Grid generation is a subtask in a numerical simulation of a flow in industrial and research environments. This embedding of the grid generation task is considered in section 2 from an informatics point of view. The design of a grid generator is the topic of section 3. Requirements to be met concern the geometrical input, the desired grid as output, the technical means to control grid resolution and grid quality, and the requirements concerning turnaround time performance for completing a grid generation task. Conceptual starting points for such designs are that flow domains are alwyas made finite if they are not, and that blocked grids offer better perspectives than unstructured grids. The construction of a blocked grid can be subdivided in a block decomposition task and a grid point distribution task. We opted here for "structured" block decompositions (blocks packed face-to-face). Degenerated blocks may be used to solve problems with connectivity relations between vertices, edges, faces, and blocks. Further, a technique for using these connectivity relations to define conventions about local coordinate systems in edges, faces and blocks is presented. Examples and

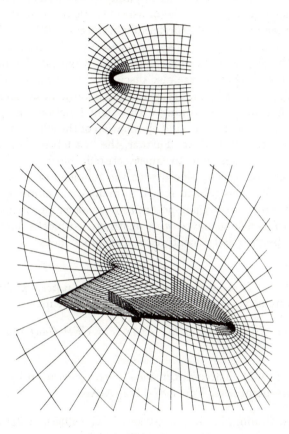

Fig. 4.5: Regular partial continuity of grid lines over block faces

experiences are reported in section 4. The example concerns a 96–blocked grid around a complex aerodynamic configuration. The chosen procedure was found effective and flexible. Slope discontinuities and stretch factors of grid lines over block faces are accepted. Experience with flow solver results gave indications about desired grid quality (measured by stretch factors, skewness angles, and slenderness (or aspect) ratios): in smooth parts of the grid, stretch factors should be preferably in the range 0.9–1.1, but skewness angles and slenderness ratios can vary between wide limits, if their variation over the grid in each block is gradual. Concepts for improvements in the presented technique are also discussed in section 4. These concern accuracy enhancement by mesh size halving in only those blocks where this is required, and mesh size tuning with weight functions in the transfinite linear interpolation technique and elimination of the elliptic technique from the grid generation procedure. Further, the "structured" block decompositions should be generalised to "unstructured" block decompositions, by allowing desired unions of blocks with the topology of a cube to become a new block. This generalization will in many cases eliminate the need of degenerated blocks with their potential flow solver problems. Better accuracy control is also achieved with unstructured blocks, and by allowing grid lines to be "partially" continuous over block faces.

References

AGARD (1984). *Large scale computing in aeronautics*. AGARD-AR-209, page 1, summary; page 23, section (e): pacing items.

AIAA (1987) *Proceedings of the AIAA 8th CFD Conference*. Honolulu, Hawaii.

Baker, T. J. (1987). Three-dimensional mesh generation by triangulation of arbitrary point sets. Paper 87-1124, AIAA.

Benek, J. A., Buning, P. G., and Steger, L. G. (1985). A 3D chimera grid embedding technique. Paper 85-1523, AIAA.

Boerstoel, J. W. (1986). Preliminary design and analysis of procedures for the numerical generation of 3d block-structured grids. Technical Report TR 86102 U, NLR.

Coleman, R. M. (1982). *Generation of boundary-fitted coordinate systems using segmented computational regions*, pages 633–652. In Thompson, 1982 edition.

Eiseman, P. R. (1982). *Automatic algebraic coordinate generation*, pages 447–464. In Thompson, 1982 edition.

Gordon, W. J. (1969). Distributive lattices and the approximation of multivariate functions. In *Approximations with special emphasis on spline functions*, pages 223–277. Academic Press, New York.

Gordon, W. J. and Thiel, L. C. (1982). *Transfinite mappings and their application to grid generation*, pages 171–192. In Thompson, 1982 edition.

Häuser, J. and Taylor, C. (1986). *Numerical grid generation in computational fluid dynamics*. Pineridge Press, Swansea, UK.

Lee, K. D., Huang, M., Yu, N. J., and Rubbert, P. E. (1980). *Grid generation for general three-dimensional configurations*, pages 355–366. In Häuser and Taylor, 1986 edition.

Rizzi, A. and Eriksson, L. E. (1983). Practical 3D mesh generation and explicit schemes to solve the Euler equations. In Habashi, W. G., editor, *Advances in computational transonics*. Pineridge Press.

Rubbert, P. E. and Lee, K. D. (1982). *Patched coordinate systems*, pages 235–252. In Thompson, 1982 edition.

Smith, R. E., Kudlinski, R. A., and Everton, E. A. (1984). Algebraic grid generation for wing-fuselage bodies. Paper 84-0002, AIAA.

Thompson, J. F. (1982). Numerical grid generation. In *Proc. Symp. Num. Grid Gen. of Curv. Coord. Syst.s and their Use in Num. Sol. of Part. Diff. Eqs.*, North-Holland, New York.

Weatherill, N. P. and Forsey, C. R. (1984). Grid generation and flow calculations for complex aircraft geometries using a multi-block scheme. Paper 84-1665, AIAA.

Weatherill, N. P., Forsey, C. R., and Shaw, J. A. (1985). Grid generation and flow calculations for complex aerodynamic shapes.

An approach to geometric and flow complexity using feature-associated mesh embedding (FAME): strategy and first results

C. M. Albone
RAE, Farnborough

1. Introduction

Flows of aeronautical interest are characterised by high Reynolds numbers, compressibility and the presence of bodies of great geometric complexity (Green 1982). As a result, such flows exhibit many complex features and these have usually been classified as geometric features (detailed body shape) or as flow features (shock waves, thin shear layers, etc.). The task of generating high quality meshes that will allow a flow algorithm to resolve all these features with sufficient accuracy for design purposes has occupied workers in CFD for many years. Accurate resolution of geometric features has conventionally been approached through the generation of a computational mesh (Thompson et al. 1985) that is aligned with the solid boundaries. Mesh generation schemes usually however make little concession to flow features, although solution-adaptive meshes (Nakahashi and Deiwert 1985; Catherall 1988) are beginning to address this problem. Accurate resolution of flow features has relied upon the careful construction of discretization schemes (Roe 1982) that keep the distortion of the feature to an acceptable level. This disparity in approach to geometric and flow features is understandable. Intolerable constraints would be placed upon a mesh that was required to be aligned to all body surfaces and also to a set of flow features whose location, or even existence, was not known in advance. Priority has in the past been given to the treatment of geometric features since these are know *a priori* and overall solution accuracy is critically affected by this treatment.

The aim of the work reported here is to unify the treatment of geometric and flow features through a more flexible approach to mesh generation. In pursuing this objective we will be making demands upon the mesh that exceed those required for mesh-generation schemes which largely ignore flow features. Since the latter schemes are already heavily constrained by geometric features alone, it is clear that a new mesh-generation strategy is required. In line with other workers (Rai 1985), we shed some constraints on the mesh by covering the flow field with several meshes. Multiblock methods (Shaw et al. 1986) adopt this approach, but they do not permit mesh overlap and usually require a fair degree of continuity at mesh interfaces, because flow calculations are perfomed as if mesh interfaces were absent. As a result, multiblock meshes, although capable of use with quite complex shapes, remain heavily constrained for practical geometries and so appear unsuitable for the unified treatment sought here. In the present approach, we allow more than one mesh to cover certain parts of the field. Where overlapping occurs, the position in a mesh hierarchy determines which mesh takes precedence. We dispense where necessary with continuity requirements (position, slope, orientation) between meshes, retaining just one condition; that the densities of meshes that occupy the same region of space are similar (up to a factor of two) in the region of overlap. This condition is satisfied primarily through the extensive use of mesh embedding. Other overlapping mesh-generation methods (Rai 1985) give this condition some consideration,but the flexibility of mesh control necessary to satisfy the condition in all the overlap regions is not always present. It is this flexibility, achieved through the use of separate feature-associated meshes together with the control of mesh density by embedding, that sets the work reported here apart from most other mesh-generation schemes and places it more in the category of unstructured-mesh methods (Löhner et al. 1986), but without the overheads usually associated with such methods.

In Section 2 we describe some elements of the strategy for feature- associated mesh embedding. The implementation of this strategy is illustrated in Section 3 for two-dimensional meshes and examples of meshes for multi-element aerofoils are shown. In Section 4 we suggest how the present approach might be generalised by proposing treatments for two further types of feature. Flow-algorithm requirements are discussed in Section 5 and a novel second-order accurate Euler algorithm is briefly described. The current status of the work programme and concluding remarks comprise Section 6.

2.   Mesh-generation strategy

In pursuit of a unified approach to geometric and flow features, we seek a classification that reflects the nature of each feature from the mesh-generation viewpoint. Here, we use the term feature in a rather wider context that is customary. The classification adopted is inspired by the nature and range of features present in high Reynolds number compressible flow past complex configurations; it is however equally valid, though less necessary, for simpler flows. We consider four types of features; these are classified according to the number, N, of directional constraints associated with the feature and are denoted as being of type N, N = 0, 1, 2, 3. Features of type 1 are associated with surfaces, those of type 2 with lines and those of type 3 with points. Type-0 features have no directional constraints associated with them. We assert that each feature merits its own mesh having a topology that is appropriate to the feature. We refer to meshes that correspond to features of type N as being type-N meshes. In the remainder of this Section and in Section 3, we concentrate upon meshes of types 0 and 1. Those of types 2 and 3 are discussed further in Section 4.

Regions of the flow field lying away from solid surfaces, shear layers and shock waves are comprised of features of type 0. Such regions could alternatively be termed 'featureless', and indeed, since we classify features according to the number of directional constraints associated with them, this term is not inappropriate. However, the numerical treatment employed for type-0 features is crucial to the accuracy of flow calculations, since these features may include regions of high (but not directionally dominated) flow gradient. Thus flow gradients are type-0 features, and, consistent with the spirit of feature-associated meshes, the meshes that we construct should reflect the magnitude of the flow gradients. There is of course nothing novel in this statement; all good mesh-generation methods seek to match mesh density with anticipated flow gradients or with actual flow gradients through solution adaption. Here though, we achieve this control of density exclusively through the use of mesh embedding (not stretching), so that a set of locally uniform embedded meshes of type 0 may be constructed to cover all type-0 features. Since these features do not have directional constraints associated with them, type-0 meshes can simply be rectangular Cartesian type as sketched in Fig. 1a for two dimensions.

Features such as body geometry, shock waves and vortex sheets in inviscid flow are associated with surfaces in space; others, such as shear layers in high Reynolds-number flow, are associated with thin regions adjacent to surfaces (which may be solid surfaces or surfaces within the fluid). All such features are characterised by the direction of the normal to the associated surface and accordingly these are denoted as being of type 1. Each type-1 feature merits its own mesh, which should be orientated according to the orientation of the associated surface. Where, for example, the surface is a solid boundary, this simplifies the task of satisfying the solid-surface boundary conditions. Each type-1 mesh could therefore be constructed with two coordinate directions in the surface and the normal to it. The distribution of mesh points within each mesh of type 1 is not a major consideration here; it could be prescribed by any of the methods currently used for conventional meshes or it could be subject to solution adaption. The novel aspect of the present approach concerns the spatial extent of each type-1 mesh in the direction normal to the associated surface. The only requirement on this is that the feature is entirely enclosed by the mesh. In inviscid flows, all type-1 features are simply surfaces and so, unless there are good reasons for doing otherwise, meshes of type 1 need only extend one mesh interval away from the surface, as sketched in Fig. 1b. For features such as thin shear layers, the need to resolve very high flow gradients normal to the associated surface will require us to use a mesh with many small intervals normal to that surface. Nevertheless, the extent in physical space again need only cover the feature, see Fig. 1c.

At the outset of a flow calculation, we cannot usually predict the nature, location or even the existence of all features of type 1. In consequence we cannot *a priori* partition all space into sets of meshes of the appropriate topology. In order to overcome this, we allow more than one mesh to cover the same region of space. Initially, a very coarse mesh of type 0 is generated to cover the whole field of interest. This is constructed ignoring any type-1 features that are present. Meshes of type 1 are then generated and embedded within (i.e., overlie) the type-0 mesh as required, in response to specified surface geometry or at a later stage to flow features that evolve during the calculation. At this stage, the densities of the (coarse) type-0 mesh and those of the type-1 meshes may be vastly different in the region of overlap. Comparability of mesh density is achieved by generating finer type-0 meshes and embedding these in the initial coarse mesh. This highly flexible embedding procedure, described in more detail in Section 3 for a two-dimensional example, may have any number of embedded levels. We consider the *complete set* of type-0 meshes as comprising the main (or background) mesh.

It is convenient to consider that the set of embedded type-0 meshes overlie each other. Thus if the original coarse type-0 mesh is defined as being at embedding level 1 and the finest mesh is at embedding level m, then the region of space covered by the level-m mesh is also covered by coarser meshes on levels $m-1$, $m-2$, ... 1. Thus certain regions of space could be covered by a type-1 mesh and by up to m type-0 meshes. It is necessary to establish a hierarchy of meshes so that where many meshes cover the same region, one is deemed to take precedence over all others, in that it is the most appropriate mesh for the feature located in that region. If part of mesh B occupies the same region of space as mesh A, and the latter is further up the hierarchy, then flow data at points on mesh B that lie in the overlap region are replaced by flow data (interpolated where appropriate) from points on mesh A. This replacement is necessary because mesh B was not constructed to cope adequately with the feature with which mesh A is associated, whereas, by design, mesh A was. Thus close to a solid surface, a type-1 mesh, which has been designed to cope with that surface, will take precedence over all type-0 meshes that occupy the same region of space, since these latter meshes were constructed with no regard for that (or any other) surface. Meshes of type 1 therefore head the hierarchy, followed by the type-0 mesh at embedding level m, then m-1, and so on with the coarsest type-0 mesh at embedding level 1 at the bottom.

Finally, we should note that each locally uniform embedded mesh of type 0 is synthesized from a set of computationally identical units called blocks. These blocks play a crucial role in the embedding procedure as shown in Section 3 for a two-dimensional example.

3.   Illustrative examples

Let us illustrate the ideas developed so far by considering a two-dimensional aerofoil in an unbounded inviscid flow.

3.1   Type-1 mesh

For this example, the only type-1 feature that we are concerned with is the aerofoil surface. From a given mesh-point distribution on the aerofoil, we construct a one-cell-thick surface-aligned mesh of type 1 by projecting normals into the field from each point. At each surface mesh point, we make the length, δ, of the normal proportional to the mean of the two adjacent surface mesh intervals, so that the constant of proportionality, K, determines the cell aspect ratio. This mesh is extended beyond the trailing edge as a C mesh (Thompson et al. 1985) but is then truncated one cell down-stream of the trailing edge. Fig. 2 shows an aerofoil together with a type-1 C mesh, for a case having 64 surface mesh intervals around the profile.

3.2   Type-0 meshes

In Fig. 3 we sketch an aerofoil with (for simplicity of illustration) only 12 surface mesh points, but now with a coarse type-0 rectangular mesh extending to the far-field boundary. For generality, the aerofoil is offset from the centre of the mesh. We consider this mesh to consist of just one block, denoted block 1 at embedding level 1. In Fig. 3, we have taken the number, n, of cells per block in each coordinate direction as 4 (although n could take any even number ≥ 4). The embedding procedure (Albone and Joyce), which is designed to achieve comparability of mesh size in the region of overlap with the type-1 mesh, proceeds as follow.

(i) Divide block 1 into four quadrants.

(ii) Taking each quadrant in turn, determine whether there are any surface mesh points in that quadrant for which $\delta < \Delta_1 d$, where Δ_1 is the mesh length in block 1 and d is a control parameter (usually $\sqrt{2}$ for two-dimensional flows).

(iii) If a quadrant contains no such points, pass to the next quadrant; otherwise refine the mesh in that *quadrant* by a factor of two in each coordinate direction, so creating a new embedded block that has the same number of cells as block 1.

(iv) Up to four such blocks may be created and these are considered to constitute the set of blocks at embedding level 2 that comprise the level-2 type-0 mesh. Each new block has a mesh length $\Delta_2 (= \frac{1}{2}\Delta_1)$.

(v) Repeat steps (i)-(iii) for each quadrant of each block on level 2, where the criterion for further embedding is that points for which $\delta < \Delta_2 d$ are found in a quadrant.

(vi) The process is repeated and embedding stops when, for all quadrants of all blocks on the current level, the criterion is not satisfied.

The result of the above procedure is shown in Fig. 4 for our simple example and, at that stage, the main-mesh cell size is locally comparable to that of the surface mesh. However, the procedure is not complete because very coarse meshes of type-0 can occur close to (but not at) the aerofoil surface. (Offsetting the aerofoil from the centre of the original square highlighted this.) Also, the ratio of mesh lengths of adjacent blocks varies enormously. The final step in the embedding procedure is to generate more blocks as necessary in order that the mesh lengths of adjacent blocks do not differ by more that a factor of two. This prevents coarse type-0 meshes from being located close to the aerofoil surface and it simplifies the transfer of flow data between the meshes. Further, it ensures that the comparability of mesh size, required where the main mesh and type-1 meshes overlap, is a property of the main mesh itself. Fig. 5 shows the final embedded structure of type-0 meshes for our example and this is constructed simply from the surface distribution of mesh points and from the three control parameters, K, n, and d, referred to earlier.

3.3 Data transfer between meshes

The main-mesh embedded structure described above may be constructed without reference to the computational molecule associated with the flow algorithm. However, transfer of data between type-0 meshes is clearly simpler if the molecule is fairly compact. Data transfer is also influenced by whether the flow variables are stored at cell centres or at cell vertices. An upwind finite-difference (cell-vertex) flow algorithm (Walkden; Albone), described briefly for the Euler equations in Section 5, is adopted in the present work. It has a five-point molecule in two dimensions (seven points in three dimensions) as shown in Fig. 6. However, in constructing procedures for the transfer of flow data we have allowed for a nine-point molecule in two dimensions (27 points in three dimensions) as shown in Fig. 7, so that the method could accommodate cell-vertex, Lax-Wendroff Euler schemes (Hall 1986). In Fig. 8, we sketch the main mesh and a type-1 mesh close to a solid boundary to illustrate the data transfer between meshes of these types, where for simplicity of explanation, we have omitted main-mesh embedding. The main mesh has not been constructed to cope with the solid surface and so at a set of main-mesh points (marked by triangles in Fig. 8) the nine-point computational molecule cannot be defined without reference to at least one main-mesh point inside the solid surface. The flow data at these points are replaced by data obtained from further up the hierarchy using linear interpolation from the four vertices of the appropriate cell of the type-1 mesh. Outer boundary conditions needed for the type-1 mesh are obtained at a set of points marked by circles using linear interpolation of flow data from the four vertices of the appropriate main-mesh cell. In this case we note that data comes from further down the hierarchy. Strong coupling of the two interpolations is avoided by ensuring that the points marked by triangles are not too close to the edge of the type-1 mesh, and this is controlled by the parameters K and d.

3.4   Other two-dimensional examples

Main and surface-aligned, type-1 meshes for one-, two-, and three-element aerofoils are shown in Figs. 9, 10, 11 respectively. In each case, the type-1 meshes have 64 surface points on each element and the values of the control parameters were $n = 4$, $K = 1.2$ and $d = \sqrt{2}$. The main-mesh embedding in these cases was driven, not only by the surface distributions of points, (as described earlier) but also by the set of points that constitute the edge of each type-1 mesh and by a fictitious set of points located two mesh intervals from the surface. This procedure ensures that the embedding level achieved at the surface is maintained over the whole of the overlap region.

4.   Generalisations

In the example of the previous section, we did not consider other type-1 flow features such as shock waves nor did we have to cope with intersecting surfaces. An overlying mesh treatment of shocks which employed shock

fitting was described in (Albone 1986). The approach was based upon
the same principles as described in this paper, but in that work, the main
mesh did not have an embedded structure. It did however require that the
shock-orientated overlying mesh extended both sides of the shock and that
it was capable of movement with respect to the main mesh. These aspects
whilst increasing program complexity a little were entirely compatible with
the principles established here for features of type 1.

The generalisations we consider in this section concern the features of
types 2 and 3 that were mentioned at the beginning of Section 2. We define
features of type 2 to be those associated with lines, examples of which may
be the line of intersection of two surfaces or a line along which the surface
normal is discontinuous. (Whilst we may sometimes view these as distinct
in aerodynamic terms, geometrically they are one and the same.) Whereas
features of type 1 are associated with a single surface and characterised by
the surface normal, features of type 2 are associated with two surfaces and
are characterised by the normals to each surface at their line of intersection.
We refer to type-2 features as edge lines (or edges) and consider this term
to embrace all lines of intersection of two surfaces irrespective of their sense
(either 'convex' or like a wing trailing edge or 'concave' like a typical wing-
fuselage junction line). The two surfaces concerned may be associated
with any geometric or flow features of type 1. We assert here that each
feature of type 2 merits its own feature- associated mesh. Such meshes are
necessary because each type-1 mesh is designed to cope with only a single
surface (one directional constraint). Therefore, close to an edge line where
two surfaces intersect (two directional constraints), neither of the type-
1 meshes associated with each surface will be suitable. Type-2 meshes,
however, are designed specifically for edge lines.

In conventional mesh-generation schemes edges are treated in a variety
of different ways according to the sign and magnitude of the discontinuity in the normal. We may view these treatments two-dimensionally by
considering planes normal to edge lines and characterising the edge by an
angle θ, $-\pi < \theta < \pi$, which measures the discontinuity in the surface normal, with θ taking positive values for edges of convex type and negative
for those of concave type. If we denote by ε a positive angle that is small
compared with π, we may identifiy three different edge-line topologies in
common use as shown in Figs. 12a-12c for $| \theta | < \varepsilon$, $| \theta \pm \pi/2 | < \varepsilon$ and
$| \theta \pm \pi | < \varepsilon$, respectively. Whilst these topologies are perfectly adequate
for the geometry concerned, their use presents difficulties where θ varies
significantly along the edge as for example in the case of a wing-fuselage
junction where topologies (a) and (b) may occur along the junction line. In
multiblock mesh-generation schemes (Shaw et al. 1986) a switch from one
topology to another along the junction line would require several blocks

covering the region local to the junction which were topologically distinct from each other and possibly from those blocks lying away from the juction. Only an O-H topology (Thompson et al. 1985) will permit a unified treatment of all three cases and this would take the form of a cylindrical-polar type mesh with its axis running along the (possibly twisted) edge line. The inclusion of edge blocks of O-H topology is of course quite feasible in some of the better multiblock schemes (Shaw et al. 1986), but it could add significantly to the constraints imposed upon the mesh and so may not be easy to implement in practise.

Here, we propose that each type-2 feature should have its own local mesh of O-H topology. Where a type-2 mesh covers a region of space that is aso covered by part of the main mesh and/or by part of a type-1 mesh, the edge mesh takes precedence over the others (and so is located further up the hierarchy), since it is locally the most appropriate mesh. In Fig. 13, we have sketched a proposed mesh structure close to a wing trailing edge as viewed along the edge for the case where the meshes each have a field penetration of two cells. This structure enables the user to specify, if necessary, a locally very fine edge mesh, with many cells of field penetration, so as to resolve the high flow gradients close to the trailing edge; such resolution is not easily achievable with conventional wing meshes. Alternatively, if the wing trailing-vortex sheet (or that from a lifting aerofoil with strong shocks) were to be treated as a type-1 feature, at least close to the wing, then, given the vortex-sheet surface, two edge lines each with their own type-2, O-H mesh could be defined. In fact, the flexibility of the structure is such that the classic swept-wing flow pattern (Albone and Hall 1980) of a triple shock wave with a vortex sheet springing from the intersection line is resolvable with shock and vortex-sheet fitting in the flow field through the use of four type-1 meshes (three shock and vortex sheet) and four type-2 edge meshes.

We complete our feature-orientated mesh hierarchy by defining the third type of feature (type 3) which is associated with three (or more) surfaces and is characterised by the normals to each surface at their point of intersection. Features of type 3 are referred to simply as points and they usually occur where two or more type-2 edge lines meet (at for example the point of intersection of the wing trailing edge and the wing-fuselage junction line or at the trailing edge of a cranked wing). They may alternatively exist at isolated points on an otherwise smooth surface (at for example the nose of a pointed body of revolution). As with types 1 and 2, type-3 features merit their own feature-associated mesh. Their topology, being appropriate to a point, should have O-O structure (Thompson et al. 1985) and so take the form of a spherical-polar-type mesh. Where a type-3 mesh covers a region of space that is also covered by part of the main mesh and/or by part of a type-1 or type-2 mesh, the type-3 mesh takes precedence over the others since locally it is the most appropriate. This is because meshes of type 2 are designed to cope with a single edge line and so they cannot cope with the extra constraint(s) that applies close to a point where two such edge lines meet. Thus type-3 meshes head the hierarchy, followed by those of type 2, then type 1, and finally with the type-0 meshes ordered according to the level of embedding.

With the full hierarchy now complete, we may summarise the transfer of flow data. Replacement of flow data due to the inappropriateness of the mesh on which those data reside takes place from meshes further up the hierarchy. Boundary data, required where a mesh terminates within the field, are supplied from meshes lower down the hierarchy.

Finally, we note that the full range of mesh types is necessary in order to deal with a configuration as simple as an isolated swept wing, even where only geometric features are considered. However, when once the hierarchical structure is established together with the housekeeping necessary to organise the transfer of flow data between meshes, configurations and flows of far greater complexity can be considered with the established structure, simply by introducing further embedded meshes of the appropriate type according to the nature of each new feature.

5. Flow solution methods

In the foregoing sections, we have made only passing reference to possible flow algorithms by noting the allowable extent of the computational molecule used for the flow calculation on the main mesh. By creating a structure of feature-associated embedded meshes, we have given ourselves the flexibility not only to have the most appropriate mesh for each feature, but to choose the flow algorithm appropriate to that feature. We could solve different equation sets on different meshes (for example the Euler equations or the Reynolds-averaged Navier–Stokes equations with various closure models). Further, for any given equation set, we could solve the equations in conservation form on one mesh and in quasi-linear form on another according to requirements. Finally, the discretization schemes adopted for any given equation set could vary from mesh to mesh. This 'zonal' approach is of course not new; however our mesh structure should enable it to be implemented simply and efficiently.

The ultimate objectives are to set up the following zonal structure for flow solution. On meshes that are associated with geometric features (solid surfaces), the Navier–Stokes equations should be solved together with no-slip boundary conditions in order to resolve the boundary layers. Free shear layers should also be covered by meshes upon which the Navier–Stokes equations are solved. Solution of the Euler equations should suffice for all other meshes including those associated with shock waves. It is intended that, where possible, shocks should be fitted as surfaces across which the flow variables are discontinuous. Where this is so, solution of the Euler equations on shock-associated meshes can be in quasi-linear form (Albone 1986). If on the other hand shock capture is preferred then the conservative form (Roe 1982) of the Euler equations should be adopted. The flow in regions of the field covered only by the main mesh should by design be smooth and in consequence the simpler, quasi-linear form of the Euler equations may be used there.

Here, we choose the split-coefficient matrix method of Chakravarthy (Chakravarthy et al. 1980) for discretization of the Euler equations in quasi-linear form. This algorithm is virtually identical to the one employed in the earlier (Albone 1986) shock-fitting work. In its first-order accurate form, this upwind algorithm is stable when combined with simple explicit forward time stepping, it is amenable to implicit or multigrid (Albone 1985) solution procedures and it naturally supports simple characteristic-based treatments of all forms of boundary conditions (Chakravarthy et al. 1980). It uses symmetric five-point and seven-point computational molecules in two and three dimensions, respectively. No added dissipative terms are required either to enforce stability or to capture shocks cleanly. (It should be remembered though that shocks so captured do not satisfy the conservation laws due to the adoption of quasi-linear form.) In common with all other first-order accurate upwind schemes, the algorithm has leading truncation error terms that take the form of products of second-order dissipative terms with Δ, the mesh interval. Whilst this property is desirable for capturing shock waves, it renders the algorithm far too dissipative for smooth regions of the flow and, in consequence, exceedingly fine meshes are necessary in order to obtain results of acceptable accuracy. Conventional second-order accurate upwind schemes have a leading truncation error term that is dispersive and of order Δ^2, followed by terms that take the form of products of fourth-order dissipative terms with Δ^3. Their accuracy in smooth regions of the flow is vastly superior to their first-order accurate counterparts, though they require modification (Yee 1987) near to shocks in order to avoid 'wiggles'. Unfortunately, this improved accuracy is accompanied by an increase in the size of the computational molecule. Use of second-order accurate upwind methods therefore leads to a more difficult treatment of boundary conditions, and in the present approach it would complicate considerably the implementation of the algorithm on the complex embedding structure of the main mesh.

Spatial second-order accuracy is obtained here through the use of two first-order calculations (Walkden; Albone). In the first part, the split-coefficient matrix method in its first-order accurate form is applied to the

Euler equations to produce a first-order accurate solution. We can then generate a forcing term at each mesh point, which is an approximation to the local truncation error (Walkden), by operating upon this solution with any second-order accurate discretization operator for the steady-flow Euler equations. In the second part, we again apply the split-coefficient matrix method in its first-order accurate form but now with the forcing term subtracted from the right-hand side of the discretized equations to obtain a second solution. It can be shown (Albone) for linear equations, and has been demonstrated (Albone) numerically for the Euler equations, that this solution is second-order accurate. This result is due to the presence of the forcing term which cancels the leading dissipative truncation error term associated with the use of the first-order accurate upwind scheme. In both parts of the solution procedure, therefore, time-marching is carried out in conjunction with first-order upwinding, although for the second part a *fixed* anti-dissipative term is present. Hence the method has the stability properties of a first-order upwind algorithm yet has spatial second-order accuracy, and, if the second-order accurate operator chosen for evaluation of the forcing term is of central-difference form, the computational molecule remains unchanged from its seven-point form in three dimensions. This algorithm was used to calculate two-dimensions flows (Albone and Joyce) using the meshes shown in Figs. 9, 10, and 11. Surface pressure distributions for the three-element aerofoil of Fig. 11, computed at an undisturbed stream Mach number of 0.2 and an angle of incidence of 20°, are shown in Fig. 14 for meshes having 64, 96 and 128 surface points on each element.

It is required that the flow algorithm on the main mesh should be capable of capturing any flow features whose location or even existence is not known in advance, such as shock waves and free shear layers (vortex sheets in inviscid flow). Features so-captured should be identifiable by a detection algorithm, although their location and strength need not be given precisely. The algorithm adopted here satisfies this requirement (Albone; Albone 1986) in respect of shock waves but its vortex-sheet-capturing capabilities have yet to be assessed. Once identified, a flow feature will be covered by its own embedded mesh of the appropriate topology (which may then move with respect to the main mesh) and the flow algorithm selected as appropriate to the feature will be solved on that mesh so optimising the resolution of that feature. Exchange of flow data between meshes will, by construction, take place in smooth regions of the flow and so interpolation between meshes can be carried out using any convenient set of dependent variables.

6.   Current work status and concluding remarks

In this paper, work already completed is described and plans for the subsequent phases are promulgated. It may therefore be helpful to separate fact from fiction by clarifying the present position.

Proof of concept in two dimensions has been limited to the generation of meshes for multi-element aerofoils (Figs. 9, 10 and 11), and the approach has proved to be very flexible and simple to apply for each new configuration (Albone and Joyce). The accuracy of the flow calculations, as exampled by Fig. 14, is comparable to that obtained from other methods. Earlier shock-fitting work (Albone 1986) in two dimensions demonstrated the utility of dynamically-overlying meshes. Three-dimensional mesh generation and flow programs, based upon the present approach, are currently being assembled. Provision for main-mesh embedding structure and for meshes of type 1 is essentially complete together with all necessary three-dimensional interpolation schemes and data-transfer procedures. Meshes of types 2 and 3 will be incorporated into the structure following successful testing of the system in its present form. Development of the flow program is in hand and, and in its initial form, the second-order accurate Euler algorithm (Albone) of Section 5 will be used for the flow calculation on all meshes.

Clearly there is a long way to go before the complete system is fully assembled and available for use 'in anger' (many of the current problems relate more to computer science than to CFD). The unified approach to complex geometric and flow features has evolved slowly over the past few years (Albone and Taylor 1985). It is based in part upon the author's many years of involvement in CFD methods for complex configurations (much of which is unfortunately not reported in the open literature), and in part upon the author's observations of the successes and failures of other workers in the field. Whilst this approach may appear to be highly ambitious, it is not considered to be incompatible with the size of the task, which, with ever increasing demands for high accuracy for ever more complex configurations and flows, represents a moving target. The flexibility being incorporated into the present approach is aimed at building a system that can readily adapt to the changing requirements to which CFD methods are continually subjected.

Figures

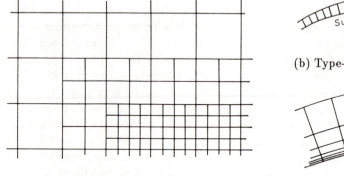

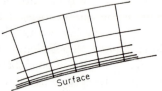

(b) Type-1 mesh: inviscid flow.

(a) Type-0 meshes.

(c) Type-1 mesh: viscous flow.

Fig. 1: Examples of Type-0 and Type-1 meshes

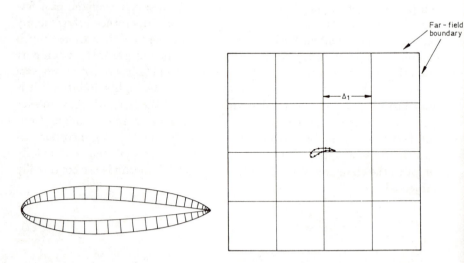

Fig. 2: Aerofoil and Type-1 mesh

Fig. 3: Aerofoil with far-field boundary and block 1 of a Type-0 mesh, containing $n \times n$ cells ($n = 4$)

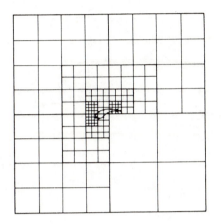 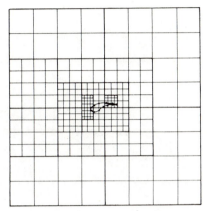

Fig. 4: First stage of embedding procedure

Fig. 5: Embedding procedure — complete

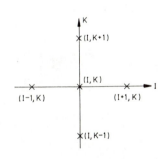 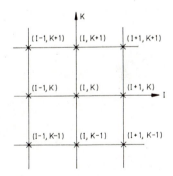

Fig. 6: Five-point computational molecule (two dimensions)

Fig. 7: Nine-point computational molecule (two dimensions)

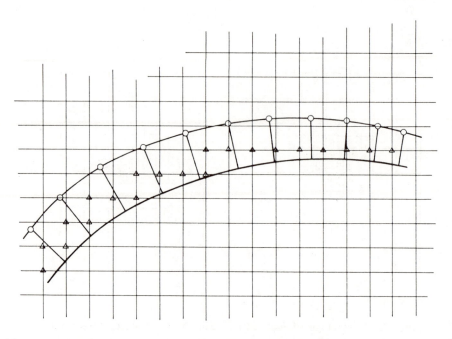

Fig. 8: Mesh points at which data transfer between main and Type-1 meshes occurs

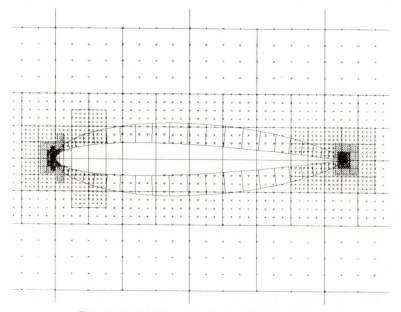

Fig. 9: Meshes for a single-element aerofoil

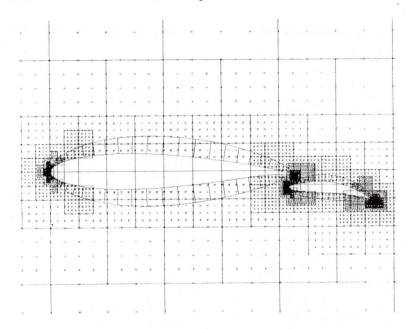

Fig. 10: Meshes for a two-element aerofoil

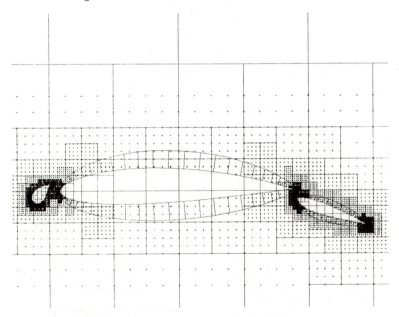

Fig. 11: Meshes for a three-element aerofoil

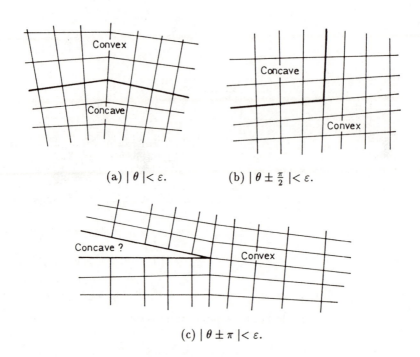

(a) $|\theta| < \varepsilon$.   (b) $|\theta \pm \frac{\pi}{2}| < \varepsilon$.

(c) $|\theta \pm \pi| < \varepsilon$.

Fig. 12: Topologies used in the treatment of edges

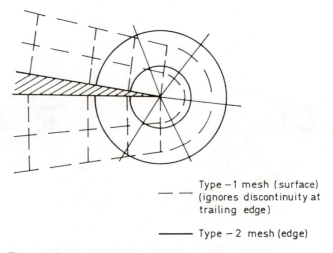

— — Type −1 mesh (surface)
(ignores discontinuity at
trailing edge)

——— Type − 2 mesh (edge)

Fig. 13: Proposed mesh topologies: wing trailing edge

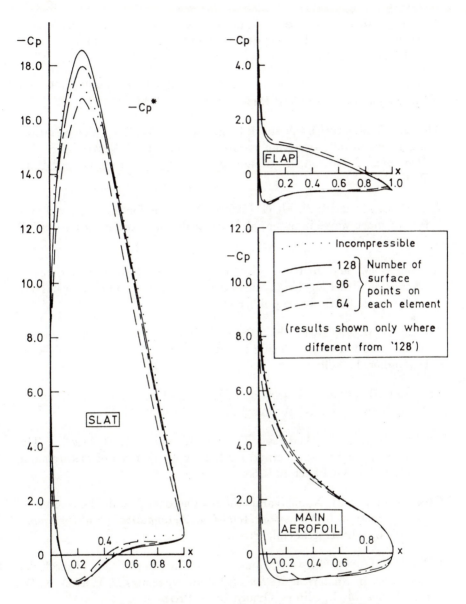

(x is measured along the chord line of each element)

Fig. 14: Three-element aerofoil $M_\infty = 0.2$, $\alpha = 20°$

References

Albone, C. M. A second-order accurate scheme for the Euler equations by deferred correction of a first-order upwind algorithm. Technical report, RAE. To be published.

Albone, C. M. (1985). An upwind, multigrid, shock-fitting scheme for the Euler equations. Technical Report TR 85004, RAE.

Albone, C. M. (1986). A shock-fitting scheme for the Euler equations using dynamically overlying meshes. In *Numerical Methods for Fluid Dynamics II*, Morton, K. W. and Baines, M. J., editors. Oxford Univ. Press.

Albone, C. M. and Hall, M. G. (1980). A scheme for the improved capture of shock waves in potential flow calculations. Technical Report TR 80128, RAE.

Albone, C. M. and Joyce, G. An overlying/embedded mesh-generation scheme and Euler algorithm for flows past multi-element aerofoils. Technical report, RAE. To be published.

Albone, C. M. and Taylor, K. (1985). Euler methods for complex geometries — are we, or you, losing the way? *CFD News*, (2/85), Walkden F., Laidler, P., editors.

Catherall, D. (1988). A solution-adaptive-grid procedure for transonic flows around aerofoils. Technical Report TR 88020, RAE.

Chakravarthy, S. R., Anderson, D. A., and Salas, M. D. (1980). The split-coefficient matrix method for hyperbolic systems of gas dynamic equations. AIAA Paper 80-0268.

Green, J. E. (1982). Numerical methods in aeronautical fluid dynamics—an introduction. In *Numerical Methods in Aeronautical Fluid Dynamics*, Roe, P. L., editor. Academic Press.

Hall, M. G. (1986). Cell-vertex multigrid schemes for solution of the Euler equations. In *Numerical Methods for Fluid Dynamics II*, Morton, K. W. and Baines, M. J., editors. Oxford Univ. Press.

Löhner, R., Morgan, K., Peraire, J., Zienkiewicz, O. C., and Kong, L. (1986). Finite-element methods in compressible flow. In *Numerical Methods for Fluid Dynamics II*, Morton, K. W. and Baines, M. J., editors. Oxford Univ. Press.

Nakahashi, K. and Deiwert, G. (1985). A self-adaptive-grid method with applications to airfoil flows. AIAA Paper 85-1525.

Rai, M. M. (1985). Navier–Stokes simulations of rotor-stator interaction using patched and overlaid grids. AIAA Paper 85-1519.

Roe, P. L. (1982). Numerical modelling of shock waves and other discontinuities. In *Numerical Methods in Aeronautical Fluid Dynamics*, Roe, P. L., editor. Academic Press.

Shaw, J. A., Forsey, C. R., Weatherill, N. P., and Rose, K. E. (1986). A block-structured mesh-generation technique for aerodynamic geometries. In *First International Conference on Numerical Grid Generation in Computational Fluid Dynamics, Landshut*. Pineridge Press.

Thompson, J. F., Warsi, Z. U. A., and Mastin, C. W. (1985). *Numerical grid generation, foundations and applications*. North-Holland.

Walkden, F. Private communication.

Yee, H. C. (1987). Upwind and symmetric shock-capturing schemes. Technical Report TM 89464, NASA.

Lax-stability vs. eigenvalue stability of spectral methods

Lloyd N. Trefethen[1]

Department of Mathematics,

Massachusetts Institute of Technology, USA.

1. Background

It is well known that spectral methods for non-periodic geometries lead to matrices that are not normal. We show that in such situations there may be a wide gap between eigenvalue stability and Lax-stability, especially for first-order problems. For example, in some cases eigenvalue analysis predicts a stability restriction $\Delta t \leq CN^{-1}$, whereas the Lax-stability restriction is $\Delta t = O(N^{-2})$. When such anomalies occur, the results of spectral calculations may be highly sensitive to rounding errors and to the smoothness of the initial and boundary data.

At this conference spectral methods were mentioned in only two or three talks, but I believe this relative obscurity is temporary. Spectral methods are nothing more than finite difference or finite element methods carried to unusually high orders of accuracy — typically by means of global trigonometric or polynomial interpolants which are differentiated to yield approximate derivatives of discrete data sequences. Since many fluid flows are smooth in at least part of the domain of interest, high-order methods are of natural utility.

It is not my purpose to present any applications of spectral methods in detail, but I will mention some representative references. One early paper of lasting interest is the report by Fornberg and Whitham (1978) on interacting solitons in the KdV equation and in other nonlinear models of water waves. Important advances in our understanding of instabilities in incompressible flows were achieved by the spectral calculations of Orszag and Kells (1980), and Orszag and Patera (1982), who showed numerically, for example, that the onset of turbulence in plane Poiseuille flow is triggered by a three-dimensional finite-amplitude instability at Reynolds number ≈ 1000.

---

[1]Supported by an IBM Faculty Development Award and an NSF Presidential Young Investigator Award

A recent paper by Fornberg (1988) illustrates the impressive capabilities of spectral methods for solving problems in elastic wave propagation, as occur in geophysics, even in the presence of discontinuous interfaces. Finally, examples of "spectral element" calculations in fluid dynamics, in which a complicated domain is subdivided into smaller domains where spectral formulas are applied, can be found in the papers of Patera and his colleagues (Maday and Patera 1988).

Recently a comprehensive monograph has appeared on this subject: *Spectral Methods in Fluid Dynamics,* by Canuto, Hussaini, Quarteroni and Zang (1988). For details and further references the reader can do no better than consult that book. Another work well worth looking at is the earlier monograph by Gottlieb and Orszag (1977).

2. The stability problem

The purpose of this paper is to discuss the numerical stability of fully discrete spectral methods for time-dependent problems with boundaries — a subject that is incompletely understood. With finite difference or finite element methods, the process of computing derivatives is local and at least approximately translation-invariant, so the matrices that arise are typically close to normal[1] and the behaviour of the process as a whole can often be approximated by Fourier techniques ("von Neumann analysis"), possibly supplemented by an investigation of wave reflection at boundaries ("GKS analysis"). For details see Richtmyer and Morton (1967), Gustafsson et al. (1972), and Trefethen (1983). By contrast, spectral differentiation is a global process, and the matrices involved may be far from normal. This paper will explore some consequences of that fact.

In particular, for spectral methods on bounded domains, Lax-stability time-step restrictions may be much tighter than the restrictions associated with various weaker definitions of stability (related, but not all identical) which go by a number of names, including "eigenvalue stability," "time-stability," "von Neumann stability," "practical stability," "s-stability," and "stability in the sense of Forsythe and Wasow." In outline, the two ideas to be contrasted here are as follows:

Lax-stability: stability for fixed t as $\Delta x, \Delta t \to 0$.

Eigenvalue stability: stability for fixed Δx and Δt as $t \to \infty$.

To make the definitions precise, one has to look at norms of operators. Let $\| \ \|$ be an appropriate norm, and let $S_{N,\Delta t}$ be the discrete solution operator

---

[1]A normal matrix is one that possesses a complete orthogonal system of eigenvectors. Equivalently, A is normal if and only if $A^H A = AA^H$, where A^H is the conjugate transpose. Symmetric, skew-symmetric, orthogonal, and circulant matrices fall in this category.

on a grid of N points with time step Δt:

$$S_{N,\Delta t} : v^n \mapsto v^{n+1}, \qquad v^n = S_{N,\Delta t}^n v^0, \tag{2.1}$$

where v^n represents the computed solution at time step n. (For an m-step discretization, v^n is replaced by $w^n = (v^n, \ldots, v^{n+1-m})^T$ and $S_{N,\Delta t}$ becomes a block companion matrix of dimension mN.) Lax-stability is defined by

Lax-stability: $\|S_{N,\Delta t}^n\| \le C$ for all N, n such that $n\Delta t \le T$

for any T and some constant $C = C(T)$, where we assume a fixed relationship between the space and time discretizations:

$$\Delta t = \Delta t(N). \tag{2.2}$$

Eigenvalue stability is defined for *fixed* N and Δt by

Eigenvalue stability: $\|S_{N,\Delta t}^n\| \le C$ for all n,

for some constant C. This is equivalent to $\rho(S_{N,\Delta t}) \le 1$, if ρ denotes the spectral radius (largest eigenvalue in absolute value), together with the condition that any eigenvalues on the unit circle should be nondefective.

For some purposes these definitions must be weakened by logarithmic or algebraic factors in n or N, but since my concern here is general phenomena rather than precise theorems, I will not worry about such details. What is important is that Lax-stability is a uniform bound for all matrices in a certain class corresponding to finer and finer meshes, whereas eigenvalue stability is a bound on the powers of a single matrix corresponding to a fixed mesh.

For the common situation in which a spectral discretization consists of a spectral differentiation operator D_N in x coupled with a standard o.d.e. formula in t, there is a further equivalent statement of eigenvalue stability:

> *Eigenvalue stability:* The eigenvalues of D_N lie within the stability region for the time-integrator.

(To be precise, we again permit only non-defective eigenvalues on the boundary.) An advantage of eigenvalue stability is that it is a relatively elementary concept, for stability regions are a familiar tool among numerical analysts (Gear 1971).

The points to be made in this paper can be summarized as follows. In comparison with finite difference and finite element methods, instabilities in spectral methods are

(A) Less well understood, and (B) More troublesome.

In particular, the stability condition for an explicit spectral method is typically both harder to predict and more restrictive than for explicit finite differences (assuming a uniform mesh). Furthermore, because a gap may arise between Lax-stability and eigenvalue stability, instabilities in spectral methods are sometimes

(C) More sensitive to rounding errors

and

(D) More sensitive to smoothness of the solution,

especially for first-order problems. The origin of (C) and (D) is the characteristic virtue of spectral methods: their high order of accuracy. For smooth problems, the truncation errors in a spectral calculation are often zero or negligible, so that any instability present is excited only by rounding errors. Consequently there are circumstances in which a Lax-unstable spectral method may give accurate answers. However, the accuracy may be destroyed by rounding errors, non-smooth data, or other perturbations such as variable coefficients or lower-order terms.

The distinctions between various notions of stability have been investigated for many years in the theoretical literature of finite difference methods; perhaps the central point of this paper is that certain spectral methods provide examples in which these distinctions are of paramount importance. A classic paper on convergence of unstable formulas with analytic initial data was written by Dahlquist (1954). The Lax stability theory appeared shortly thereafter in the survey by Lax and Richtmyer (1956). The sensitivity to perturbations of certain eigenvalue stable but Lax-unstable formulas was explored in an important paper by Kreiss (1962), which unfortunately has had limited influence because it was written in German. Discussions of various weaker definitions of stability can be found in Strang (1960) and Gottlieb and Orszag (1977), among others. The book by Richtmyer and Morton (1967) remains an excellent source on many of these topics.

3.   2nd-order differentiation

Consider the model problem

$$u_t = u_{xx}, \quad x \in [-1, 1], \quad u(\pm 1, t) = 0, \tag{3.1}$$

and for concreteness let

$$1 = x_0 > x_1 > \cdots > x_{N-1} > x_N = -1 \tag{3.2}$$

be the *Chebyshev extreme points* (= *Gauss-Lobatto-Chebyshev points*) defined by $x_j = \cos(j\pi/N)$. Like most grids for nonperiodic spectral methods, this grid has spacing $O(N^{-1})$ in the interior but $O(N^{-2})$ near the

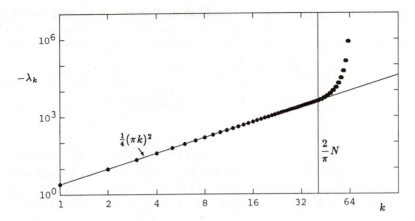

Fig. 1: Eigenvalues of the second-order spectral differentiation matrix for Chebyshev extreme points, $N = 64$ (log-log scale). The eigenvalues with $k > 2N/\pi$ occur in approximate pairs

boundary; such nonuniformity is essential if one is to avoid the Runge phenomenon of wild oscillations (Trefethen and Weideman 1989). If $v = (v_1, \ldots, v_{N-1})^T$ denotes a vector of data at positions $x_1, \ldots, x_{N-1}$, then the second-order spectral differentiation matrix D_N is the $(N-1) \times (N-1)$ matrix

$$D_N : v \mapsto w \qquad (3.3)$$

defined implicitly as follows (explicit matrix entries can be derived from the first-order matrix given by Canuto et al. (1988), p. 69:

(1) Interpolate v by a polynomial p of degree N with $p(\pm 1) = 0$;
(2) Set $w_j = (D_N v)_j = p''(x_j)$, $1 \leq j \leq N - 1$.

This differentiation process is global and, for "smooth" vectors v, highly accurate. A semidiscrete spectral approximation to (3.1) is now provided by the system of $N - 1$ ordinary differential equations

$$v_t = D_N v, \qquad (3.4)$$

and if the time derivative is replaced by a linear multistep or Runge–Kutta formula, the approximation becomes fully discrete.

The eigenvalues of D_N are real and negative — and as illustrated in Fig. 1 for $N = 64$, some of them are huge. A proportion $2/\pi$ are of order $O(N^2)$, and approximate closely the eigenvalues $\frac{1}{4}(\pi k)^2$ of the associated differential operator; the corresponding eigenvectors are very nearly the sines and cosines one gets for the latter, represented by at least two points per wavelength throughout the spatial grid. But the remaining "outlier" eigenvalues are of order $O(N^4)$, with the largest being about $0.0474N^4$ (Ouazzani et al. 1982, Trefethen and Weideman 1989). A rigorous upper bound as $N \to \infty$ is $\sqrt{11/4725}N^4 \approx 0.0482N^4$.

Because the largest eigenvalues are so big, any explicit time integration of (3.4) will be subject to an eigenvalue stability restriction

$$\Delta t \leq CN^{-4} \tag{3.5}$$

for some constant C that depends on how much of the negative real axis is contained in the stability region for the time-integrator. For Runge–Kutta formulas of orders 1–4 the constants are $C \approx 42, 42, 53, 59$, respectively; for the corresponding Adams–Bashforth formulas they are $C \approx 42, 21, 11, 6.3$.

Although the eigenvalues of D_N are real, D_N is not symmetric, or normal; but it is close to normal. One measure of this is a comparison between the spectral radius (largest eigenvalue) and the 2-norm (largest singular value):

$$\rho(D_N) \approx 0.0474N^4, \qquad \|D_N\| \approx 0.0483N^4. \tag{3.6}$$

Another is the size of the commutator:

$$\frac{\|D_N^T D_N - D_N D_N^T\|}{\|D_N^T D_N\|} \approx 0.163. \tag{3.7}$$

(These are empirical results for large N, based on the 2-norm.) A more useful measure of closeness to normality is the modest size of the condition number $\kappa(V) = \|V\| \|V^{-1}\|$ of V, the matrix of normalized eigenvectors of D_N:

$$\kappa(V) \approx 1.72, \ 4.06, \ 11.31 \quad \text{for } N = 8, \ 32, \ 128. \tag{3.8}$$

(If D_N were normal $\kappa(V)$ would be 1.) So far as I am aware, because D_N is close to normal, important differences do not arise in this second-order problem between Lax-stability and eigenvalue stability, and the same condition (3.5) is at least approximately valid for both. This conclusion appears to extend also to other Chebyshev and Legendre meshes.

Here is a summary of the stability of second-order spectral differentiation in terms of statements (A)–(D) of the last section. First, (A) stability restrictions for explicit formulas are not as easy to predict as with finite differences, but reasonable estimates can be obtained in some cases by looking at coefficients of characteristic polynomials (Weideman and Trefethen 1988). Second, (B) these stability limits are extremely restrictive. Third, the matrices are close enough to normal that Lax-stability and eigenvalue stability correspond closely, and there is no exaggerated sensitivity to (C) rounding errors or (D) smoothness of the solution.

Because the stability limits for explicit second-order spectral formulas are so restrictive, every effort should be made to employ implicit schemes instead. The efficient implementation of implicit spectral methods is a topic of active current research, and impressive results have been achieved by iterative methods with suitably chosen preconditioners; see Chapter 5 of Canuto et al. (1988).

4. 1st-order differentiation in Chebyshev points

Now consider the first-order problem

$$u_t = u_x, \quad x \in [-1, 1], \quad u(1, t) = 0, \qquad (4.1)$$

again on the Chebyshev grid (3.2); to make the problem well-posed only the inflow boundary condition has been specified. We now deal with a vector $v = (v_1, \ldots, v_N)^T$ of data values at $x_1, \ldots, x_N$, and the first-order spectral differentiation matrix

$$D_N : v \mapsto w \qquad (4.2)$$

is an $N \times N$ matrix defined implicitly as follows (for explicit entries see Canuto et al. (1988), p. 69):

(1) Interpolate v by a polynomial p of degree N with $p(1) = 0$;
(2) Set $w_j = (D_N v)_j = p'(x_j)$, $1 \le j \le N$.

The eigenvalues of D_N are complex, and they are plotted in Fig. 2 for $N = 32$. As in the second-order case of the last section, most of them (about 82%) are of reasonable size, $O(N)$, but the remainder are much larger: $O(N^2)$. (Details can be found in the recent paper by Dubiner (1987); the exact proportion of "outlier" eigenvalues approaches $1/2 - 1/\pi \approx 0.1817$ as $N \to \infty$.) Therefore, any explicit time integration of the spectral semidiscretization of (4.1) will be subject to an eigenvalue stability restriction

$$\Delta t \le C N^{-2} \qquad (4.3)$$

for some C that depends on the stability region for the time-integrator.

As pointed out first by Trefethen and Trummer (1987), the eigenvalues of D_N are highly sensitive to small perturbations such as rounding errors. To illustrate this, Fig. 2 shows eigenvalues computed in 16-digit precision together with eigenvalues of the same matrix after rounding each entry to 8 digits (relative to the largest element of the matrix). This figure shows that D_N must be far from normal, since the eigenvalues of normal matrices

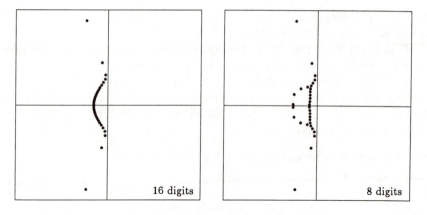

Fig. 2: Eigenvalues in the complex plane of the first-order spectral differentiation matrix for Chebyshev extreme points, N=32. The region shown is the square bounded by $\pm 100 \pm 100i$

are well-conditioned functions of the matrix entries. In analogy to (3.6)–(3.8), here are some measures of non-normality: the spectral radius and 2-norm differ considerably,

$$\rho(D_N) \approx 0.0886N^2, \qquad \|D_N\| \approx 0.5498N^2, \qquad (4.4)$$

the commutator is larger than before,

$$\frac{\|D_N^T D_N - D_N D_N^T\|}{\|D_N^T D_N\|} \approx 0.870, \qquad (4.5)$$

and the condition number of the matrix of eigenvectors is huge:

$$\kappa(V) \approx 5.8e2, \ 7.1e5, \ 1.1e12 \ \text{ for } \ N = 8, \ 16, \ 32. \qquad (4.6)$$

What are the consequences of the non-normality of D_N? Are Lax-stability and eigenvalue stability quite distinct for this problem? So far as I am aware the answer is no, except possibly if an unusual time-integration formula is chosen. Experiments indicate that Lax-stability as well as eigenvalue stability is essentially determined by the need to fit the largest of the eigenvalues in a stability region; Fig. 2 suggests that the outlier eigenvalues, after all, are insensitive to perturbations. Since space is limited, therefore, we shall leave this example without further discussion and turn to a problem in which the difference between Lax-stability and eigenvalue stability is pronounced.

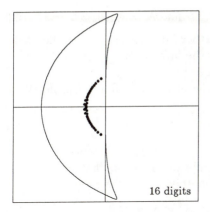 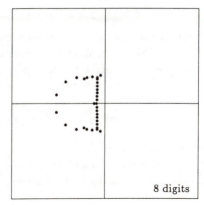

16 digits   8 digits

Fig. 3: Eigenvalues in the complex plane of the first-order spectral differentiation matrix for Legendre points, $N = 32$ (same scale as in Fig. 2). The stability region plotted on the left applies to Fig. 7, below

5.   1st-order differentiation in Legendre points

We now consider the most interesting example of this paper: the same problem as in the last section, but with $x_1, \ldots, x_N$ replaced by the zeros of the Legendre polynomial $P_N(x)$. These Legendre points are not far different from Chebyshev points, but according to a discovery of Dubiner (1987), the eigenvalues of the corresponding matrix D_N are much smaller than before: $O(N)$ rather than $O(N^2)$. Therefore should it not be possible to replace (4.3) by a more favorable stability restriction $\Delta t \leq C N^{-1}$?

Following upon Dubiner's theoretical work, Tal-Ezer (1986) carried out experiments that indicated that this optimistic expectation is indeed justified, at least for certain scalar problems in one dimension. However, our own view is that eigenvalue analysis can be misleading, and the optimism must be qualified. Legendre spectral methods with large time steps may be sensitive to non-smooth data and to other perturbations of the problem such as the introduction of variable coefficients or lower-order terms. This sensitivity is of just the kind Kreiss warned of in 1962.

To begin the numerical illustrations, Fig. 3 repeats Fig. 2 for Legendre points. The prediction that the eigenvalues are $O(N)$ is clearly valid, in exact arithmetic, but as in the Chebyshev case, they are highly sensitive to perturbations. Therefore D_N must again be far from normal, and our measures of normality come out as follows:

$$\rho(D_N) \approx N, \qquad \|D_N\| \approx N^2, \tag{5.1}$$

$$\frac{\|D_N^T D_N - D_N D_N^T\|}{\|D_N^T D_N\|} \approx 1, \tag{5.2}$$

$$\kappa(V) \approx 8.5e3, \ 3.1e8, \ 2.5e17 \ \text{for} \ N = 8, \ 16, \ 32. \tag{5.3}$$

Evidently the matrices D_N are even farther from normal than in the Chebyshev case. What is more important, the large outlier eigenvalues are no longer present to mask the non-normality.

Let us now restrict attention to the 3rd-order Adams–Bashforth formula

$$v^{n+1} = v^n + \frac{\Delta t}{12} D_N(23v^n - 16v^{n-1} + 5v^{n-2}). \tag{5.4}$$

For eigenvalue stability, the stability region of the time-integrator must enclose the eigenvalues of D_N, and computations show that the condition for this in exact arithmetic is approximately

$$\textit{Eigenvalue stability:} \quad \Delta t \le 0.734 N^{-1}. \tag{5.5}$$

In the presence of rounding errors, however, Fig. 3 suggests that this condition would have to be tightened, as is confirmed by experiments (Trefethen and Trummer 1987). It follows that (5.5) cannot be enough to ensure Lax-stability, for rounding errors are small perturbations, and thus by definition (together with the discrete Duhamel principle), Lax-stability would entail insensitivity to them.

Instead, I conjecture that the Lax-stability restriction for this model problem is approximately

$$\textit{Lax-stability:} \quad \Delta t = O(N^{-2}). \tag{5.6}$$

The use of the notation $O(N^{-2})$ rather than CN^{-2} is deliberate: (5.6) means that the spectral method is Lax-stable as $N \to \infty$ and $\Delta t(N) \to 0$ if and only if there exists *any* constant a such that

$$\Delta t \le aN^{-2} \tag{5.7}$$

for all N. Equation (5.6) may not be the exact condition for Lax-stability of this model problem, since rigorous analysis of the stability of spectral methods is typically complicated by small factors, as mentioned in Section 2; in any case an exact result may depend on the choice of norm. But I believe it is at least close to correct. The "only if" half of the statement is proved in the next section.

For empirical justification of (5.6), consider the solution operator $S_{N,\Delta t}$: $(v^n, v^{n-1}, v^{n-2})^T \mapsto (v^{n+1}, v^n, v^{n-1})^T$ for the Adams–Bashforth formula (5.4), which is a matrix of dimension $3N$:

$$S_{N,\Delta t} = \begin{pmatrix} I & 0 & 0 \\ I & 0 & 0 \\ 0 & I & 0 \end{pmatrix} + \frac{\Delta t}{12} \begin{pmatrix} 23D_N & -16D_N & 5D_N \\ 0 & 0 & 0 \\ 0 & 0 & 0 \end{pmatrix}. \tag{5.8}$$

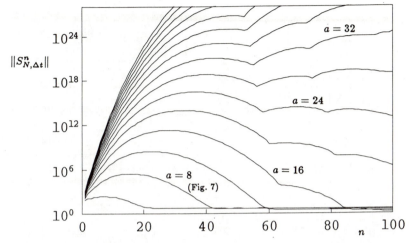

Fig. 4: Legendre grid with $N = 32$: $\|S_{N,\Delta t}^n\|$ as a function of n for $\Delta t = aN^{-2}$, various a. $S_{N,\Delta t}$ is power-bounded approximately for $a \leq 23.5$

Fig. 4 shows numerically computed norms $\|S_{N,\Delta t}^n\|$ as functions of n with $N = 32$ for various values a in (5.7). (From now on $\| \ \|$ denotes $\| \ \|_\infty$, which can be computed much faster than $\| \ \|_2$.) For $a \leq 0.734 \times 32 \approx 23.5$ (see (5.5)), $S_{N,\Delta t}$ is power-bounded and thus eigenvalue stable, but the figure shows that tremendous growth of the powers $S_{N,\Delta t}^n$ takes place for much smaller a, before the eventual decay.

Now for any N and Δt, let $\gamma_{N,\Delta t}$ be the power-boundedness constant

$$\gamma_{N,\Delta t} = \sup_n \|S_{N,\Delta t}^n\|, \tag{5.9}$$

or in other words, the maximum value of one of the curves in a plot like Fig. 4. If $\rho(S_{N,\Delta t}) > 1$ (ρ being again the spectral radius), or if $\rho(S_{N,\Delta t}) = 1$ with a defective eigenvalue on the unit circle, then $\gamma_{N,\Delta t} = \infty$ and the formula is eigenvalue unstable; otherwise it is eigenvalue stable. Fig. 5 shows $\gamma_{N,\Delta t}$ as a function of N for various values of a in (5.7). (To suppress the distracting irregularity that would otherwise result from the discreteness of N, the corner of each curve has been shifted slightly to the position corresponding to (5.5).) Each curve is eventually approximately constant, no matter how large a is, but the constants grow exponentially with a. Clearly Lax-stability is in jeopardy if (5.7) does not hold for some a.

A different view of the same data is provided by Fig. 6, which is a rough contour plot, on a log-log scale, of $\gamma_{N,\Delta t}$ as a function of N and Δt^{-1}. Below the line $\Delta t = 0.734N^{-1}$, with slope 1, the prediction $\gamma_{N,\Delta t} = \infty$ is verified. The interesting behaviour lies above this line, where $\gamma_{N,\Delta t}$ is finite and we

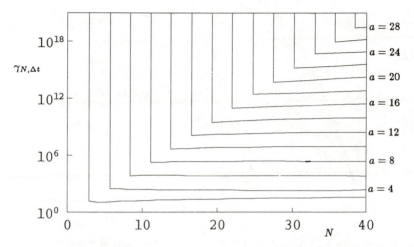

Fig. 5: Legendre grid: power-boundedness constants $\gamma_{N,\Delta t}$ (5.9) for $\Delta t = aN^{-2}$, various a. The solid dot marks the computation of Fig. 7

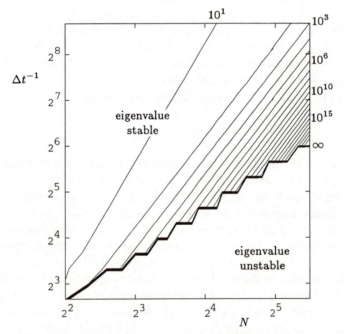

Fig. 6: Legendre grid: contour plot of the power-boundedness constants $\gamma_{N,\Delta t}$ (5.9). The solid dot marks the computation of Fig. 7

see a set of straight parallel contours with slope 2. For Lax-stability it is permissible to traverse this slope at a constant altitude, however high, and evidently (5.7) does just that. But any relationship of Δt to N that violates (5.7) as $N \to \infty$ will lead to an ascent of the slope rather than a traversal: Lax-instability.

After several pages devoted to norms of operators, our final figure represents an actual spectral calculation selected to show that Lax-instability may reveal itself in some experiments but not others. Fig. 7 is based on computations with $N = 32$ and $\Delta t = aN^{-2}$, $a = 8$, for which the Adams–Bashforth formula is eigenvalue stable even in low-precision arithmetic; the stability region was shown in Fig. 3. However, Figs. 4–6 indicate that $\gamma_{N,\Delta t} \approx 10^{5.4}$, so the potential for unstable growth is present. But in Fig. 7a the initial function is $u(x,0) = \cos^4(\pi x/2)$; when extended by zeros outside $[-1,1]$, this function is three times continuously differentiable, which is evidently smooth enough that the computation is successful, but in Fig. 7b the initial function is $\cos^2(\pi x/2)$, whose extension outside $[-1,1]$ is only once continuously differentiable, and now errors appear at the boundary that grow considerably before dying away. The computation is unsuccessful, at least if transient phenomena are of importance.

Fig. 7b represents a mild, borderline example chosen to make the plots interesting; in general the instability may be far worse. With $a = 12$ or 16, for example, the error at the boundary grows to 3.1×10^3 and 2.5×10^6, respectively, before eventually decaying away. Trefethen and Trummer (1987) used this same initial data with $a = 16$, but examined the results only at $t = 1$ and failed to notice the instability — an indication that errors appearing in at least some unstable spectral calculations may indeed be transient.

6. Conclusion; a pseudo CFL condition

Although many spectral calculations exhibit none of the stability problems highlighted in the example of the last section, that example is by no means unique. For example, another model problem with similar properties (but easier to analyze) is the equation $u_t = -xu_x$ on a Chebyshev or Legendre grid in $[-1,1]$ with no boundary conditions (Solomonoff and Turkel 1988, Trefethen and Trummer 1987). Since the papers in the literature that address the question of stability mainly restrict their examples to constant coefficients and analytic initial data, the limitations of eigenvalue analysis have not received as much notice as they deserve.

This paper has presented experiments; what is needed now is theory. A starting point would be to prove that the example of the last section is indeed Lax-stable if and only if (5.6) holds (perhaps in a slightly strengthened form). But deeper and more general questions suggest themselves

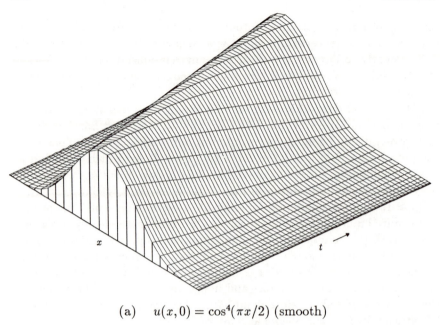

(a) $u(x,0) = \cos^4(\pi x/2)$ (smooth)

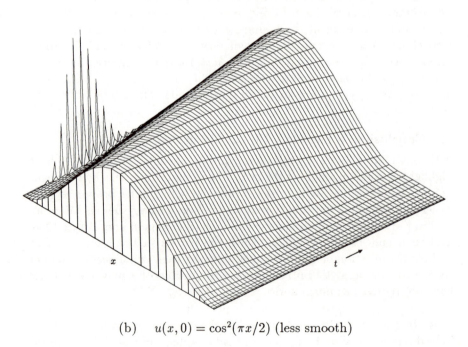

(b) $u(x,0) = \cos^2(\pi x/2)$ (less smooth)

Fig. 7: Legendre grid: two calculations for $0 \leq t \leq 0.5$ with $N = 32$, $\Delta t = 8N^{-2}$

too. What general techniques can be devised for predicting stability limits of fully discrete spectral methods? When is Lax-stability needed in practice? For steady-state rather than time-accurate calculations, under what circumstances is it safe to settle for less than Lax-stability and ignore transient blow-ups like that of Fig. 7b?

I will close with an observation that goes half-way towards justifying (5.6). That condition must be *necessary* for Lax-stability, for the simple reason that D_N contains elements of order $O(N^2)$, which implies that $S_{N,\Delta t}$ contains elements of order $O(\Delta t N^2)$; if (5.6) is violated, $S_{N,\Delta t}$ cannot itself be bounded uniformly in N, to say nothing of its powers. This argument generalizes to a "pseudo CFL condition" for explicit spectral methods: small gaps between mesh points imply correspondingly tight stability restrictions. (The usual CFL argument, based on domains of dependence, is vacuous for spectral methods because the spectral differentiation process is global.) Here is one example of how this assertion can be made into a theorem:

THEOREM. *Consider the pseudospectral approximation to (4.1) based on any grid (3.2) and any consistent, explicit time-integration formula, and define* $\Delta x = x_{N-1} - x_N$. *A necessary condition for Lax-stability of this approximation, in any matrix norm subordinate to a vector norm, is*

$$\Delta t = O(\Delta x) \text{ as } N \to \infty.$$

Proof. Let $p(x) = (x - x_0)\cdots(x - x_{N-1})/(x_N - x_0)\cdots(x_N - x_{N-1})$ be the polynomial interpolant to the values $0, \ldots, 0, 1$ at $x_0, \ldots, x_{N-1}, x_N$, respectively. Then $(D_N)_{NN} = p'(x_N) = \sum_{j=0}^{N-1}(x_N - x_j)^{-1} < -1/\Delta x$, and thus $S_{N,\Delta t}$ contains an element of order at least $\Delta t/\Delta x$ as $N \to \infty$. For Lax-stability this element must be $O(1)$ as $N \to \infty$, hence the condition $\Delta t = O(\Delta x)$. ∎

In later work, I hope to make the observations of this paper more general and precise. Many fundamental problems of numerical analysis, not only in the area of spectral methods, depend upon a proper treatment of functions of non-normal matrices. Meanwhile, I remind readers that despite all of this attention to pathologies, spectral methods can be extraordinarily accurate; we shall see more of them in the years ahead.

7. Acknowledgments

My work on spectral methods has profited from collaboration with Manfred Trummer, André Weideman, and most recently Andrew Stuart and Satish Reddy. I am grateful to these people for their assistance, and also to Yvon Maday, Gil Strang, and Hillel Tal-Ezer. Thanks also to Louis Howell for help with graphics.

References

Canuto, C., Hussaini, M. Y., Quarteroni, A. and Zang, T. (1988). *Spectral Methods in Fluid Dynamics*. Springer–Verlag, New York.

Dahlquist, G. (1954). Convergence and stability for a hyperbolic difference equation with analytic initial-values. *Mathematica Scandinavica*, 2:91–102.

Dubiner, M. (1987). Asymptotic analysis of spectral methods. *Journal of Scientific Computing*, 2:3–31.

Fornberg, B. and Whitham, G. B. (1978). A numerical and theoretical study of certain nonlinear wave phenomena. *Philosophical Transactions of the Royal Society of London*, A289:373–404.

Fornberg, B. (1988). The pseudospectral method: accurate representation of interfaces in elastic wave calculations. *Geophysics*, 53:625–637.

Gear, C. W. (1971). *Numerical Initial Value Problems in Ordinary Differential Equations*. Prentice–Hall, Englewood Cliffs, New Jersey.

Gottlieb, D. and Orszag, S. A. (1977). *Numerical Analysis of Spectral Methods: Theory and Applications*. SIAM, Philadelphia.

Gustafsson, B., Kreiss, H.-O., and Sundström, A. (1972). Stability theory of difference approximations for mixed initial boundary value problems. II *Mathematics of Computation*, 26:649–686.

Kreiss, H.-O. (1962). Über die Stabilitätsdefinition für Differenzengleichungen die partielle Differentialgleichungen approximieren. *BIT*, 2:153–181.

Lax, P. D. and Richtmyer, R. D. (1956). Survey of the stability of linear finite difference equations. *Communications on Pure and Applied Mathematics*, 9:267–293.

Maday, Y. and Patera, A. T. (1988). Spectral element methods for the incompressible Navier-Stokes equations. In *State of the art surveys in computational mechanics*, Noor, A. K., Editor, ASME.

Orszag, S. A. and Kells, L. C. (1980). Transition to turbulence in plane Poiseuille flow and plane Couette flow. *Journal of Fluid Mechanics*, 96:159–205.

Orszag, S. A. and Patera, A. T. (1982). Secondary instability of wall-bounded shear flows. *Journal of Fluid Mechanics* 128:347–385.

Ouazzani, J., Peyret, R. and Zakaria, A. (1985). Stability of collocation-Chebyshev schemes with application to the Navier-Stokes equations. In *Proceedings 6th GAMM Conference on Numerical Methods in Fluid Mechanics*, Rues, D. and Kordulla, W., Editors, Vieweg.

Richtmyer, R. D. and Morton, K. W. (1967). *Difference Methods for Initial-value Problems* (2nd edn). Wiley–Interscience, New York.

Solomonoff, A. and Turkel, E. (1988). Global collocation methods for approximation and the solution of partial differential equations. *Journal of Computational Physics*, to appear.

Strang, W. G. (1960). Difference methods for mixed boundary-value problems. *Duke Mathematical Journal* 27 : 221–232.

Tal-Ezer, H. (1986). A pseudospectral Legendre method for hyperbolic equations with an improved stability condition. *Journal of Computational Physics* 67 : 145–172.

Trefethen, L. N. (1983). Group velocity interpretation of the stability theory of Gustafsson, Kreiss, and Sundström. *Journal of Computational Physics*, 199–217.

Trefethen, L. N. and Trummer, M. R. (1987). An instability phenomenon in spectral methods. *SIAM Journal of Numerical Analysis*, 24 : 1008–1023.

Trefethen, L. N. and Weideman, J. A. C. (1989). Two results on interpolation in equally spaced points. Submitted to *Journal of Approximation Theory*.

Weideman, J. A. C. and Trefethen, L. N. (1988). The eigenvalues of second-order spectral differentiation matrices. *SIAM Journal of Numerical Analysis*, to appear.

Acceleration of compressible Navier–Stokes calculations

M. O. Bristeau   R. Glowinski

INRIA, France   *University of Houston, USA*

B. Mantel, J. Périaux and G. Rogé

AMD/BA, France

1. Introduction and synopsis

Operator splitting techniques applied to the numerical simulation of unsteady compressible viscous flows lead to the solution of generalized (linear) Stokes subproblems and of nonlinear steady systems. These problems involve a quite large number of variables requiring efficient solution algorithms. In this paper, we discuss the solution of the above linear subproblems by iterative techniques, involving a preconditioning step by a suitable boundary operator. The solution of the nonlinear (advection) subproblems is achieved through an iterative method generalizing the so-called GMRES algorithm. Another issue to be briefly discussed here is the compatibility of the velocity and density finite element approximations. Numerical results obtained from the simulation of complex flows originating from Aerospace Engineering illustrate the possibilities of the above methods.

The numerical solution of complicated two and three dimensional viscous flows modeled by the Navier–Stokes equations can be achieved by operator splitting methods like those discussed in e.g. (Yanenko 1971; Marchuk 1975; LeVeque and Oliger 1983; Glowinski 1984; Bristeau et al. 1987). See also the references therein. The resulting subproblems to be solved at each time step are still non-trivial ones and therefore their numerical solution requires the use of advanced computational methods (mostly iterative); such methods are precisely discussed in this paper together with compatibility problems associated to velocity and density finite element discretizations.

2. The Navier–Stokes equations modeling compressible viscous flows

We consider the case of external flows around bodies with the notation of Fig. 1.

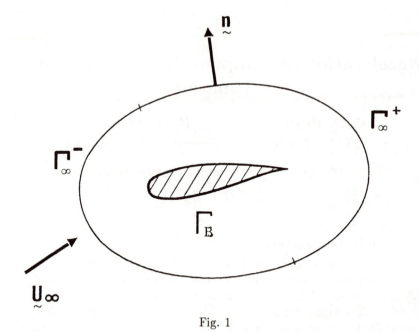

Fig. 1

We can use for the modeling of such flows the following (non-conservative) system of partial differential equations

$$\frac{\partial \rho}{\partial t} + \mathbf{\nabla} \cdot (\rho \mathbf{u}) = 0, \tag{2.1}$$

$$\rho \frac{\partial \mathbf{u}}{\partial t} + \rho(\mathbf{u} \cdot \mathbf{\nabla})\mathbf{u} + (\gamma - 1)\mathbf{\nabla}\rho T = \frac{1}{\text{Re}}[\mathbf{\nabla}^2\mathbf{u} + \frac{1}{3}\mathbf{\nabla}(\mathbf{\nabla} \cdot \mathbf{u})], \tag{2.2}$$

$$\rho \frac{\partial T}{\partial t} + \rho \mathbf{u} \cdot \mathbf{\nabla} T + (\gamma - 1)\rho T\mathbf{\nabla} \cdot \mathbf{u} = \frac{1}{\text{Re}}[\frac{\gamma}{Pr}\mathbf{\nabla}^2 T + F(\mathbf{\nabla}\mathbf{u})], \tag{2.3}$$

where the pressure p is given by the perfect gas law

$$p = (\gamma - 1)\rho T, \tag{2.4}$$

and where $\rho, \mathbf{u}$ $(= \{u_i\}_{i=1}^N, N = 2, 3)$, T are the non-dimensionalized density, velocity and temperature, respectively, with (if $N = 2$)

$$F(\mathbf{\nabla}\mathbf{u}) = \frac{4}{3}\left\{|\frac{\partial u_1}{\partial x_1}|^2 + |\frac{\partial u_2}{\partial x_2}|^2 - \frac{\partial u_1}{\partial x_1}\frac{\partial u_2}{\partial x_2}\right\} + |\frac{\partial u_1}{\partial x_2} + \frac{\partial u_2}{\partial x_1}|^2. \tag{2.5}$$

In (2.2), (2.3) Re, Pr and γ are the Reynolds number, the Prandtl number, and the ratio of specific heats, respectively, ($\gamma = 1.4$ for air flows).

With Γ_∞ as in Fig. 1 and

$$\Gamma_\infty^- = \{x \mid x \in \Gamma_\infty, \mathbf{u}_\infty \cdot \mathbf{n}_\infty < 0\}, \quad \Gamma_\infty^+ = \{x \mid x \in \Gamma_\infty, \mathbf{u}_\infty \cdot \mathbf{n}_\infty \geq 0\} \quad (2.6)$$

($\mathbf{n}_\infty$: unit outward normal vector at Γ_∞), we prescribe on the inflow boundary Γ_∞^-

$$\mathbf{u} = \mathbf{u}_\infty, \quad \rho = 1, \quad T = 1/\gamma(\gamma - 1)M_\infty^2, \quad (2.7)$$

while on the outflow boundary Γ_∞^+, we prescribe Neumann and other "natural" boundary conditions (see Sec. 4). On the "wall" ∂B, we shall use the following Dirichlet boundary conditions

$$\mathbf{u} = \mathbf{0} \text{ and } T = T_\infty(1 + \frac{\gamma - 1}{2}M_\infty^2). \quad (2.8)$$

In (2.7), (2.8), $\mathbf{u}, T_\infty, M_\infty$, denote the free stream velocity, temperature, and Mach number, respectively. Finally, we shall also prescribe the following initial conditions:

$$\rho(x, 0) = \rho_0(x), \quad \mathbf{u}(x, 0) = \mathbf{u}_0(x), \quad T(x, 0) = T_0(x). \quad (2.9)$$

In order to render the above system of equations closer to the incompressible flow model, we introduce the following new dependent variable

$$\sigma = \ln\rho \quad (\sigma : \text{logarithmic density}). \quad (2.10)$$

With this variable, the Navier–Stokes equations become

$$\frac{\partial \sigma}{\partial t} + \mathbf{V} \cdot \mathbf{u} + \mathbf{u} \cdot \mathbf{V}\sigma = 0, \quad (2.11)$$

$$\frac{\partial \mathbf{u}}{\partial t} + (\mathbf{u} \cdot \mathbf{V})\mathbf{u} + (\gamma - 1)(T\mathbf{V}\sigma + \mathbf{V}T) = \frac{e^{-\sigma}}{Re}(\mathbf{V}^2\mathbf{u} + \frac{1}{3}\mathbf{V}(\mathbf{V} \cdot \mathbf{u})), \quad (2.12)$$

$$\frac{\partial T}{\partial t} + \mathbf{u} \cdot \mathbf{V}T + (\gamma - 1)T\mathbf{V} \cdot \mathbf{u} = \frac{e^{-\sigma}}{Re}(\frac{\gamma}{Pr}\mathbf{V}^2T + F(\mathbf{V} \cdot \mathbf{u})). \quad (2.13)$$

Equations (2.12), (2.13) can also be written as follows:

$$\frac{\partial \mathbf{u}}{\partial t} - \mu\mathbf{V}^2\mathbf{u} + \beta\mathbf{V}\sigma = \Psi(\sigma, \mathbf{u}, T), \quad (2.14)$$

$$\frac{\partial T}{\partial t} - \pi\mathbf{V}^2T = \chi(\sigma, \mathbf{u}, T), \quad (2.15)$$

with:

(a) δ: a mean value of the *reciprocal of the density* ($\delta = 1$ is a possible value),

(b) $\nu = \dfrac{1}{\mathrm{Re}}, \quad \mu = \nu\delta, \quad \pi = \dfrac{\gamma\nu\delta}{Pr}$,

(c) $\beta = (\gamma - 1)T_B = \dfrac{1}{\gamma}(\dfrac{\gamma - 1}{2} + \dfrac{1}{M_\infty^2})$,

(d) $\boldsymbol{\Psi}(\sigma, \mathbf{u}, T) = -(\gamma - 1)[\boldsymbol{\nabla}T + (T - T_B)\boldsymbol{\nabla}\sigma] - (\mathbf{u} \cdot \boldsymbol{\nabla})\mathbf{u}$
$\quad + \nu e^{-\sigma}(\boldsymbol{\nabla}^2\mathbf{u} + \dfrac{1}{3}\boldsymbol{\nabla}(\boldsymbol{\nabla} \cdot \mathbf{u}))$,

(e) $\chi(\sigma, \mathbf{u}, T) = -(\gamma - 1)T\boldsymbol{\nabla} \cdot \mathbf{u} - \mathbf{u} \cdot \boldsymbol{\nabla}T + \dfrac{\gamma\nu}{Pr}(e^{-\sigma} - \delta)\boldsymbol{\nabla}^2 T + \nu e^{-\sigma}F(\boldsymbol{\nabla}\mathbf{u})$.

3.   Time discretization by operator splitting methods

Using time discretization by operator splitting methods, it is possible to decouple, as shown in, e.g. (Bristeau et al. 1987), the various operators involved in the above model. We obtain therefore the following scheme (where $\Delta t(> 0)$ is the time discretization step, and where $\theta \in (0, 1/2)$):

$$\sigma^o = \sigma_o = \ln\rho_o, \quad \mathbf{u}^o = \mathbf{u}_o, \quad T^o = T_o; \qquad (3.1)$$

then for $n \geq 0$, starting from $\{\sigma^n, \mathbf{u}^n, T^n\}$ we solve

$$\dfrac{\sigma^{n+\theta} - \sigma^n}{\theta\Delta t} + \boldsymbol{\nabla} \cdot \mathbf{u}^{n+\theta} = -\mathbf{u}^n \cdot \boldsymbol{\nabla}\sigma^n, \qquad (3.2)_1$$

$$\dfrac{\mathbf{u}^{n+\theta} - \mathbf{u}^n}{\theta\Delta t} - a\mu\boldsymbol{\nabla}^2\mathbf{u}^{n+\theta} + \beta\boldsymbol{\nabla}\sigma^{n+\theta} = \boldsymbol{\Psi}(\sigma^n, \mathbf{u}^n, T^n) + b\mu\boldsymbol{\nabla}^2\mathbf{u}^n, \qquad (3.2)_2$$

$$\dfrac{T^{n+\theta} - T^n}{\theta\Delta t} - a\pi\boldsymbol{\nabla}^2 T^{n+\theta} = \chi(\sigma^n, \mathbf{u}^n, T^n), \qquad (3.2)_3$$

$$\dfrac{\sigma^{n+1-\theta} - \sigma^{n+\theta}}{(1 - 2\theta)\Delta t} + \mathbf{u}^{n+1-\theta} \cdot \boldsymbol{\nabla}\sigma^{n+1-\theta} = -\boldsymbol{\nabla} \cdot \mathbf{u}^{n+\theta} \qquad (3.3)_1$$

$$\dfrac{\mathbf{u}^{n+1-\theta} - \mathbf{u}^{n+\theta}}{(1 - 2\theta)\Delta t} - b\mu\boldsymbol{\nabla}^2\mathbf{u}^{n+1-\theta} - \boldsymbol{\Psi}(\sigma^{n+1-\theta}, \mathbf{u}^{n+1-\theta}, T^{n+1-\theta})$$
$$= a\mu\boldsymbol{\nabla}^2\mathbf{u}^{n+\theta} - \beta\boldsymbol{\nabla}\sigma^{n+\theta} \qquad (3.3)_2$$

$$\dfrac{T^{n+1-\theta} - T^{n+\theta}}{(1 - 2\theta)\Delta t} - b\pi\boldsymbol{\nabla}^2 T^{n+1-\theta} - \chi(\sigma^{n+1-\theta}, \mathbf{u}^{n+1-\theta}, T^{n+1-\theta})$$
$$= a\pi\boldsymbol{\nabla}^2 T^{n+\theta} \qquad (3.3)_3$$

$$\frac{\sigma^{n+1} - \sigma^{n+1-\theta}}{\theta \Delta t} + \boldsymbol{\nabla} \cdot \mathbf{u}^{n+1} = -\mathbf{u}^{n+1-\theta} \cdot \boldsymbol{\nabla} \sigma^{n+1-\theta}, \tag{3.4}_1$$

$$\frac{\mathbf{u}^{n+1} - \mathbf{u}^{n+1-\theta}}{\theta \Delta t} - a\mu \boldsymbol{\nabla}^2 \mathbf{u}^{n+1} + \beta \boldsymbol{\nabla} \sigma^{n+1}$$
$$= \boldsymbol{\Psi}(\sigma^{n+1-\theta}, \mathbf{u}^{n+1-\theta}, T^{n+1-\theta}) + b\mu \boldsymbol{\nabla}^2 \mathbf{u}^{n+1-\theta}, \tag{3.4}_2$$

$$\frac{T^{n+1} - T^{n+1-\theta}}{\theta \Delta t} - a\pi \boldsymbol{\nabla}^2 T^{n+1}$$
$$= \chi(\sigma^{n+1-\theta}, \mathbf{u}^{n+1-\theta}, T^{n+1-\theta}) + b\pi \boldsymbol{\nabla}^2 T^{n+1-\theta}, \tag{3.4}_3$$

with $0 < a, b < 1$, $a + b = 1$ satisfying (cf. (Bristeau et al. 1987))

$$a = (1 - 2\theta)/(1 - \theta), \quad b = \theta/(1 - \theta). \tag{3.5}$$

The above scheme is a θ-scheme close to those used for the incompressible Navier–Stokes equations and described in, e.g., (Bristeau et al. 1987).

4. Accelerated solution of the generalized Stokes subproblems

Each time step requires the solution of the two systems of coupled equations $(3.2)_1$, $(3.2)_2$, and $(3.4)_1$, $(3.4)_2$ (there is no difficulty for solving $(3.2)_3$ and $(3.4)_3$). Indeed systems $(3.2)_1$, $(3.2)_2$ and $(3.4)_1$, $(3.4)_2$ are particular cases of the generalized Stokes problem

$$\alpha \sigma + \boldsymbol{\nabla} \cdot \mathbf{u} = g \text{ in } \Omega, \tag{4.1}_1$$

$$\alpha \mathbf{u} - a\mu \boldsymbol{\nabla}^2 \mathbf{u} + \beta \boldsymbol{\nabla} \sigma = \mathbf{f} \text{ in } \Omega, \tag{4.1}_2$$

with the associated boundary conditions

$$\mathbf{u} = \mathbf{0} \text{ on } \Gamma_B, \quad \mathbf{u} = \mathbf{u}_\infty \text{ on } \Gamma_\infty^-, \quad a\mu \frac{\partial \mathbf{u}}{\partial \mathbf{n}} - \beta \sigma \mathbf{n} = \mathbf{0} \text{ on } \Gamma_\infty^+, \tag{4.1}_3$$

where, in (4.1), (4.2), α is a it positive parameter ($\alpha \sim 1/\Delta t$).

After an appropriate space discretization (to be discussed in Sec. 6), problem (4.1) will lead to a linear system which can be solved by direct methods; however, for a large number of variables, which is the case of three-dimensional flow problems, the lack of symmetry and positive definitess of the matrix associated to the above system makes the solution process quite costly. We should use therefore iterative methods taking advantage of the saddle point structure of problem (4.1). Such methods are described in e.g. (Glowinski 1984; Bristeau et al. 1987). Here we shall present conjugate gradient techniques for solving (4.1) in which the master unknown is the trace of σ on Γ. This method takes advantage of the

Helmholtz decomposition of the generalized Stokes problem (4.1) and uses as preconditioner a boundary operator quite easy to implement. Let's take thus the divergence of both sides in equation $(4.1)_2$; we obtain then:

$$\alpha \mathbf{\nabla} \cdot \mathbf{u} - a\mu\Delta(\mathbf{\nabla} \cdot \mathbf{u}) + \beta\Delta\sigma = \mathbf{\nabla} \cdot \mathbf{f}; \qquad (4.2)$$

on the other hand, it follows from $(4.1)_1$ that

$$\mathbf{\nabla} \cdot \mathbf{u} = g - \alpha\sigma. \qquad (4.3)$$

Combining (4.2), (4.3) we obtain then

$$\alpha^2\sigma - (\beta + \alpha a\mu)\Delta\sigma = \alpha g - \mathbf{\nabla} \cdot \mathbf{f} - a\mu\Delta g. \qquad (4.4)$$

To the elliptic equation (4.4) we associate the Dirichlet boundary condition

$$\sigma = \lambda \text{ on } \Gamma. \qquad (4.5)$$

Assuming that λ is known, we should compute σ from (4.4), and then $\mathbf{u}$ from $(4.1)_2$, $(4.1)_3$; we shall adjust λ in such a way that $(4.1)_1$ will hold. To reach this goal, it is convenient to introduce ψ solution of the following elliptic problem:

$$\alpha^2\psi - \zeta\Delta\psi = \alpha\sigma + \mathbf{\nabla} \cdot \mathbf{u} - g \text{ in } \Omega, \qquad (4.6)_1$$

$$\psi = 0 \text{ on } \Gamma, \qquad (4.6)_2$$

with $\zeta = \beta + \alpha a\mu$.
We should easily check that ψ satisfies the following biharmonic equation

$$\alpha^2\psi - (\zeta + \alpha a\mu)\Delta\psi + \alpha a\mu\zeta\Delta^2\psi = 0; \qquad (4.7)$$

if we can show that $\psi \equiv 0$, then equation $(4.1)_1$ will hold. A sufficient condition to reach such a goal is to select λ such that

$$\frac{\partial\psi}{\partial\mathbf{n}} = 0 \text{ on } \Gamma. \qquad (4.8)$$

To show the feasibility of this approach we introduce the linear operator A defined as follows: the function λ being known on Γ, we solve the following elliptic problems:

$$\begin{cases} \alpha^2\sigma_\lambda - \zeta\Delta\sigma_\lambda = 0 & \text{in } \Omega, \\ \sigma_\lambda = \lambda & \text{on } \Gamma, \end{cases} \qquad (4.9)$$

$$\begin{cases} \alpha\mathbf{u}_\lambda - a\mu\Delta\mathbf{u}_\lambda = -\beta\mathbf{\nabla}\sigma_\lambda & \text{in } \Omega, \\ \mathbf{u}_\lambda = 0 & \text{on } \Gamma_B \cup \Gamma_\infty^-, \\ a\mu\dfrac{\partial\mathbf{u}_\lambda}{\partial\mathbf{n}} - \beta\mathbf{n}\sigma_\lambda = \mathbf{0} & \text{on } \Gamma_\infty^+, \end{cases} \tag{4.10}$$

$$\begin{cases} \alpha^2\psi_\lambda - \zeta\Delta\psi_\lambda = \alpha\sigma_\lambda + \mathbf{\nabla}\cdot\mathbf{u}_\lambda & \text{in } \Omega, \\ \psi_\lambda = 0 & \text{on } \Gamma, \end{cases} \tag{4.11}$$

We define then operator A by

$$A\lambda = -\dfrac{\partial\psi_\lambda}{\partial\mathbf{n}} \mid_\Gamma. \tag{4.12}$$

Since the following relation holds,

$$\begin{cases} < A\lambda_1, \lambda_2 >= \dfrac{\alpha}{\zeta} \int \sigma_1\sigma_2 \, dx + \dfrac{1}{\beta\zeta} \int_\Omega [a\mu\mathbf{u}_1\cdot\mathbf{u}_2 + a\mu\mathbf{\nabla}\mathbf{u}_1\cdot\mathbf{\nabla}\mathbf{u}_2]dx, \\ \forall\lambda_1, \lambda_2, \end{cases} \tag{4.13}$$

where $\sigma_1, \mathbf{u}_1$ (resp. $\sigma_2, \mathbf{u}_2$) are the solutions of (4.9)–(4.10) corresponding to λ_1 (resp. λ_2), the operator A is self adjoint; similarly combining (4.13) with (4.10), (4.11), we can easily show that A is strongly elliptic.
To apply the above result to the solution of problem (4.1), let's introduce $\sigma_o, \mathbf{u}_o$, and ψ_o solutions of

$$\begin{cases} \alpha^2\sigma_o - \zeta\Delta\sigma_o = \alpha g - \mathbf{\nabla}\cdot f - a\mu\Delta g & \text{in } \Omega, \\ \sigma_o = 0 & \text{on } \Gamma, \end{cases} \tag{4.14}$$

$$\begin{cases} \alpha\mathbf{u}_o - a\mu\Delta\mathbf{u}_o = \mathbf{f} - \beta\mathbf{\nabla}\sigma_o & \text{in } \Omega, \\ \mathbf{u}_o = 0 \text{ on } \Gamma_B, \mathbf{u}_o = \mathbf{u}_\infty & \text{on } \Gamma_\infty^-, \\ \dfrac{\partial\mathbf{u}_o}{\partial\mathbf{n}} = 0 & \text{on } \Gamma_\infty^+, \end{cases} \tag{4.15}$$

$$\begin{cases} \alpha^2\psi_o - \zeta\Delta\psi_o = \alpha\sigma_o + \mathbf{\nabla}\cdot\mathbf{u}_o - g & \text{in } \Omega, \\ \psi_o = 0 & \text{on } \Gamma, \end{cases} \tag{4.16}$$

respectively. Combining (4.1) and (4.14)-(4.16) we obtain that the trace λ of σ on Γ satisfies the following (pseudodifferential) equation

$$A\lambda = \dfrac{\partial\psi_o}{\partial\mathbf{n}} \mid_\Gamma. \tag{4.17}$$

The boundary problem (4.17) can be solved by a quasi-direct method in the sense of (Glowinski 1984, Chapter 7), (Glowinski and Pironneau

1979); since this approach would require the construction of a full matrix approximating A, which is very costly in dimension three, we shall focus in this paper on preconditioned conjugate gradient solution methods. Using classical conjugate gradient notation let's define the preconditioning operator S as follows:

$$S^{-1} = B, \tag{4.18}$$

$$Bz = c\phi_{z|\Gamma} + 2z + \frac{1}{c}\frac{\partial\theta_z}{\partial\mathbf{n}}\Big|_\Gamma, \tag{4.19}$$

with ϕ_z and θ_z solutions of:

$$\alpha^2\phi_z - \zeta\Delta\phi_z = 0, \text{ in } \Omega, \frac{\partial\phi_z}{\partial\mathbf{n}} = z \text{ on } \Gamma, \tag{4.20}$$

$$\alpha^2\theta_z - \zeta\Delta\theta_z = 0, \text{ in } \Omega, \theta_z = z \text{ on } \Gamma, \tag{4.21}$$

and

$$c = \frac{\alpha}{\sqrt{\zeta}}\{\frac{1}{2}[(1 + \frac{\zeta}{\alpha a\mu})(1 + \sqrt{\frac{\alpha a\mu}{\zeta}})]^{1/2} - 1\}, \tag{4.22}$$

where

$$\zeta = \beta + \alpha a\mu. \tag{4.23}$$

Indeed parameter c occurs quite naturally from a Fourier analysis of operator A, cf. (Bégue et al. 1988) for details.

The corresponding conjugate gradient algorithm for solving (4.1) via (4.17) is given by

Step 0: Initialization

$$\lambda^o \text{ is given}; \tag{4.24}$$

solve then

$$\begin{cases} \alpha^2\sigma^o - \zeta\Delta\sigma^o = \alpha g - \mathbf{\nabla}\cdot f - a\mu\Delta g & \text{in } \Omega, \\ \sigma^o = \lambda^o & \text{on } \Gamma, \end{cases} \tag{4.25}$$

$$\begin{cases} \alpha\mathbf{u}^o - a\mu\Delta\mathbf{u}^o = \mathbf{f} - \beta\mathbf{\nabla}\sigma^o & \text{in } \Omega, \\ \mathbf{u}^o = \mathbf{0} \text{ on } \Gamma_B, \mathbf{u}^o = \mathbf{u}_\infty & \text{on } \Gamma_\infty^-, \\ a\mu\frac{\partial\mathbf{u}^o}{\partial\mathbf{n}} - \beta\mathbf{n}\sigma^o = \mathbf{0} & \text{on } \Gamma_\infty^+, \end{cases} \tag{4.26}$$

$$\begin{cases} \alpha^2\psi^o - \zeta\Delta\psi^o = \alpha\sigma^o + \mathbf{\nabla}\cdot\mathbf{u}^o - g & \text{in } \Omega, \\ \psi^o = 0 & \text{on } \Gamma, \end{cases} \tag{4.27}$$

and define

$$r^o = -\frac{\partial \psi^o}{\partial \mathbf{n}} \big|_\Gamma .\tag{4.28}$$

The preconditioning is achieved through

$$\alpha^2 \phi^o - \zeta \Delta \phi^o = 0, \quad \text{in } \Omega, \frac{\partial \phi^o}{\partial \mathbf{n}} = r^o \text{ on } \Gamma,\tag{4.29}$$

$$\alpha^2 \theta^o - \zeta \Delta \theta^o = 0, \quad \text{in } \Omega, \theta^o = r^o \text{ on } \Gamma,\tag{4.30}$$

$$g^o = c\phi^o|_\Gamma + 2r^o + \frac{1}{c}\frac{\partial \theta^o}{\partial \mathbf{n}}\Big|_\Gamma .\tag{4.31}$$

then we set

$$w^o = g^o.\tag{4.32}$$

Next, for $n \geq 0$, with $\lambda^n, \sigma^n, \mathbf{u}^n, \psi^n, r^n, g^n, w^n$ known, compute $\lambda^{n+1}, \sigma^{n+1}$, $\mathbf{u}^{n+1}, \psi^{n+1}, r^{n+1}, g^{n+1}, w^{n+1}$ as follows:

Step 1: Descent

$$\begin{cases} \alpha^2 \bar{\sigma}^n - \zeta \Delta \bar{\sigma}^n = 0 & \text{in } \Omega, \\ \bar{\sigma}^n = w^n & \text{on } \Gamma, \end{cases}\tag{4.33}$$

$$\begin{cases} \alpha \bar{\mathbf{u}}^n - a\mu \Delta \bar{\mathbf{u}}^n = \mathbf{f} - \beta \mathbf{\nabla} \bar{\sigma}^n & \text{in } \Omega, \\ \bar{\mathbf{u}}^n = 0 & \text{on } \Gamma^-_\infty, \end{cases}\tag{4.34}$$

$$\begin{cases} \alpha^2 \bar{\psi}^n - \zeta \Delta \bar{\psi}^n = \mathbf{\nabla} \cdot \bar{\mathbf{u}}^n + \alpha \bar{\sigma}^n & \text{in } \Omega, \\ \bar{\psi}^n = 0 & \text{on } \Gamma, \end{cases}\tag{4.35}$$

$$\rho_n = -\frac{\int_\Gamma r^n g^n \, d\Gamma}{\int_\Gamma \frac{\partial \bar{\psi}^n}{\partial n} w^n \, d\Gamma}.\tag{4.36}$$

We define then $\sigma^{n+1}, \mathbf{u}^{n+1}, \psi^{n+1}, r^{n+1}$ by

$$\lambda^{n+1} = \lambda^n - \rho_n w^n,\tag{4.37}$$

$$\sigma^{n+1} = \sigma^n - \rho_n \bar{\sigma}^n,\tag{4.38}$$

$$\mathbf{u}^{n+1} = \mathbf{u}^n - \rho_n \bar{\mathbf{u}}^n,\tag{4.39}$$

$$\psi^{n+1} = \psi^n - \rho_n \bar{\psi}^n,\tag{4.40}$$

$$r^{n+1} = r^n - \rho_n \frac{\partial \bar{\psi}^n}{\partial \mathbf{n}},\tag{4.41}$$

Step 2: Construction of the new descent direction

Solve

$$\alpha^2 \phi^{n+1} - \zeta \Delta \phi^{n+1} = 0, \text{ in } \Omega, \frac{\partial \phi^{n+1}}{\partial \mathbf{n}} = r^{n+1} \text{ on } \Gamma, \qquad (4.42)$$

$$\alpha^2 \theta^{n+1} - \zeta \Delta \theta^{n+1} = 0, \text{ in } \Omega, \theta^{n+1} = r^{n+1} \text{ on } \Gamma, \qquad (4.43)$$

and define

$$g^{n+1} = c\phi^{n+1}|\Gamma + 2r^{n+1} + \frac{1}{c}\frac{\partial \theta^{n+1}}{\partial n}_{|\Gamma}; \qquad (4.44)$$

compute then

$$\gamma_n = \frac{\int_\Gamma r^{n+1} g^{n+1} d\Gamma}{\int_\Gamma r^n g^n d\Gamma} \qquad (4.45)$$

$$w^{n+1} = g^{n+1} + \gamma_n w^n. \qquad (4.46)$$

Do $n = n + 1$ and go to (4.33).

5.   Solutions of the nonlinear subproblems by GMRES type algorithms

5.1   *Generalities*

Among the numerical methods which can be used for solving large nonlinear problems, let's mention nonlinear least squares methods since they have been sucessfully applied – coupled to conjugate gradient algorithms – to the solution of complicated problems arising from Fluid Dynamics. See e.g. (Glowinski 1984; Bristeau et al. 1987; Bristeau et al. 1985; Glowinski et al. 1985) for such applications. One of the drawbacks of the above methods is that they require an accurate knowledge of the derivatives of the least squares cost function, which for some problems may be by itself a fairly costly operation (the compressible Navier–Stokes equations seem to belong to that class of hard problems). More recently, variations of the above methods which do not require the knowledge of derivative have been introduced by several investigators. Among those methods the GMRES one has shown a promising potential to solve those complicated nonlinear problems. See (Saad and Schultz 1983; Wigton et al. 1985) for the theory and C.F.D. applications of GMRES. In the following section, we shall describe in a general Hilbert space framework a natural generalization of the GMRES algorithm. Actually these methods have been applied to the compressible Euler equations. See e.g. (Wigton et al. 1985; Mallet et al. 1988).

5.2   The GMRES algorithm for nonlinear problems

Applying operator splitting methods to the Navier–Stokes equations leads to the solution of nonlinear systems of partial differential equations, which in turn reduce to nonlinear finite dimensional systems once they have been approximated by either finite element, finite difference or spectral methods. All these problems can be formulated as follows:

Let V be a real Hilbert space for the scalar product $(.,.)$ and the corresponding norm $||.||$. We denote by V' the dual space of V, by $< .,. >$ the duality pairing between V' and V and finally by S the duality isomorphism between V and V' i.e. the isomorphism from V onto V' such that

$$< Sv, w >= (v, w), \forall v, w \in V,$$
$$< Sv, w >=< Sw, v >, \forall v, w \in V,$$
$$< f, S^{-1}g >= (f, g)_*, \forall f, g \in V'.$$

Now with F a (possibly nonlinear) operator from V into V', we consider the problem

$$F(u) = 0. \tag{5.1}$$

5.2.1   Description of the GMRES algorithm for solving (5.1)

$$u^o \in V, \text{given}; \tag{5.2}$$

then for $n \geq 0$, u^n being known we obtain u^{n+1} as follows:

$$r_1^n = S^{-1}F(u^n), \tag{5.3}$$

$$w_1^n = r_1^n / ||r_1^n||. \tag{5.4}$$

Then for $j = 2, ..., k$, we compute r_j^n and w_j^n by

$$r_j^n = S^{-1}DF(u^n; w_{j-1}^n) - \sum_{i=1}^{j-1} b_{ij-1}^n w_i^n, \tag{5.5}$$

$$w_j^n = r_j^n / ||r_j^n||; \tag{5.6}$$

in (5.5), $DF(u^n; w)$ is defined by either

$$DF(u^n; w) = F'(u^n).w \tag{5.7}_1$$

(where $F'(u^n)$ is the differential of F at u^n, or if the calculation of F' is too costly

$$DF(u^n; w) = \frac{F(u^n + \varepsilon w) - F(u^n)}{\varepsilon}. \qquad (5.7)$$

with ε sufficiently small; we define then $b_{i\ell}^n$ by

$$b_{i\ell}^n = < DF(u^n; w_\ell^n), w_i^n > (= (S^{-1}DF(u^n; w_\ell^n), w_i^n)). \qquad (5.8)$$

Next, we solve

$$\text{Find } \mathbf{a}^n = \{a_j^n\}_{j=1}^k \in \mathbb{R}^k \text{ such that}$$

$$\forall \mathbf{b} = \{b_j\}_{j=1}^k \in \mathbb{R}^k \text{ we have}$$

$$\left\| F(u^n + \sum_{j=1}^k a_j^n w_j^n) \right\|_* \leq \left\| F(u^n + \sum_{j=1}^k b_j w_j^n) \right\|_*, \qquad (5.9)$$

and obtain u^{n+1} by

$$u^{n+1} = u^n + \sum_{j=1}^k a_j^n w_j^n. \qquad (5.10)$$

Do $n = n + 1$ and go to (5.3).

In algorithm (5.2)–(5.10), k is the dimension of the so-called Krylov space.

Remark 5.1: We should easily prove that

$$(w_j^n, w_\ell^n) = 0, \quad \forall 1 \leq \ell, j \leq k, j \neq \ell. \qquad (5.11)$$

Remark 5.2: To compute $DF(u^n; w)$, we can use instead of $(5.7)_2$ the second order accurate discrete derivative

$$DF(u^n; w) = \frac{F(u^n + \varepsilon w) - F(u^n - \varepsilon w)}{2\varepsilon}. \qquad (5.12)$$

Remark 5.3: Norm $\|.\|_*$ in (5.9) satisfies the following relations

$$\|f\|_* = \|S^{-1}f\|, \quad \forall f \in V', \qquad (5.13)$$

$$\|f\|_* = < f, S^{-1}f >^{1/2}, \quad \forall f \in V', \qquad (5.14)$$

indeed we have used (5.13) to evaluate the various norms $\|.\|_*$ occuring in (5.9).

Remark 5.4: In order to evaluate $\mathbf{a}^n$ via the solution of (5.9), it is sufficient to approximate, in the neighborhood of $\mathbf{b} = 0$ the functional

$$\mathbf{b} \to \|F(u^n + \sum_{j=1}^{k} b_j w_j^n)\|_*^2 \tag{5.15}$$

by the *quadratic* one defined by

$$\mathbf{b} \to \|F(u^n) + \sum_{j=1}^{k} b_j DF(u^n; w_j^n)\|_*^2. \tag{5.16}$$

Minimizing the functional (5.16) is clearly equivalent to solving a linear system whose matrix $k \times k$ is symmetric and positive definite (this approach is equivalent to taking for $\mathbf{a}^n$ the first iterate provided by Newton's method applied to the solution of (5.9) and initialized with $\mathbf{b}=\mathbf{0}$).

5.3 Application to the solution of the nonlinear problems (3.3)

It follows from Sec. 3 that at each time step of the θ-scheme we have to solve a nonlinear system of the following type

$$\left\{ \begin{array}{l} \alpha\sigma + \mathbf{u} \cdot \boldsymbol{\nabla}\sigma = g, \\ \alpha\mathbf{u} - b\mu\Delta\mathbf{u} - \Psi(\sigma, \mathbf{u}, T) = \mathbf{f}, \\ \alpha T - b\pi\Delta T - \chi(\sigma, \mathbf{u}, T) = h, \end{array} \right. \tag{5.17}$$

with appropriate boundary conditions. In this context, we shall use the GMRES algorithm of Sec. 5.2. taking as preconditioner the elliptic operator associated to the following product norm

$$\{\phi, \mathbf{v}, \theta\} \to \{\alpha \int_\Omega \phi^2 dx + \alpha A \int \mathbf{v}^2 dx + b\mu A \int_\Omega |\boldsymbol{\nabla}\mathbf{v}|^2 dx$$
$$+ \alpha B \int_\Omega \theta^2 dx + b\pi B \int_\Omega |\boldsymbol{\nabla}\theta|^2 dx\}^{1/2},$$

with $A, B > 0$, $0 < b < 1$; cf. (Bristeau et al.) for more details.

6. Comments of the compatibility of the velocity and density finite element approximations

It is well known that in the incompressible Navier–Stokes equations pressure and velocity cannot be approximated independently, cf. (Girault and Raviart 1986) and the references therein.

Typical choices are for example:

i) *continuous* and piecewise *linear* approximation for the *pressure* associated to a *continuous* and piecewise *quadratic* approximation for the velocity.

ii) continuous and piecewise linear approximation for the *pressure*
associated to a *continuous* and piecewise *linear* approximation
for the velocity, on a grid *twice finer* that the pressure grid. See
(Glowinski 1984, Chapter 7) and Sec. 5 (Bristeau et al. 1987)
for more details.

iii) the so-called Arnold-Brezzi Fortin mini-element; cf. (Arnold
et al. 1984) and Sec. 5, (Bristeau et al. 1987).

Concerning now the compressible Navier–Stokes equations, there has
been a natural tendancy to approximate velocity and density (or logarithmic density) using the same finite element spaces for both quantities.
Indeed such a strategy can produce spurious oscillations particularly for
density. Once more a most efficient cure is again to use approximations
of type (i), (ii), (iii), with the role of the pressure played by the density
(or the logarithmic one); as in the incompressible case, approximations of
type (ii) and (iii) are particularly attractive and therefore have been used
for the subsequent calculations.

7. Numerical Experiments

We discuss in this section the results associated to the numerical simulation of compressible viscous flow inside and around air intakes and after
bodies. We shall present first the performances associated to the iterative
generalized Stokes solver discussed in Section 4. For these calculations, we
have used continuous and piecewise linear approximations for the temperature T, coupled to an approximation of $\{\mathbf{u}, \sigma\}$ of type (iii) in the sense
of Sec. 6, the test problem being associated to the 2D air intake of Fig.
2, Fig. 3(a), and Fig. 3(b) shows the variation of the residual norm versus the iteration number and the CPU time, respectively. On Fig. 4(a)
and Fig. 4(b) we observe that the speeds of convergence increase with the
ration $\frac{Re}{\Delta t}$. We observe with interest (cf. Fig. 5) the dramatic increase in
computational cost associated to the quasi-direct solution in the sense of
(Bristeau et al. 1987), Sec. 12 of problem (4.17).

The above test problem has been also used as a benchmark for comparisons between the least square conjugate gradient algorithm described
in (Bristeau et al. 1987) and the GMRES algorithm of Sec. 6.

Acknowledgements

The authors would like to thank P. Brown, Q. V. Dinh, P. Perrier, O. Pironneau and Y. Saad for fruitful discussions. They thank also Mrs. C. Demars
for the typing of the manuscript. The support of DRET (Grant No. 8634)
is also acknowledged.

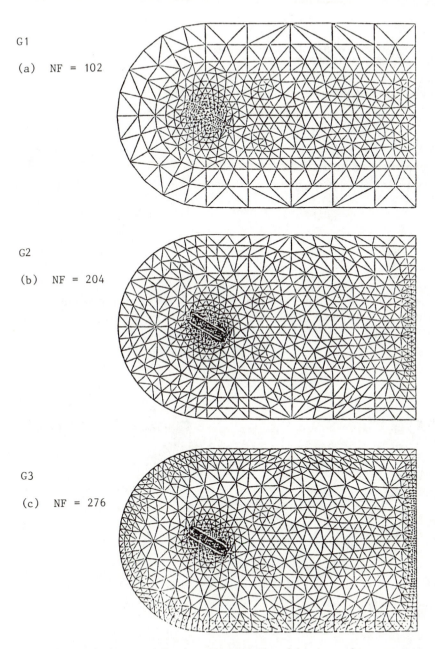

G1

(a)   NF = 102

G2

(b)   NF = 204

G3

(c)   NF = 276

Fig. 2: Triangulations used for the approximation of density and temperature.
NF: Boundary nodes

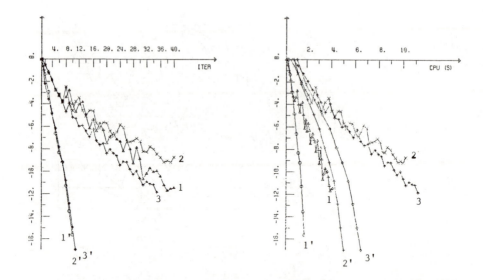

Fig. 3: Generalized Stokes problem. Influence of the number of boundary nodes. On G1 $\Delta t = 0.1$, Re $= 10000$; on G2 2, 2′; on G3 1, 2, 3 without preconditioner, 1′, 2′, 3′ with preconditioner

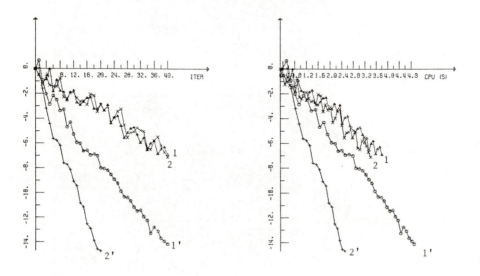

Fig. 4: Influence of the Reynolds number, 1, 1′ Re $= 500$; 2, 2′ Re $= 10000$

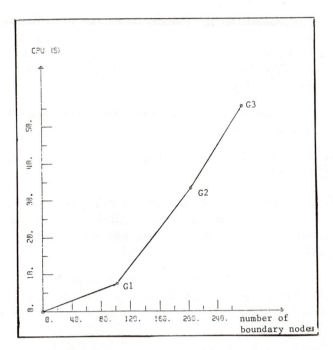

Fig. 5: Generalized Stokes problem. Quasi-direct solution. Influence of the number of boundary nodes on the computational cost of the boundary operator

References

Arnold D.N., Brezzi F., and Fortin M. (1984). A stable finite element for the Stokes equations, *Calcolo*, 21, 337.

Bègue C., Bristeau M.O., Glowinski R., Mantel B., Périaux J. and Rogé G. (1988). Accélération de la convergence dans le calcul des écoulements de fluides visqueux. 1re partie: cas incompressible et cas compressible non conservatif. Rapport INRIA, Rocquencourt, France, no 861.

Bristeau M.O., Glowinski R., Mantel B., Périaux J., Rogé and G. Terrasson G. Domain Decomposition. Finite Element and Iterative Methods for Fluid Flow Simulation, *Proceedings of the 1987 CISME Conference on Nonlinear Mechanics*, Udine, Italy (to appear).

Bristeau M.O., Glowinski R. and Périaux J. (1987). Numerical Methods for the Navier–Stokes Equations. Applications to the Simulation of Compressible and Incompressible Viscous Flows. *Computer Physics Reports*, 6, North-Holland, Amsterdam, 73-187.

Bristeau M.O., Glowinski R., Périaux J. Perrier P., Pironneau P. and Poirier G. (1985). On numerical solution of nonlinear problems in Fluid Dynamics by least squares and finite element methods (II). Application to transonic flow simulation, *Computing Methods in Applied Mechanics and Engineering, 51*, pp. 363-394.

Girault V. and Raviart P.A. (1986). *Finite Element Methods for Navier–Stokes Equations*, Springer-Verlag, Berlin.

Glowinski R. (1984). *Numerical Methods for Nonlinear Variational Problems*, Springer-Verlag, New-York.

Glowinski R., Keller H.B., and Reinhart L. (1985). Continuation conjugate gradient methods for the least squares solution of nonlinear boundary value problems, *SIAM Journal on Scientific and Statistical Computing, 4*, pp. 793-832.

Glowinski R., Pironneau O. (1979). On numerical methods for the Stokes problem, Chapter 13 of *Energy Methods in Finite Element Analysis*, R. Glowinski, E.Y. Rodin, O.C. Zienkiewicz eds., J. Wiley, Chichester, pp. 243-264.

LeVeque R. and Oliger J. (1983). Numerical methods based on additive splittings for hyperbolic partial differential equations, *Math. Computing, 40*, 469.

Mallet M., Périaux J., Stoufflet B. (1988). On fast Euler and Navier–Stokes Solvers, *Proceedings of the Seventh GAMM-Conference on Numerical Methods in Fluid Mechanics*, M. Deville Ed., Notes on Numerical Fluid Mechanics, Vol. 20, Vieweg.

Marchuk, G.I. (1975). *Methods of Numerical Mathematics*. Springer-Verlag, New-York.

Saad Y., Schultz M.H. (1983). GMRES: a Generalized Minimal Residual Algorithm for Solving Nonsymmetric Linear Systems, *Research Report, Yale University*, YALEU/DCS/RR-254.

Wigton L.B., Yu N.J. and Yound D.P. (1985). GMRES Accceleration of Computational Fluid Dynamics Codes, *AIAA 7th Computational Fluid Dynamics Conference*, Cincinnati, Ohio, July 1985, Paper 85-1494, pp. 67-74.

Yanenko, N.N. (1971). *The method of fractional steps*. Springer, New-York.

Steady incompressible and compressible solution of Navier–Stokes equations by rotational correction

F. El Dabaghi

INRIA, France

1. Introduction

In this work, we propose a new formulation derived from the Helmholtz decomposition in order to compute a steady solution of the Navier–Stokes equations governing an incompressible or compressible viscous flow at moderate Reynolds numbers. The main idea is based upon the fact that the general aspect of a viscous flow governed by the Navier–Stokes equations is essentially given by the Euler equations. Indeed let us consider a viscous flow, the vorticity created by the viscosity effects limited to a small area covering the boundary layers developed on the adherence frontiers, and everywhere else the flow can be considered as an inviscid one. This fundamental remark combined to our recent works (El Dabaghi 1984; El Dabaghi and Pironneau 1986; El Dabaghi et al. 1987) on the Helmholtz decomposition of a velocity vector field in a potential part plus a rotational one

$$\mathbf{u} = \nabla\phi + \mathbf{v}, \quad \nabla.\mathbf{v} = 0 \quad (\mathbf{v} = \nabla \times \mathbf{\Psi}) \tag{1.1}$$

leads to the following iterative method to compute the velocity $\mathbf{u}$ of a viscous flow in two steps: The first one consists in determining the potential velocity $\nabla\phi$ by solving the continuity equation with $\mathbf{v}$ given. The second step , more complicated, computes the rotational velocity $\mathbf{v}$ by solving the momentum equations with the characteristic method (Pironneau 1980) combined with an Uzawa formulation.

For simplicity, in the compressible case, we will restrict the analysis to iso-entropic and iso-enthalpic solutions which means that the flow is supposed close to the subsonic regime. On the other hand since the numerical scheme is based on an unsteady approach to reach the steady state, we neglect in this preliminary work the acoustic effects due to the time derivative of the density. In the next section, we recall briefly the status of the problem and its practical motivation and then we describe the general $(\phi - \mathbf{v})$ formulation of the compressible Navier–Stokes equations by noticing that the incompressible case is a particular one. In the last section, a brief comment on the different methods of solution used in the ($\phi - \mathbf{v}$) algorithm derived from the continuous case and then the numerical simulation of 2-D external flows around an NACA0012 profile and a cylinder are presented in order to establish the validity of the above methodology.

2. ($\phi - \mathbf{v}$) formulation

2.1 Notation

Ω bounded domain of R^3

$\Gamma = \Gamma_\infty \cup \Gamma_c$ boundary of Ω

Γ_∞ upstream and downstream boundary

Γ_c adherence or body boundary

$\mathbf{n}$ outward unit normal defined on Γ

$\mathbf{u}$ velocity

$\mathbf{v}$ rotational velocity

$\boldsymbol{\Psi}$ stream vector (or vector potential)

ϕ scalar potential

ρ density

p pressure

S modified entropy

H total enthalpy

ω vorticity

$\gamma = c_p/c_v$, specific heat ratio (≈ 1.4 for air)

η kinematic viscosity

Re Reynolds number ($Re = \mathbf{u}_\infty L/\eta$ where L is the characteristic length of the obstacle)

$\nabla.$ divergence

∇ gradient

Δ Laplacian

$\nabla\times$ curl

$\mathbf{u}.\mathbf{v}$ inner product between $\mathbf{u}$ and $\mathbf{v}$

$(\mathbf{u}.\mathbf{v})$ inner product between $\mathbf{u}$ and $\mathbf{v}$ in L^2, $\|.\|$ the associated norm

$H^1(\Omega) = \{w \in L^2(\Omega) : \nabla w \in L^2(\Omega)\}, \|w\|_{1,\Omega} = (\|w\|_{0,\Omega}^2 + \|\nabla w\|_{0,\Omega}^2)^{\frac{1}{2}}$ the associated norm

$H^1(\Omega)/\mathbf{R} = H^1(\Omega)$ quotiented by the constants

$H_0^1(\Omega) = H^1(\Omega)$ with zero traces on Γ

2.2 Status of the problem

In a finite elements environment, and without spending time describing their merits and their flexibilty in industrial applications, solving the equations involved in fluid mechanics is still today a challenge for all industrial research sections and an important goal for numerical analysts. Actually their joint efforts have provided accurate predictions for inviscid flow; these studies have taken two general directions, the first based on the potential approach and its rotational correction and the second one by solving directly the Euler equations with a conservative scheme or not, inspired essentially by finite difference solvers. It goes without saying that theoretical results obtained have to be neglected compared to the numerical ones. For the Navier–Stokes equations, the theoretical situation is not much better but in the last few years the numerical simulation of viscous flows greatly increased due to the relative simple extension of direct Euler solvers, conservative or not, to compute a viscous flow and undoubtedly to the evolution of super-computers. This extension reveals the same disadvantages as for an inviscid flow: it is very expensive in CPU time for transonic or rotational subsonic regimes. This implies a comeback for the potential approach with rotational correction which is the aim of this work. This method claims to be cheaper and more adequate to describe the boundary layers and its wake.

2.3 New ($\phi - \mathbf{v}$) formulation

As announced in the introduction, this new approach consists in superposing the potential part of the flow on its rotational one in order to obtain the global velocity. This choice was guided by some theoretical considerations: if we suppose that the velocity $\mathbf{u}$ exists then all the problems generated by this methodology are well-posed. On the other hand, our expertise in the different numerical tools involved enable us to treat this formulation. We recall briefly the compressible Navier–Stokes equations by neglecting the time derivative of the density

$$\nabla.(\rho\mathbf{u}) = 0 \quad \text{(continuity)} \quad (2.1)$$

$$\rho(\mathbf{u}_t + \mathbf{u}.\nabla\mathbf{u}) - \eta(\Delta\mathbf{u} + \frac{1}{3}\nabla\nabla.\mathbf{u}) + \nabla p = 0 \quad \text{(momentum)} \quad (2.2)$$

$$p = C\rho^\gamma \quad \text{(ideal state gas law)} (2.3)$$

where C is an aerodynamic constant ($C = \frac{M_\infty^2}{\gamma}$, M_∞ being the Mach number at infinity). Then by putting (1.1) in (2.1) we obtain a non-linear equation known as the transonic rotational equation

$$\nabla.(\rho\nabla\phi) = -\nabla\rho.\mathbf{v} \quad (2.4)$$

where the non-linearity is due to the ρ formula obtained under the isoenthalpic hypothesis

$$\rho = \left[\frac{\gamma - 1}{C\gamma S}(H - \frac{1}{2}\mathbf{u}^2)\right]^{\frac{1}{\gamma-1}}. \quad (2.5)$$

We add to (2.4) the classical Neumann boundary conditions

$$\rho \nabla \phi . \mathbf{n} = \begin{cases} \rho_\infty \mathbf{u}_\infty . \mathbf{n} & \text{on } \Gamma_\infty \\ 0 & \text{on } \Gamma_c. \end{cases} \quad (2.6)$$

Substituting (1.1) in (2.2) leads to

$$\rho(\mathbf{v}_t + \nabla \phi_t + \mathbf{u} . \nabla(\mathbf{v} + \nabla \phi)) - \eta(\Delta(\mathbf{v} + \nabla \phi) + \frac{1}{3} \nabla \nabla . (\mathbf{v} + \nabla \phi)) + \nabla p = 0 \quad (2.7)$$

and by using the fact that $\mathbf{v}$ is incompressible, we have

$$\begin{aligned}
\rho(\mathbf{v}_t + \nabla \phi_t + \mathbf{u} . \nabla(\mathbf{v} + \nabla \phi)) - \eta \Delta \mathbf{v} + \nabla q &= 0 \\
\nabla . \mathbf{v} &= 0 \quad (2.8) \\
q &= p - \frac{4}{3} \eta \Delta \phi.
\end{aligned}$$

We observe that we do not use here the state gas law and an important phenomena is kept homogeneous under a conservative form in the equation, that is the convective part of the flow. We add to this system the following boundary conditions

$$\mathbf{v} = \begin{cases} \mathbf{u}_\infty - \nabla \phi & \text{on } \Gamma_\infty \\ -\nabla \phi & \text{on } \Gamma_c. \end{cases} \quad (2.9)$$

The time discretization of (2.8), (2.9) leads to a well-posed quasi-Stokes problem.

2.4 Method of solution

The variational formulation of (2.4), (2.6) is solved by a $\mathbf{P}^1$ FEM associated with a least squares method (Glowinski 1984; Poirier 1981). For the problem (2.8), (2.9) we use an iterative method consisting of solving a quasi-Stokes problem at each time step by using an Uzawa algorithm; the convective part of the flow is computed by the characteristic method. Indeed the total derivative of $\mathbf{u}$ is approximated to the first order by

$$\mathbf{u}_t + \mathbf{u} . \nabla \mathbf{u} = \frac{d\mathbf{u}}{dt} \approx \frac{\mathbf{u}(\mathbf{x}, t + \Delta t) - \mathbf{u}(\mathbf{x}', t)}{\Delta t} \quad (2.10)$$

where $\mathbf{x}'$ is the point at time t, occupying the point $\mathbf{x}$ at time $(t + \Delta t)$ and it is the solution of the differential equation

$$\frac{d\mathbf{X}(\mathbf{x}, t + \Delta t, \tau)}{dt} = \mathbf{u}(\mathbf{X}(\mathbf{x}, t + \Delta t, \tau), \tau) \qquad t \leq \tau \leq t + \Delta t \quad (2.11)$$

with

$$\mathbf{X}(\mathbf{x}, t + \Delta t, t) = \mathbf{x}' \quad , \quad \mathbf{X}(\mathbf{x}, t + \Delta t, t + \Delta t) = \mathbf{x} \quad (2.12)$$

and by using the following notations

$$\mathbf{u}^{n+1} = \mathbf{u}(\mathbf{x}, t^n + \Delta t), \quad \mathbf{u}^n = \mathbf{u}(\mathbf{x}, t^n), \quad \mathbf{X}(\mathbf{x}, t^n + \Delta t, t^n) = \mathbf{X}^n \quad (2.13)$$

we rewrite (2.10) as follows

$$\frac{d\mathbf{u}}{dt} \approx \frac{(\mathbf{u}^{n+1} - \mathbf{u}^n \circ \mathbf{X}^n)}{\Delta t}. \tag{2.14}$$

Then by replacing (2.14) in (2.8) and by using (1.1), we obtain a $(\mathbf{v}, q)$ quasi-Stokes system to be solved at each time step:

$$\rho^n \frac{\mathbf{v}^{n+1}}{\Delta t} - \eta \Delta \mathbf{v}^{n+1} + \nabla q^{n+1} = \rho^n \frac{((\mathbf{v}^n + \nabla \phi^n) \circ \mathbf{X}^n + \nabla \phi^n)}{\Delta t} \tag{2.15}$$

where $\mathbf{X}^n$ is calculated by (2.11), (2.12), ρ^n and $\nabla \phi^n$ are obtained by (2.4), (2.5), (2.6).

The variational form of (2.15) associated to (2.8), (2.9) is:
Find $\mathbf{v}^{n+1} \in \mathbf{V}^N$ $(N = 2, 3)$ and $q^{n+1} \in L_0^2(\Omega)$ such that $\forall \mathbf{w} \in (H_0^1(\Omega))^N$ and $\forall z \in L_0^2(\Omega)$

$$\frac{\rho^n}{\Delta t}(\mathbf{v}^{n+1}, \mathbf{w}) \; + \; \eta(\nabla \mathbf{v}^{n+1}, \nabla \mathbf{w}) - (q^{n+1}, \nabla.\mathbf{w}) =$$

$$\frac{\rho^n}{\Delta t}((\mathbf{v}^n + \nabla \phi^n) \circ \mathbf{X}^n + \nabla \phi^n, \mathbf{w}) \tag{2.16}$$

$$(\nabla.\mathbf{v}^{n+1}, z) = 0 \tag{2.17}$$

where

$$L_0^2 = \{z \in L^2(\Omega) : \int_\Omega z = 0\}$$

$$\mathbf{V}^N = \left\{ \mathbf{v} \in (H_0^1(\Omega))^N : \mathbf{v} = \left\{ \begin{array}{ll} \mathbf{u}_\infty - \nabla \phi^n & \text{on } \Gamma_\infty \\ -\nabla \phi^n & \text{on } \Gamma_c \end{array} \right\} \right\}.$$

The problem (2.16), (2.17) is discretized by a FEM $\mathbf{P}^1$-Bubble for $\mathbf{v}^{n+1}$ and $\mathbf{P}^1$ for q^{n+1} associated with a Conjugate Gradient-Uzawa. The incompressible case can be found easily by taking ρ^n constant in (2.15), (2.16), and (2.17). Now let us describe the algorithm for solution:
Do $n = 0$ to M and initialise $\mathbf{v}^0 = \mathbf{0}$

1. Solve (2.4) - (2.6). $\Longrightarrow \phi^n, \rho^n$

2. Solve (2.11), (2.12). $\Longrightarrow \mathbf{X}^n, \mathbf{F}^n$

3. Solve (by Uzawa)

$$\frac{\rho^n \mathbf{v}^{n+1}}{\Delta t} - \eta \Delta \mathbf{v}^{n+1} + \nabla q^{n+1} = \mathbf{F}^n$$

$$\nabla \cdot \mathbf{v}^{n+1} = 0 \qquad\qquad \implies \mathbf{v}^{n+1}, q^{n+1}$$

where

$$\mathbf{F}^n = \rho^n \frac{(\mathbf{v}^n + \nabla \phi^n) \circ \mathbf{X}^n + \nabla \phi^n}{\Delta t}.$$

4. Calculate ρ^{n+1} by using the state law.
   Go to step 1 until convergence.

End Do.

We remark that we use a very effective preconditioner $(\eta I d^{-1} - \frac{\rho}{\Delta t} \Delta^{-1})$ due to Cahouet and Chabard (1981).

3. Numerical results

At the moment this approach is incomplete; indeed the iso-entropic hypothesis decreases the physical meaning of the solutions obtained but as a first guess this formulation is validated. On Fig. 1a the triangulation around a NACA0012 is presented and on Fig. 1b an enlargement of this is shown around the body. The flow simulated is at Re 500 with an upstream infinity Mach number of 0.85 and a time step $\Delta t = 0.8$. Fig. 1c represents the iso-Mach lines of the full potential solution. Then on Fig. 1d we show the iso-Mach lines of the velocity potential part at convergence to steady state and on Fig. 1e we present the iso-Mach lines of the total velocity and we note the rotational correction influence. Another simulation was done around a cylinder (Figs. 2a, 2b) at Re 50 with an upstream infinity Mach number of 0.5 and a time step $\Delta t = 0.8$. Fig. 2c represents the iso-Mach lines of the full potential solution. Then on Fig. 2d we show the iso-Mach lines of the velocity potential part at convergence to steady state and on Fig. 2e we present the iso-Mach lines of the total velocity and we note the rotational correction influence is more efficient in this case because of its low Re number and undoubtedly its subsonic nature. To conclude we can say for the moment that the quality of the boundary layers and its wake obtained by simulation is reasonable under the hypothesis and in the near future we will develop this new method with the complete Navier–Stokes equations.

Acknowledgements
This work was partly supported by STPA.

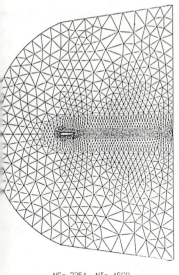

NS= 2354 NT= 4560

Fig. 1a

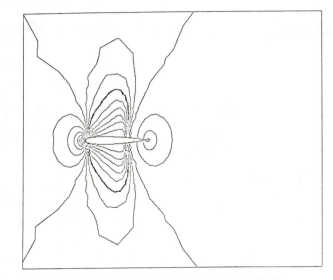

NAVIER-STOKES M= 0.85 REYNOLDS 500 DT=.0 ITER 1 POTENTIEL

Fig. 1c

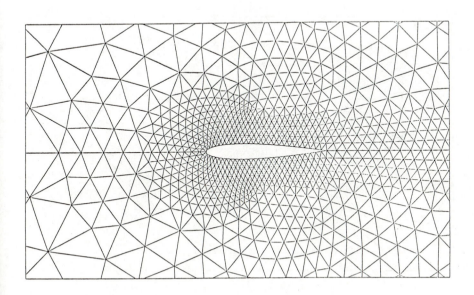

NS= 2354 NT= 4560

Fig. 1b

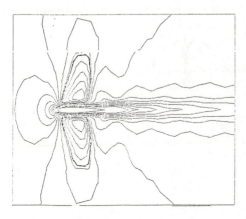

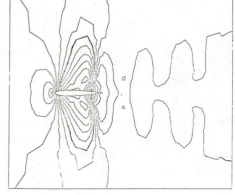

NAVIER-STOKES M= 0.85 REYNOLDS 500 DT=.0 ITER 25 CORRECTION   NAVIER-STOKES M= 0.85 REYNOLDS 500 DT=.0 ITER 25 POTENTIEL

Fig. 1d   Fig. 1e

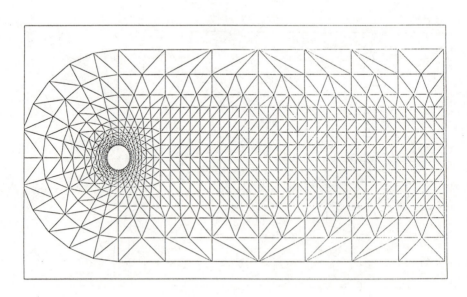

NS= 664   NT= 1260

Fig. 2a

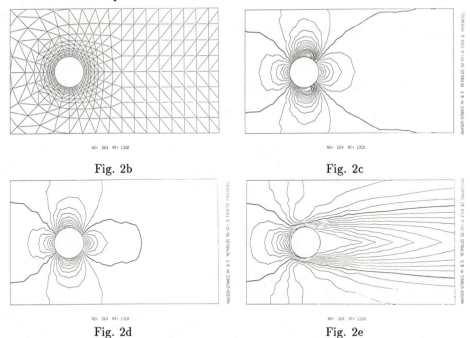

Fig. 2b

Fig. 2c

Fig. 2d

Fig. 2e

References

Cahouet, J. and Chabard, J. P. (1987). Some fast 3-D finite element solvers for generalised Stokes problems *EDF Report* HE/41/87.03

El Dabaghi, F. (1984). Thèse de 3ème cycle, Université Paris XIII.

El Dabaghi, F. and Pironneau O. (1986). Stream vectors in three dimensional aerodynamics. *Numer. Math.* 48:561–589

El Dabaghi, F., Periaux, J., Pironneau O., and Poirier, G. (1987). 2-D/3-D finite element solution of steady Euler Equations for transonic lifting flow by stream vector correction. *Int. J. for Num. Meth. in Fluids* 7.

Glowinski, R. (1984). *Numerical Methods for Non-linear Variational Problems.* Springer Verlag.

Pironneau, O. (1980). On the transport diffusion algorithm and its applications to the Navier–Stokes equations. *Numer. Math.* 8:309–332

Poirier, G. (1981). Thèse de 3ème cycle, Université Paris VI.

Comparison of implicit methods for the compressible Navier–Stokes equations

Y. Marx and J. Piquet

UA 1217 - ENSM

Nantes, France

1. Introduction

The solution of the compressible Navier–Stokes equations still requires much computing time even on super computers, unless acceleration techniques are used. In this work several implicit methods are tested on a 2D configuration. The combination of implicit methods and of multigrid techniques is presented. The efficiency of the overall procedure is demonstrated as steady states on a 321×65 mesh can be computed in about 1 min.

2. Numerical procedure

The Navier–Stokes equations written in their two dimensional conservation form are

$$\frac{\partial U}{\partial t} + \frac{\partial F'}{\partial x} + \frac{\partial G'}{\partial y} = \frac{\partial}{\partial x}(N_x \frac{\partial \tilde{U}}{\partial x} + M_y \frac{\partial \tilde{U}}{\partial y}) + \frac{\partial}{\partial y}(N_x \frac{\partial \tilde{U}}{\partial y} + M_y \frac{\partial \tilde{U}}{\partial x}) \quad (2.1)$$

where U and $\tilde{U}$ correspond respectively to the conservative variables and to the convective variables, F', G' are the advective fluxes and N_x, N_y, M_x, M_y are the diffusion matrices.

These equations are solved with a two step procedure. An explicit approximation step is first performed: it rests on a Godunov or MUSCL approach (van Leer 1977) where the Osher (1982) scheme is employed for solving the Riemann problems. This explicit approximation is then corrected with an implicit correction and a multigrid operator in order to accelerate the convergence towards the steady state.

Three implicit correction schemes corresponding to different space approximations of the following implicit operator have been tested:

$$\frac{\delta U}{\Delta t} + \frac{\partial}{\partial x}[A\delta U] + \frac{\partial}{\partial y}[B\delta U]$$

$$- \frac{\partial}{\partial x}[N_x\frac{\partial}{\partial x}\delta\tilde{U} + M_x\frac{\partial}{\partial y}\delta\tilde{U}] - \frac{\partial}{\partial y}[N_y\frac{\partial}{\partial y}\delta\tilde{U} + M_y\frac{\partial}{\partial x}\delta\tilde{U}]$$

$$= \frac{\Delta U}{\Delta t} \tag{2.2}$$

where $\delta U = U^{n+1} - U^n$, ΔU is the explicit increment A, B are the jacobians of F', G'. For each implicit scheme, a centered approximation is used for the diffusion and a first order upwind approximation for the advection.

In the Coakley (1981) scheme, the N_x, N_y matrices are replaced by their spectral radius ν,

$$\nu = \rho^{-1}\max(\mu, \lambda + 2\mu, \gamma\mu/\delta), \qquad \delta \text{ is the Prandtl number} \tag{2.3}$$

while the matrices of the cross derivatives are set equal to zero. A factorization of the two dimensional implicit operator is also performed and the resulting system is "diagonalized" in the sense of Chaussee and Pulliam (1980). As a result, only scalar tridiagonal systems have to be solved.

In order to see the influence of the factorization, an unfactorized version of the Coakley scheme has been developed in (Marx and Piquet 1988b). As the A and B matrices have different eigenvectors, the matrix of the unfactored Coakley cannot be efficiently inverted with a direct procedure. A line block Gauss-Seidel relaxation is used for that purpose.

The only simplification made in the third implicit scheme consists in taking all the matrices out of the space derivatives. Therefore the influence of the implicit treatment of the diffusive terms can be studied. A vectorizable line red-black Gauss-Seidel relaxation is used instead of the line block Gauss-Seidel relaxation. This scheme which is described more extensively in (Marx and Piquet 1988a) is now referred to as the "full" implicit scheme.

3. Results

The properties of the implicit schemes are now studied on the GAMM test case (Bristeau and Glowinski 1987) The performances of the Coakley corrections are presented in Figs. 1 and 2 for the first order scheme. An obvious mesh dependency is obtained. Moreover, a sudden change of the convergence slope is found with the factored Coakley scheme at high CFL number. As demonstrated in (Marx and Piquet 1988b), this corresponds to the bad damping properties of the factored Coakley scheme when large CFL numbers are used. If the CFL number is related to the range of dominating frequencies — calculated with a FFT analysis — a significant improvement of the convergence (Fig. 3) is obtained. The sensitivity of the convergence rate to the viscous coefficient (2.3) is demonstrated in Fig. 4. One can notice that the convergence decreases as the diffusion number is increased by putting more points in the viscous layer or by enlarging the coefficient, ν. A mesh-independent convergence rate is found with the "full" implicit scheme which rests on a better approximation of the diffusive terms (Fig. 5). A further reduction in the computing time can be observed (Fig. 6) if the "full" implicit scheme is used as a relaxation scheme in a multigrid procedure, (Marx and Piquet 1988a).

Problems related to the non linearities of the slope limiters appear when the second order scheme is employed. On a case without shock, the machine accuracy can be reached with the full implicit scheme if underrelaxation is performed, whereas no improvement is found when underrelaxing the factored Coakley scheme (Figs. 7 and 8). The multigrid technique leads again to a significant reduction in the computing time (Fig. 9).

The importance of using a continuous slope limiter is demonstrated in Fig. 10. It is found that with a discontinuous slope limiter (Harten-Osher) (Harten and Osher 1985), the convergence ends with a limit cycle. Such a phenomenon is not obtained with the C^1 slope limiter of Van Albada. It can also be noticed that, in both cases, freezing in time the slope limiters results in a quadratic convergence. With regard to the convergence, the Van Albada limiter seems superior to the Harten-Osher reconstruction, but looking at the computing field, a "kick" in the shock is observed on a coarse grid (Fig. 11) with the Van Albada limiter. This "kick" disappears on finer grids (Fig. 12). Therefore, despite its poor convergence features, the Harten-Osher reconstruction is used in the multigrid computations. Comparing the computed field obtained ofter 1100 iterations on a single grid and those obtained with the multigrid procedure, one can conclude that only a few cycles are needed and a great reduction of the computing time is achieved.

Table 1 sums up the computing time per iteration and per grid point for the implicit schemes. Because the line Gauss-Seidel relaxation does not vectorize, the unfactored Coakley scheme is more costly than the initial (factored) Coakley scheme. The "full" implicit scheme does not vectorize as well as the factored Coakley scheme on the Cray2, but on a VP200 it is only 25% more expensive than the factored Coakley scheme. As its convergence rate is faster and as it can be used in a multigrid procedure, the "full" implicit scheme is by far the most efficient implicit scheme.

4. Conclusion

A comparison of different implicit schemes for the compressible Navier–Stokes equations has been presented and the importance of the treatment - at the implicit step - of the diffusive terms has been demonstrated. The effectiveness of using the implicit scheme as a relaxation scheme in a multigrid procedure has been also pointed out.

Acknowledgments

Computations have been performed on the VP200 (CIRCE) and, with a donation by the Scientific Committee of the CCVR, on the Cray2 (CCVR).

Tables and figures

Schemes	Cray2	VP200
Factored Coakley	19 s	23 s
Unfactored Coakley	60 s	106 s
"Full" implicit	52 s	29 s

Table 1: CPU times for the three implicit schemes per grid point, 220×30 grid.

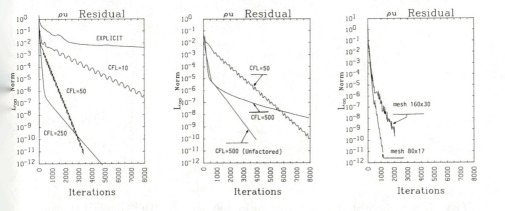

Figs. 1, 2, 3: First Order Coakley schemes, Re = 100: Fig. 1, 80×17 mesh; Fig. 2, 160×30 mesh; Fig. 3, Factored scheme, variable CFL number

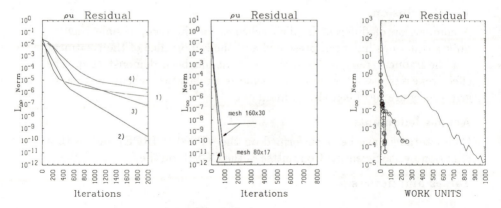

Figs. 4, 5, 6: Fig. 4. Influence of (1), factored Coakley scheme, 160 × 30 grid; (2), unfactored Coakley scheme, 160 × 30 grid; (3), unfactored Coakley scheme, 220 × 30 grid; (4) unfactored Coakley scheme, 220 × 30 grid, 2ν. Fig. 5. "Full" implicit scheme: CFL = 250 with 80 × 17 grid; CFL = 500 with 160 × 30 grid. Fig. 6. FMG method, 3 grids: o; single grid, CFL = 4500, 321 × 65 grid: —

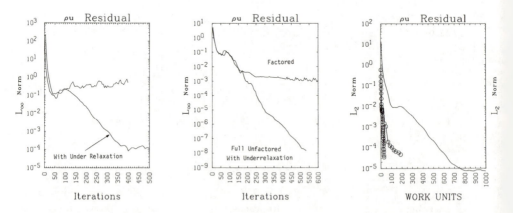

Figs. 7, 8, 9: Second order schemes; Re = 100. Fig. 7. "Full" implicit scheme with and without relaxation, 321 × 65 grid. Fig. 8. "Full" implicit scheme and Factored Coakley scheme, 80 × 17 grid. Fig. 9. FMG method, 3 grids: o; single grid, CFL = 4500, 321 × 65 grid: —

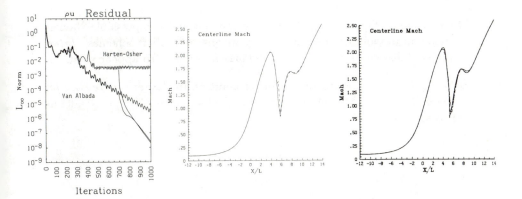

Figs. 10, 11, 12: Second order schemes; Re = 400. Fig. 10. Influence of the limiters, 80 × 17 grid. Fig. 11. Centerline Mach number, 80 × 17 grid; Harten and Osher: —; Van Albada: - - - -. Fig. 12. Centerline Mach number. Harten and Osher, 321 × 65 grid: —; Van Albada, 160 × 30 grid: - - - -; Harten and Osher, 80 × 17 grid: — - —

References

Bristeau, M. O. and Glowinski, R. (1987). Numerical simulation of compressible Navier-Stokes equations. In *Notes on Numerical Fluid Dynamics, Vol.18*, Periaux, J. and Viviand, H., editors. GAMM Workshop, Vieweg Verlag.

Chaussee, D. S. and Pulliam, T. H. (1980). A diagonal form of an implicit approximate factorization algorithm with application to a two dimensional inlet. AIAA Paper 80-0067.

Coakley, T. J. (1981). Numerical methods for gas dynamics combining characteristic and conservation concepts. AIAA Paper 81-1257.

Harten, A. and Osher, S. (1985). Uniformly high-order accurate non-oscillatory schemes I. NASA CR 175768.

van Leer, B. (1977). Towards the ultimate conservative finite difference scheme. IV a new approach to numerical convection. *Journal of Computational Physics*, 23:276–299.

Marx, Y. and Piquet, J. (1988a). Towards multigrid acceleration of 2D compressible Navier-Stokes finite volume implicit schemes. In *Robust Multigrid Methods*, Hackbush, W., editor. Vieweg Verlag. To appear soon.

Marx, Y. and Piquet, J. (1988b). Two-dimensional compressible Navier-Stokes finite volume computation by means of implicit schemes. *International Journal for Numerical Methods in Fluids*. To appear soon.

Osher, S. (1982). Shock modelling in aeronautics. In *Numerical Methods for Fluid Dynamics*, Morton and Baines, editors, pages 179–217. Academic Press.

Implicit finite difference methods for computing discontinuous atmospheric flows.

M. J. P. Cullen

Meteorological Office, Bracknell

1. Introduction

An important mathematical model for representing discontinuous atmospheric flow is given by first calculating an equilibrium velocity field from the requirement that the horizontal pressure gradient and frictional forces balance the Coriolis acceleration due to the Earth's rotation, and that the vertical pressure gradient is hydrostatic. In the absence of friction this requires the horizontal velicity to be geostrophic. The momentum in the equation of motion is then approximated by its equilibrium value, but the trajectory is not approximated. In the absence of friction this leads to the well-known geostrophic momentum approximation, Hoskins (1982), Salmon (1985).

Cullen and Purser (1984) showed that the frictionless equations possess generalised solutions which may contain discontinuities. A number of subsequent studies, e.g. Cullen et al. (1987), have shown that these solutions are useful simplified models of a number of atmospheric flows of direct relevance to weather forecasting, such as the deflection of flow round large mountain barriers and the intensification of depressions due to convective mass transfer. In these cases the equations can be solved exactly for piecewise constant data by using a Lagrangian method.

In this paper we include the friction. A finite difference method is used to solve the equations, which still admit discontinuous solutions. The method has to be at least partly implicit, since the trajectory is determined implicitly. A predictor-corrector method is used. The velocity field from the previous timestep is used as a first guess, and the evolution equations solved. The residual in the balance of forces, which should be zero, is calculated and used to generate a correction to the velocity field. The structure of the method is similar to that of the pressure correction method for the incompressible Navier-Stokes equations.

2. Model problem

The equations are for flow in a two-dimensional atmospheric cross-section. The coordinates are x and σ where σ is pressure divided by surface pressure. The upper and lower boundaries are then $\sigma = 0, 1$. The balance of forces in the x direction is given by:

$$\partial\phi/\partial x + C_P\sigma^\kappa\theta\partial\pi/\partial x - fv = F_u, \qquad (2.1)$$

where ϕ is the geopotential, C_P the specific heat of air at constant pressure, κ/C_P where R is the gas constant, $\pi = (p_*/p_0)^\kappa$ where p_* is the surface pressure and p_0 a reference pressure, f is the Coriolis parameter and F_u the frictional force. There is assumed to be no pressure gradient in the y direction, so that the balance in that direction is trivial. The hydrostatic relation is

$$\partial\phi/\partial\sigma = -RT/\sigma \qquad (2.2)$$

where T is the temperature. The evolution equations for the y momentum and potential temperature $\theta = T/(\pi\sigma^\kappa)$ are

$$Dv/Dt + fu = F_v \qquad (2.3)$$
$$D\theta/Dt = H \qquad (2.4)$$

where F_v is the frictional force in the y direction and H is the heat source. The continuity equation is

$$\partial p_*/\partial t + \partial(p_*u)/\partial x + \partial(p_*\dot\sigma)/\partial\sigma = 0 \qquad (2.5)$$

where $\dot\sigma$ is the vertical velocity.

In the case without friction, the generalised solutions are constructed at each time as an incompressible rearrangement of θ and $(v + fx)$. There is a unique rearrangement which satisfies (2.1) and (2.2) and the dynamical stability condition

$$(\partial(v + fx)/\partial x)(\partial\theta/\partial\sigma) - (\partial\theta/\partial x)(\partial v/\partial\sigma) \le 0. \qquad (2.6)$$

When forcing terms such as H in (2.4) are included, the solution evolves in time as a sequence of rearrangements. If the rearrangements are smooth, they can be represented as advection by a smooth velocity field (u, v). In general the rearrangement will not be smooth and may, for instance, require fluid to detach from the boundary. In the presence of friction, the structure of the solution is more complicated and the problem must be solved as an initial value problem. If the term H in (2.4) implies heating from below, the condition (2.6) will be weakened and considerable vertical rearrangement of the fluid will occur. If the vertical stability is completely destroyed by the heating, convective overturning will result. This cannot be represented as advection by a smooth velocity field and requires separate numerical treatment.

The frictional forces and heat fluxes are calculated as functions of the local Richardson number using the turbulence model employed in the U.K.

operational weather forecasting models. In the problem illustrated here the cross-section is half land and half sea. The land is assumed to be rougher than the sea, and is heated and cooled on a diurnal cycle. The sea temperature is assumed constant. A sea-breeze circulation is set up. During the day the hydrostatic pressure difference between land and sea implied by the heating is partly balanced by friction and there is only limited penetration of the sea breeze inland. During the evening the air becomes more stable, the friction is reduced, and the sea breeze accelerates. Later in the night the land becomes colder than the sea and a shallow land breeze develops.

3. Numerical method

A finite difference method is used. The variables are stored on a staggered grid, with u and v held at the same points, and θ and $\dot{\sigma}$ held together at points staggered apart in both x and σ. $\dot{\sigma}$ is held at the upper and lower boundaries, where it is set to zero. p_* is held instead of θ at the lower boundary points. The scheme is described in more detail in Cullen (1987).

A predictor-corrector method is used. In the predictor step the variables v, θ and p_* are stepped forward in time using equations (2.3) to (2.5) and the current values of the mass-weighted velocity field $(p_* u, p_* \dot{\sigma})$. An implicit single step method is used, with slight backward weighting (0.55). The heating and friction increments are then added, these are also calculated using an implicit method. The fields must then be adjusted to ensure satisfaction of (2.6). It has not yet been found practicable to use a two-dimensional adjustment, and therefore successive sweeps of the data are made in the x and σ directions to ensure that $\partial\theta/\partial\sigma < 0$ and $\partial(v + fx)/\partial x > 0$. These adjustments cover the cases where there is convective overturning not representable by a finite difference approximation to equations (2.3) to (2.5) .

The correction step updates the mass-weighted velocity fields by calculating a streamfunction correction $\Delta\psi$ and a vertical mean momentum correction $\Delta\overline{U}$, with $\overline{U} = p_* \overline{u}$. The correction equations are derived by using (2.3) to (2.5) to convert the velocity corrections into corrections to v, θ and p_*, and then substituting into (2.1) and (2.2) to remove the residuals. This gives

$$\kappa\delta_x p_*^{-1}\delta_x\Delta\overline{U} - f^2 p_*^{-1}(\Delta\overline{U} - \delta_\sigma\psi)_* + C_* j(\Delta\overline{U} - \delta_\sigma\psi)_j = R_* \quad (3.1)$$

$$-\kappa C_P \sigma^{\kappa-1}\pi\delta_x(\overline{\delta_\sigma\,\theta}^\sigma\delta_x\psi) + f^2\delta_\sigma(p_*^{-1}\delta_\sigma\psi) + \delta_\sigma C_{ij}(\Delta\overline{U} - \delta_\sigma\psi)_j = R \quad (3.2)$$

In these equations, the suffix $*$ denotes lower boundary values and δ a finite difference in the direction specified. The friction term F_u at level i is calculated in terms of values of u at model levels as the sum $C_{ij}u_j$, the details are given in Bell and Dickinson (1987). In deriving these equations, not all the terms in (2.3) to (2.5) were used. The selection was done to make (3.1) and (3.2) as elliptic as possible, and in particular to ensure that the one-dimensional conditions $\partial\theta/\partial\sigma < 0$ and $\partial(v + fx)/\partial x > 0$ are sufficient for ellipticity. This procedure is found to be needed to ensure stability where the solutions are discontinuous, and is analogous to the under-relaxation often used in the pressure correction method.

This scheme was used to integrate the equations on a 50x12 grid, with unequal vertical spacing giving higher resolution near the lower boundary. The horizontal grid-length was 4km and the timestep 1 hour. The correction step was performed three times for each predictor step, this was found to give adequate accuracy. The results are compared with a Lagrangian method developed by Chynoweth (1987) in which all effects can be included except the F_u term. By calculating this term *a posteriori* the accuracy of the finite difference calculation can be assessed. They are also compared with an explicit model where the Du/Dt term is included on the left hand side of (2.1). This allows non-equilibrium wave motions to develop, and illustrates the difficulty of computing real motions which are close to non-smooth equilibrium state.

4.   Results

Figures A-D show:

A: Lagrangian solution after 12 hours. The elements were originally in a regular rectangular array.

B: Finite difference θ distribution at the same time.

C: Finite difference u field.

D: The u field from the explicit model.

Both the Lagrangian and finite difference solutions show the low level sea-breeze circulation with a deeper and weaker return flow above. The air originally at the coast has moved 35km inland in the Lagrangian solution, element 96 has, however, moved 70km at the surface leading to the formation of a sharp front. In the finite difference solution the greatest lateral displacement is about 40km at this stage, though after 18 hours, when the friction is weaker, it reaches about 80km. The velocity field shown in C indicates strong surface convergence near the coast, consistent with the strong front formed in the Lagrangian model. The temperature contrast shown in B is weaker because of the effect of the heating.

$\sigma = 0$

$\sigma = 1$

$\longrightarrow x$

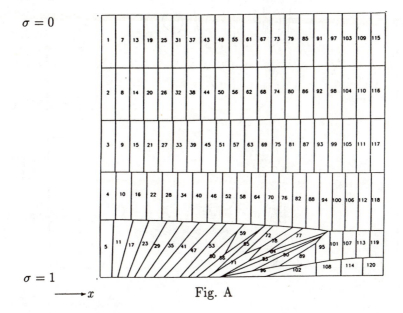

Fig. A

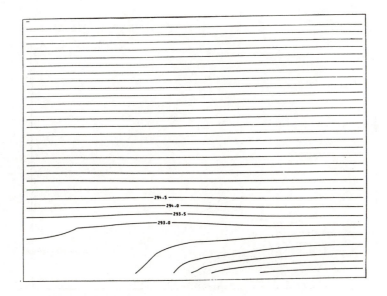

Fig. B

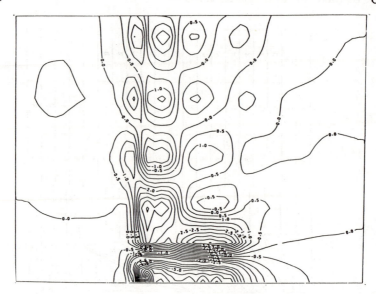

Fig. C

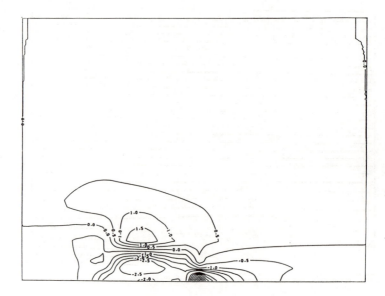

Fig. D

The solution of the apparently more general explicit equations is shown in D. The sea breeze circulation is much stronger and sets up a wave train in the vertical. In this model the air tends to overshoot its equilibrium position and oscillate about it. These oscillations have a larger horizontal scale and a small vertical scale. They can be damped by increasing the vertical diffusion, but this also tends to destroy the details of the main circulation. In reality small scale turbulent entrainment helps to damp such oscillations though some wave motions are excited. The wave response cannot be treated correctly within the context of an operational forecast model and the implicit computation of the equilibrium velocity field presented here provides a way of avoiding the problem.

References

Bell, R. and Dickinson, A. (1987). The Meteorological Office operational numerical weather prediction system. *Met. Office Scientific Paper* 41, HMSO.

Chynoweth, S. (1987). PhD thesis, Department of Maths., University of Reading.

Cullen, M. (1987). Implicit finite difference methods for modelling discontinuous atmospheric flows. *U.K. Met. Office, Met O 11 Tech. Note* 259. Submitted for publication.

Cullen, M. et al. (1987). Modelling the quasi-equilibrium dynamics of the atmosphere. *Quart. J. Roy. Meteor. Soc.*, 113:735–757.

Cullen, M. and Purser, R. (1984). An extended Lagrangian model of semi-goestrophic frontogenesis. *J. Atmos. Sci.*, 41:1477–1497.

Hoskins, B. (1982). The mathematical theory of frontogenesis. *Ann. Rev. Fluid Mech.*, 14:131–151.

Salmon, R. (1985). New equations for nearly geostrophic flow. *J. Fluid Mech.*, 153:461–477.

Numerical simulation of unsteady flows using the MUSCL approach

P. Guillen, M. Borrel and J. L. Montagne

Aerodynamics Department,

ONERA, France

1. Introduction

Simulation of blast wave interaction with a stationary object is a challenging problem for the computational fluid dynamics approach. On the one hand numerical simulation can complement experiments which could be very expensive and difficult to conduct. On the other hand, it involves a lot of theoretical aspects concerning the treatment of sharp discontinuities on the grounds both of accuracy and of robustness. Many authors, for instance Kutler & Fernquist (1980), Kurylo et al. (1986), Mark & Kutler (1983) have treated this flow problem past various kinds of objects using a viscous or an inviscid model. Most of them used an explicit Mac Cormack scheme which could present some weaknesses in low pressure cases. More recently Glaz et al. (1984) used an Eulerian second order Godunov scheme defined in Colella & Woodward (1984) and Colella & Glaz (1983) to compute the shock interaction problem with a wedge. Similar computations have been made by Rai (1983), Moon & Yee (1987), and Young & Yee (1987). The latter two used TVD schemes of the Harten-Yee form. In this study, we have considered schemes obtained with Van Leer's MUSCL approach. Basically, with the finite volume formulation we use, this approach leads to introducing linear distributions in each volume cell, while for stability reasons limiters must be used in the vicinity of strong shocks. The advance in time is known through the evaluation of perfect fluid fluxes accross interfaces between adjacent cells. This can be done using several techniques, such as the ones of Godunov, Roe, Van Leer, Osher, Steger-Warming or Vijayasundaram. Many of these flux formulations have been compared for Euler or for Navier–Stokes steady applications (Anderson et al. 1986; Van Leer et al. 1987). In this paper we intend to study the effects of different limiters and different upwinded flux evaluations when incorporated in the MUSCL approach to compute unsteady cases. After describing the numerical method in a first part, we apply it to the blast wave interaction problem with a wedge for different geometric and aerodynamic conditions for which experimental data are available.

2. The numerical method

2.1 *Formulation*

The unsteady 2D Euler equations are written in conservation form

$$W_t + F_x + G_y = 0$$

where

$$W = \begin{bmatrix} \rho \\ \rho u \\ \rho v \\ e \end{bmatrix}, \quad F = \begin{bmatrix} \rho u \\ p + \rho u^2 \\ \rho u v \\ (p+e)u \end{bmatrix}, \quad G = \begin{bmatrix} \rho v \\ \rho u v \\ p + \rho v^2 \\ (p+e)v \end{bmatrix}$$

with the perfect gas assumption

$$p = (\gamma - 1)\left(e - \frac{1}{2}\rho(u^2 + v^2)\right)$$

They are solved with a finite volume approach. The grid used has an i-j structure and the flow variables are located at the nodes. A volume control Ω_{ij} is attached to the node (see Fig. 1). At time t^n we know the average of the conservative variables in Ω_{ij}

$$\overline{W}_{ij}^n = \frac{1}{area(\Omega_{ij})} \int_{\Omega_{ij}} W(x,y,t^n)dxdy$$

The numerical scheme which has a second order accuracy in space and in time includes the 3 following stages :

1. a linear distribution is introduced in each control volume

$$U_{i+\theta,j+\lambda}^n(x,y) = \overline{U}_{ij}^n + \theta g_{ij}^i + \lambda g_{ij}^j$$

where $U = P^{-1}(W)$ is a set of variables conveniently chosen and g the slopes in the two grid directions.

The second order of accuracy in time is then achieved by a predictor-corrector scheme.

2. predictor

$$\overline{W}_{ij}^{n+\frac{1}{2}} = \overline{W}_{ij}^n - \frac{\Delta t}{2 \times area(\Omega_{ij})} \sum^{IJ} |\partial\Omega_{IJ}| \left\{ F[W_{IJ}]n_x^{IJ} + G[W_{IJ}]n_y^{IJ} \right\}$$

where IJ describes the midpoints of the sides of the cell $\partial\Omega_{ij}$ with length $|\partial\Omega_{IJ}|$ and (n_x^{IJ}, n_y^{IJ}) the outward unit normal at these points.

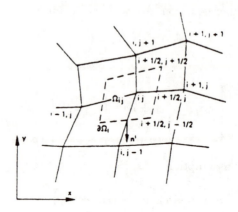

Fig. 1 Notations in the computational space

3. corrector

$$\overline{W}_{ij}^{n+1} = \overline{W}_{ij}^{n} - \frac{\Delta t}{area(\Omega_{ij})} \sum^{IJ} \mid \partial \Omega_{IJ} \mid FL \left[W_{IJ}^{n+\frac{1}{2}}, W_{ad}^{n+\frac{1}{2}} \right]$$

where W_{ad} represents the value in the adjacent cell and $W_{IJ}^{n+\frac{1}{2}}$ is defined by

$$W_{i+\theta,j+\lambda}^{n+\frac{1}{2}}(x,y) = P \left(\overline{U}_{ij}^{n+\frac{1}{2}} + \theta g_{ij}^{i} + \lambda g_{ij}^{j} \right)$$

$FL(w_1, w_2)$ represents the numerical expression of fluxes in the direction normal to the side separating w_1 and w_2

Clearly many variants of this method can be defined according to choices made for the evaluation of the slopes and of the fluxes.

2.2 Limiters

The computation of slopes depends mainly on two aspects:
Firstly, the set of variables used (ie. the choice of the biunivoque function P): in our numerical tests two sets were tried, the conservative variables and the (ρ, u, v, p) set. Another attractive set that has been considered by other authors consists of characteristics variables. Our 1D numerical experiments show that the second set is preferable to the first one, and will be the only one considered here.

We shall denote

$$\alpha^i_{i+\frac{1}{2},j} = P^{-1}(W_{i+1,j}) - P^{-1}(W_{i,j})$$

Secondly, the so-called limiter function can be more or less stringent. Many choices have been proposed in the literature and we have tested four of them. They are reviewed in the order of decreasing dissipation. The formulae are presented for the calculation of g^i_{ij} as a function of $a = \alpha^i_{i-\frac{1}{2},j}$ and $b = \alpha^i_{i+\frac{1}{2},j}$.

2.2.1 *Roe minmod formulation*
(Roe 1985)

$$g^i_{i,j} = minmod(b,a)$$

where the minmod function of a series of values is the smallest of them in absolute value if they have the same sign , else 0.

2.2.2 *Van Albada limiter*
(Van Leer 1977)

$$g^i_{i,j} = \frac{[b(a^2 + \delta_2) + a(b^2 + \delta_2)]}{(a^2 + b^2 + 2\delta_2)}$$

In this limiter, δ_2 is a small parameter

2.2.3 *Van Leer limiter*
(Van Leer 1974)

$$g^i_{i,j} = \frac{(ab+ \mid ab \mid)}{(a+b)}$$

2.2.4 *Roe superbee formulation*
(Roe 1985)

$$g^i_{i,j} = S.\max\{0, \min(2|a|, S.b), \min(|a|, 2S.b)\}; \qquad S = sign(a)$$

Similarly one can calculate $g^j_{i,j}$ for the other grid direction with these limiters.

2.3 *Fluxes*
The fluxes at an interface are evaluated by an upwinding technique between the two states w_1 and w_2 on either part of the the interface, in the normal direction. A recent survey of some upwind fluxes is made in Van Leer et al. (1987), so we will describe them only briefly hereafter. In what follows we call $F(U)$ the flux in the normal direction and A its jacobian matrix.

2.3.1 Godunov fluxes

The flux $FL(w_1, w_2)$ is found by solving the Riemann problem at the interface exactly and so it takes into account the non linear interactions between w_1 and w_2. This is done by an iterative method and is quite expensive in CPU time, which is why other formulations are considered.

2.3.2 Osher fluxes

$$FL(w_1, w_2) = \frac{1}{2}(F(w_1) + F(w_2) - \int_{w_1}^{w^2} |A(U)|dU)$$

where the integral is carried out along a path piecewise parallel to the eigenvectors of the jacobian matrix A. In our numerical experiments we use the path defined in the Osher-Solomon (1982) paper.

2.3.3 Roe fluxes

$$FL(w_1, w_2) = \frac{1}{2}\left\{(F(w_1) + F(w_2) - R.diag\psi(\lambda).R^{-1}(w_2 - w_1)\right\}$$

where the λ's are the eigenvalues of $A(\overline{w})$ and R the corresponding matrix of eigenvectors, $\overline{w}$ is the Roe mean value (Roe 1981) of w_1 and w_2 determined to solve exactly discontinuities. Also

$$\psi(\lambda) = \left\{ \begin{array}{ll} |\lambda| & \text{if } \lambda > \delta \\ \frac{(\lambda+\delta)}{2\delta^2} & \text{if } \lambda \leq \delta \end{array} \right.$$

is an entropic correction to $|\lambda|$ needed for steady cases. This correction is less necessary for unsteady cases.

2.3.4 Van Leer fluxes

(Van Leer 1982)

This is a flux splitting constructed with low order polynomials of the normal speed u which gives continuous eigenvalues at $u = 0$ and $u = +c$ and $u = -c$. Formulations are given in the reference.

2.3.5 Vijayasundaram fluxes

(Vijayasundaram 1983)

$$FL(w_1, w_2) = A^+(\overline{w}).w_1 + A^-(\overline{w}).w_2$$

where $A^\pm$ is the matrix obtain by cancelling the negative (respectively positive) eigenvalues of A. Originally $\overline{w}$ was the arithmetic average, but in this study we prefer to choose the Roe mean value.

2.3.6 Steger-Warming fluxes

(Steger and Warming 1981)

$$FL(w_1, w_2) = A^+(w_1).w_1 + A^-(w_2).w_2$$

For 1D problems the three first fluxes are well known to be able to distinguish all kinds of discontinuites (both contacts and shocks) and capture them within 1 or 2 points for steady cases. For 2D problems, these properties hold only if the discontinuities are parallel to interfaces, which is obviously not the case generally in our fixed grid unsteady computations. In the numerical experiments we consider two different kinds of grids.

2.4 Boundary conditions

The treatment of boundary conditions is the same as the one described in Borrel & Montagne (1985). In the present paper the flux version of the boundary treatment was chosen.

3. Numerical experiments

3.1 The blast wave interaction over a wedge

This problem is accurately described in Deschambault & Glass (1983), Glaz et al. (1984) and the reader is invited to consult these papers for details. We just recall that there are four kinds of reflections over a wedge according to the angle and the shock Mach number : single, double, complex, and regular. Since we wanted to use a perfect fluid solver, comparisons have to be made for cases with a low shock Mach number if the gas considered is air, or a not too high one ($M_s < 7$) in argon. In these conditions we avoid real gases and vibrational nonequilibrium effects. The four cases selected are the following ones for which experimental data are available in Deschambault & Glass (1983), Glaz et al. (1984) :

Reflexion type	M_s	wedge angle	Gas
Regular	2.05	60°	*Argon*
Single Mach	2.03	27°	*Air*
Complex Mach	5.07	30°	*Argon*
Double Mach	7.1	49°	*Argon*

The double Mach reflection case, which includes in its solution four

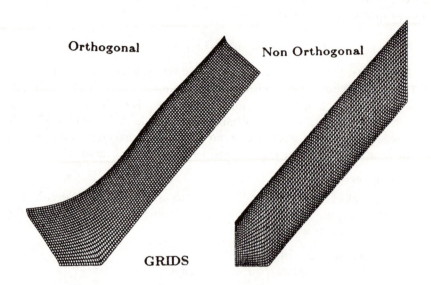

Fig. 2 Computational grids

shocks and two contact discontinuities, was retained to test the different versions of our scheme over several grids. It is a well known case which has been computed by several authors, Rai (1983), Young & Yee (1987) and Deschambault & Glass (1983), many of them using the self similar property of this application to solve it with a quasi stationary method (Rai 1983; Deschambault and Glass 1983).

3.2   Test conditions
3.2.1   Grids

Two types of grid were considered, as can be seen in Fig 2, a smooth orthogonal grid (a) and a smooth non orthogonal one (b). The first type of grid was obtained with the Steger-Sorenson GRAPE program (Steger and Sorenson 1979) which solves the Poisson equation, the second type by an optimisation method of Carcaillet (1985). We notice that one family of grid lines is aligned with eventual mach stems for the (a) case, and with the incident shock for the (b) case. For the double Mach reflection case we also consider coarse grids (200×50) and fine grids (about 400×100)

3.2.2   Initialisation

Knowing the shock Mach number, the Rankine-Hugoniot relations give the post-shock state. As noted by Glaz et al. (1984) the formation, from the initial data, of a travelling shock wave structure which diffuses over 3 or 4 points results in a starting error which appears as a low amplitude wave behind the main shock wave. This starting error is eliminated after about 50 time steps, by uniformisation of the post-shock region ,when the shock wave has nearly reached the wedge.

3.2.3   Boundary conditions

The inflow and outflow boundary values are constant and imposed to the post and pre shock values. The two other boundaries are the upper boundary and the wall. For the upper boundary it is very important to impose a null gradient in the direction parallel to the travelling shock, as our treatment consists in extrapolating values on a fictitious grid line outside the boundary, and as this extrapolation is made on the mesh indices, the easiest way for us was to make the top of the grid lines parallel to the shock. For the solid surface no special treatment was used. Results are discussed according to the direction of lines originating on the wall in the following sections.

3.3   Numerical results

3.3.1   Validation

The different cases were first computed with the van Leer fluxes and the minmod limiter. For all isocontours displayed except those concerning the single Mach reflection we chose 30 equally spaced isovalues.

Regular reflection : we use a 379×99 point non-orthogonal grid. Isobar and isopycnic lines are displayed on Fig. 3. The same test case has been computed and compared with experimental results in Glaz et al. (1984). Both numerical results are very similar and quite close to the experimental ones.

Single Mach reflection : the numerical isopycnics shown in Fig. 4 correspond to those found in Glaz et al. (1984). Our results are very similar to those of Glaz et al. which are shifted by about one fringe from experimental results. We use a 288×100 orthogonal grid. The slip contact line can be deduced by comparing isobar and isopycnic contours.

Complex Mach reflection : this was computed over a 313×200 grid. Experimental results can be found in Deschambault & Glass (1983). Direct comparison shows that the reflection shape and the contact discontinuity location agree very well. Density values around the triple point were also found to be the same (Fig. 5).

Double Mach reflection : discrepancies were found in Glaz et al. (1984) between experimental and computational results and were attributed to viscous effects. Our results correspond to the computational ones of Glaz et al. (1984). The first Mach stem is well captured as well as the first

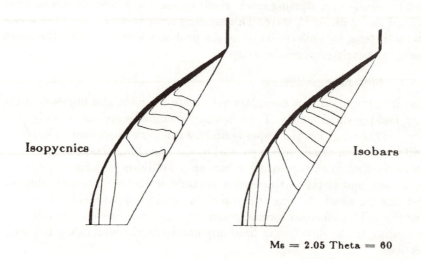

Isopycnics

Isobars

Ms = 2.05 Theta = 60

Fig. 3 Regular reflection

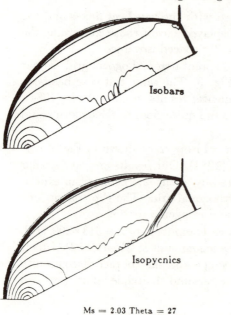

Isobars

Isopycnics

Ms = 2.03 Theta = 27

Fig. 4 Single Mach reflection

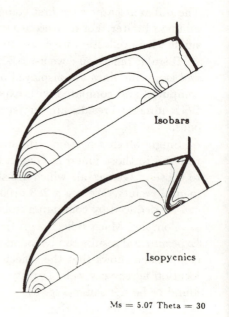

Isobars

Isopycnics

Ms = 5.07 Theta = 30

Fig. 5 Complex Mach reflection

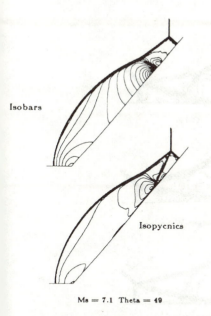

Fig. 6 Double Mach reflection

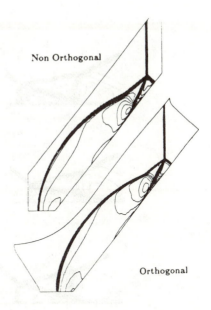

Fig. 7 Grid effects

contact discontinuity, the second stem is computed with some diffculties and the second slip line which is experimentally very weak is not captured at all. For this test case we use a 552×124 non orthogonal grid (Fig. 6).

3.3.2 Grid effects

The DMR case was selected to see the influence of some parameters. In Fig. 7 one can see the results for an orthogonal and a non-orthogonal grid. The grids were chosen so that they include roughly the same number of significant cells (about 180×40). The differences between the two results can be explained easily: if discontinuities are parallel to a family of grid lines the resolution is much better than if they are not. For instance, the higher part of the reflection shock and the two Mach stems have a better resolution in the orthogonal case, and it is the opposite for the lower part of the reflected shock and the slip line (especially when it intersects the wall). Thus we can conclude that an adaptive strategy is necessary.

3.3.3 Flux effects

The double Mach stem region is displayed on Fig. 8 for eight different types of fluxes. The grid considered here is orthogonal with 236×52 points. The thickness of the second Mach stem and the behaviour of the main slip line where it encounters the wall are good criteria to reveal differences in

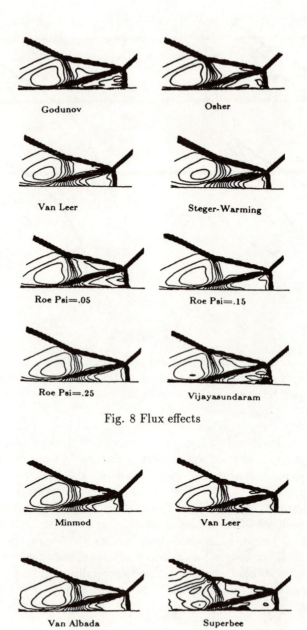

Godunov

Osher

Van Leer

Steger-Warming

Roe Psi=.05

Roe Psi=.15

Roe Psi=.25

Vijayasundaram

Fig. 8 Flux effects

Minmod

Van Leer

Van Albada

Superbee

Fig. 9 Limiter effects

accuracy of the computations. It can be seen that the best resolution is obtained, as might be expected, with Godunov fluxes; Osher evaluation is very close, but van Leer fluxes spread discontinuities much more, and the Steger-Warming fluxes are the worst from this point of view. Vijayasundaram fluxes give quite good results according to these two criteria but create a kind of numerical instability at the foot of the first Mach stem. Roe fluxes with a small parameter δ are very good except for a small instability at the first Mach stem too. It should be noticed that this small instability disappears when we add some diffusion (a δ increase) but at the same time the discontinuities spread out.

3.3.4   Limiter effects

On Fig. 9,one can see the result for the tested limiters. Superbee seems to give very sharp discontinuities but is somewhat unstable elsewhere. Van Leer and van Albada limiters are quite similar and give better results than the minmod formulation.

4.   Conclusion

The use of the MUSCL approach in numerical simulations of unsteady shock wave interaction with wedges seems valid for practical engineering problems. According to the degree of accuracy needed, we may choose different fluxes. Godunov or Osher formulations are interesting for accurate computations. Van Leer fluxes give less accurate results but with significant cpu time savings. The use of adaptive grids should improve greatly the accuracy.

References

Anderson, W. K., Thomas, J. L., and Leer, B. V. (1986). A comparison of finite volume flux vector splittings for Euler equations. *AIAA Journal*, 24:1453–1460.

Borrel, M. and Montagne, J. L. (1985). Study of a non centered scheme with application to aerodynamics. AIAA Paper 851497.

Carcaillet, R. (1985). Generation and optimisation of three dimensional computational grids. Master's thesis, Austin, Texas.

Colella, P. and Glaz, H. M. (1983). Efficient solution algorithms for the Riemann problem for real gases. Report 15776, LBL.

Colella, P. and Woodward, J. (1984). The piecewise-parabolic method (ppm) for gas-dynamical simulations. *Journal of Computational Physics*, 54:174–201.

Deschambault, R. L. and Glass, I. I. (1983). Non stationary oblique shock wave reflections. *Journal of Fluid Mechanics*, 131:27–57.

Glaz, H. M., Colella, P., Glass, I. I., and Deschambault, R. L. (1984). A numerical study of oblique shock wave reflections with experimental comparisons. Report 18156, LBL.

Kurylo, J., Hancock, S., and Kivity, K. (1986). Viscous flow calculations of shock diffraction and drag loads on arched structures. AIAA Paper 860031.

Kutler, P. and Fernquist, A. R. (1980). Computation of blast wave encounter with military targets. Report 80-82, Flow simulation, Inc.

Mark, A. and Kutler, P. (1983). Computation of shock wave-target interaction. AIAA Paper 830039.

Moon, J. and Yee, H. C. (1987). Numerical simulation by TVD schemes of complex shock reflections from airfoils. AIAA Paper 870350.

Osher, S. and Solomon, F. (1982). Upwind difference schemes for hyperbolic systems of conservation laws. *Mathematics of Computation*, 38(158):339–374.

Rai, M. M. (1983). A conservative treatment of zonal boundaries for Euler equations calculation. AIAA Paper 840169.

Roe, P. L. (1981). Approximate Riemann solvers, parameter vectors, and difference schemes. *Journal of Computational Physics*, 43:357–372.

Roe, P. L. (1985). Some contributions to the modelling of discontinuous flows. *Lectures in Applied Mathematics*, 22.

Steger, J. and Sorenson, R. L. (1979). Automatic mesh point clustering near a boundary in grid generation with elliptic partial differential equation. *Journal of Computational Physics*, 33:405–410.

Steger, J. L. and Warming, R. F. (1981). Flux vector splitting of the inviscid gasdynamic equations with applications to finite difference methods. *Journal of Computational Physicis*, 40(263-293).

Van Leer, B. (1974). II. Monotonicity and conservation combined in a second order scheme. *Journal of Computational Physics*, 14:361–370.

Van Leer, B. (1977). A second order sequel to Godunov method. *Journal of Computational Physics*, 23:276–299.

Van Leer, B. (1982). Flux vector splitting for the Euler equations. ICASE Report 82-30.

Van Leer, B., Thomas, J. L., Roe, P. L., and Newsome, R. (1987). Comparison of numerical flux formulation for the Euler and NS equations. AIAA Paper 871104.

Vijayasundaram, G. (1983). *Resolution numerique des equations d'Euler pour des ecoulements transsoniques avec un schema de Godunov en Elements finis*. PhD thesis, Paris.

Young, V. and Yee, H. C. (1987). Numerical imulation of shock wave diffraction by TVD schemes. AIAA Paper 870112.

Computation of viscous separated flow using a particle method

J. M. R. Graham

Dept. of Aeronautics, Imperial College, London.

1. Intoduction

The basic discrete vortex method provides a method of solution for the unsteady incompressible Euler equations. Usually a cut-off radius or vortex blob representation (Leonard 1980) is used to provide stability. The advantage of the method is that because it is Lagrangian with small numerical diffusion, sharp structures such as vortex sheets remain well defined in the flow field. The velocity field is calculated by a grid-free method using the Biot–Savart integral for the interaction between every pair of vortices. A faster alternative for a field containing large numbers of vortices is the Cloud-in-cell method (Christiansen 1973), in which the velocity is obtained from a finite difference solution of the Poisson equation for the stream- function on a superposed grid. Viscous versions of vortex methods have been developed using diffusing vortex blobs (Chorin 1973) and the random walk method (Smith and Stansby 1987) which approximates the solution of the diffusion equation with increasing accuracy as the number of vortices is increased. The numerical method described in the present paper is for 2-dimensional flow and is based on the Cloud-in-cell method but the viscous diffusion is carried out by finite differences on the fixed mesh. It is therefore more like a finite difference method for the Navier–Stokes equations in vorticity/ streamfunction form carried out on a fixed mesh except for the convection part which is modelled by Lagrangian moving particles. Methods of this type for general convection-diffusion problems have previously been discussed quite extensively (e.g. (Farmer and Norman 1986; Raviart 1986)).

2. Method

The two-dimensional Navier–Stokes equations in vorticity/streamfunction (ω/ψ) form are

$$\omega_t + \psi_y \omega_x - \psi_x \omega_y = \nu(\omega_{xx} + \omega_{yy}) \tag{2.1}$$

and

$$\psi_{xx} + \psi_{yy} = -\omega \tag{2.2}$$

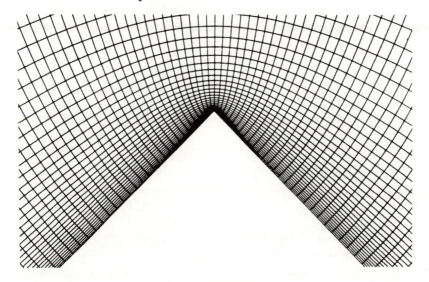

Fig. 1: Computational mesh in physical plane

where ν is the kinematic viscosity.

For many of the cases of flow past bodies it is convenient to transform the flow field to a rectangular domain (ξ, η) with the transformed boundary of the body lying along one or more boundaries, say $\eta = 0$. Using a conformal transformation $(x, y) \mapsto (\xi, \eta)$, equations (2.1) and (2.2) become:

$$J\omega_t + \psi_\eta \omega_\xi - \psi_\xi \omega_\eta = \nu(\omega_{\xi\xi} + \omega_{\eta\eta}) \tag{2.3}$$

and

$$\psi_{\xi\xi} + \psi_{\eta\eta} = -J\omega \tag{2.4}$$

where J is the Jacobian $\partial(x, y)/\partial(\xi, \eta)$.

Equation (2.4) is solved at each time step on a grid which is uniform in the ξ direction so that Fourier transforms may be used, but stretched in the η direction so that the meshes adjacent to the wall are small enough to resolve the boundary-layer (Fig. 1). After taking a fast discrete Fourier transform in the ξ direction and using central differences, equation (2.4) results in a tridiagonal set of equations for the transform of ψ on the $\eta =$ constant grid lines.

The same grid is used to solve (2.3) by a split step method. The vorticity is represented by a distribution of moving points:

$$\omega(\xi, \eta, t) = \sum_j \Gamma_j \delta(\xi - \xi_j(t)) \delta(\eta - \eta_j(t)) \tag{2.5}$$

where δ is the Dirac function. The convection part (left hand side) of (2.3) is then satisfied by integrating:

$$\frac{d\xi_j}{dt} = \frac{1}{J}\psi_\eta(\xi_j, \eta_j) \quad \text{and} \quad \frac{d\eta_j}{dt} = \frac{1}{J}\psi_\xi(\xi_j, \eta_j) \qquad (2.6)$$

where the derivatives of ψ are calculated at ξ_j, η_j by central differences and interpolation form the mesh.

The diffusion part of equation (2.3) is solved by a semi-implicit central difference scheme on the mesh:

$$(1 - \frac{\alpha\nu\Delta t}{J}\Delta_\eta^2)\omega^{(n+1)} = (1 + \frac{\nu\Delta t}{J}\Delta_\xi^2 + (1-\alpha)\frac{\nu\Delta t}{J}\Delta_\eta^2)\omega^{(n)} \qquad (2.7)$$

where Δt is the time step, Δ_ξ and Δ_η are central difference operators. $\alpha(0 < \alpha < 1)$ controls the degree of implicitness in the η direction.

In the inviscid Cloud-in-cell method the mesh vorticity remains equivalent to the point distribution $\sum \Gamma_j \delta(\xi - \xi_j)(\eta - \eta_j)$. This is no longer so when viscous diffusion occurs. It is therefore necessary to reinterpolate the new $\omega^{(n+1)}$ field from the fixed mesh back onto the moving point distribution prior to the next convection sub-step. It is very important to minimise the numerical diffusion in high Reynolds number computations for which the viscous diffusion is small. Therefore in this method the difference field $[\omega^{(n+1)} - \omega^{(n)}]$, which is $O(\nu\Delta t)$, is transferred onto the original field $\sum \Gamma_j \delta(\xi - \xi_j)(\eta - \eta_j)$, so that as $\nu\Delta t \to 0$ the inviscid method will be recovered exactly.

Reinterpolation is carried out so that the first three moments of the vorticity field $\int\int \omega \, dxdy$, $\int\int x\omega \, dxdy$ and $\int\int y\omega \, dxdy$ are conserved.

Vorticity enters the computational domain through the boundaries either by convection from the exterior or by the action of the pressure gradient at solid boundaries. In the cases presented here periodic boundary conditions have been specified so that the only vorticity convected through open boundaries is that which has just been lost at the opposing end. A second order boundary condition was used for the vorticity at the wall ($\eta = 0$), setting the velocity at the wall to zero:

$$J_0\omega_0 = -0.5J_1\omega_1 - 3\psi_1/\eta_1^2 \qquad (2.8)$$

Subscripts 0 and 1 refer to the wall and first mesh node out.

The boundary condition for the streamfunction sets $\psi = 0$ on the wall and is computed for the outer boundary by the Biot–Savart integral of the vorticity field and its image. This is an expensive process and is only carried out exactly for a few far boundary points.

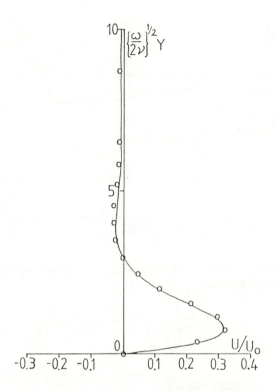

Fig. 2: Velocity distribution through an oscillatory laminar boundary layer. — theory; o computation

3. Results

(1) *Stokes boundary-layer.*
This flow is a test of the diffusion algorithm since convection normal to the wall and streamwise gradients are zero. Fig. 2 shows a comparison at one time of results computed using the method (Cozens 1987) with the exact solution for the oscillatory boundary-layer. Similar agreement is obtained at other phases of the flow cycle. These results were taken at the end of 2 complete cycles from start-up.

(2) *Impulsively started flow past a right-angle corner.*
This is a separating flow which, if the corner is sharp, has an inviscid self-similar solution (Pullin 1978). Making a transformation in terms of a

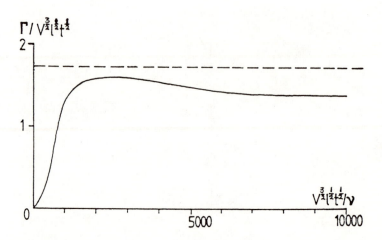

Fig. 3: Variation of circulation with time in separated flow round a right angle edge. - - inviscid; — viscous

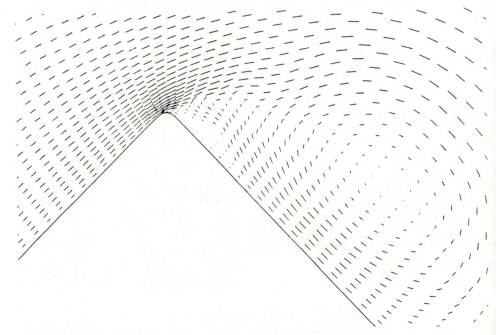

Fig. 4: Velocity vector plot for separated flow round a right angle edge

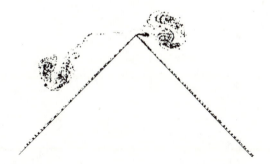

Fig. 5: Vortex plot for oscillatory flow round a right angle edge

length scale l: $X = \left(\frac{ut}{l}\right)^{-3/4} \frac{x}{l}$, $Y = \left(\frac{ut}{l}\right)^{-3/4} \frac{y}{l}$, $\Omega = t\omega$, $\Psi = \left(\frac{ut}{l}\right)^{-3/4} \frac{\psi t}{l^2}$ converts equations (2.1) and (2.2) into:

$$-\Omega + (\Psi_y - \frac{3}{4}x)\Omega_x - (\Psi_x + \frac{3}{4}y)\Omega_y = -\frac{R}{2}\Omega_R + R^{-1}(\Omega_{xx} + \Omega_{yy}) \quad (3.1)$$

and

$$\Psi_{xx} + \Psi_{yy} = -\Omega \quad (3.2)$$

from which it is clear that the inviscid sharp-edged self-similar solution should emerge from the viscous initial solution (including rounded corner cases), as time $\to \infty$ or the Reynolds number $R = \left(\left(\frac{ut}{l}\right)^{3/2} \frac{l^2}{\nu t}\right) \to \infty$. Fig. 3 shows a plot of the non-dimensionalised total circulation shed from a slightly rounded edge into the separating vortex as a function of R (or time) (Cozens 1987). After initial rapid growth towards the inviscid value the apparent large time asymptotic behaviour falls well below this value. This is probably because of the formation of secondary separation which is visible in Fig. 4 and which is not accounted for in the inviscid solution.

(3) Oscillatory flow past edges.
Fig. 5 shows the vortex formation which occurs under sinusoidal oscillatory flow past an edge sharp enough to cause separation (Cozens 1987). A pair of vortices are formed each cycle and then convect away in their own self-induced velocity field. For larger ratios of edge radius to flow amplitude a more complicated pattern ensues, the vortices are weaker and eventually shedding ceases altogether.

(4) Impulsively started steady incident flow past a circular cylinder.
The method has been used to model a uniform incident flow U past a cir-

Fig. 6: Vortex plot for steady flow past a circular cylinder at a Reynolds number of 250

cular cylinder over a range of Reynolds numbers from 100 to 10000. Fig. 6 shows a plot of the vortex particle positions at a non-dimensional time $Ut/d = 40$, where d is the cylinder diameter, at a Reynolds number of 250. The regular form of the Von-Karman vortex sheet in this Reynolds number range is predicted. The computed Strouhal number (0.175) and mean drag coefficient (1.4) are in reasonable agreement with other computations and experimentally measured values. At higher Reynolds numbers the shed vortex sheets become increasingly unstable and long two-dimensional filaments are formed. The predicted drag coefficient is also too large due to the failure of a two-dimensional laminar model to represent the growth of turbulence in the wake.

4. Conclusions

The two-dimensional method of solving the unsteady incompressible Navier–Stokes equations based on a vortex particle method presented above appears to give results at least as good as those given by conventional finite difference schemes for laminar flows. Like most methods representation of turbulence in unsteady separated wakes is inadequate.

References

Chorin, A. (1973). Numerical study of slightly viscous flow. *Journal of Fluid Mechanics*, 57 : 785.

Christiansen, J. (1973). Numerical simulation of hydrodynamics by a method of point vortices. *Journal of Computational Physics*, 13 : 363.

Cozens, P. (1987). *Numerical modelling of the roll damping of ships due to vortex shedding*. Ph.D. thesis, London University.

Farmer, C. and Norman, R. (1986). The implementation of moving point methods for connection diffusion equations. In *Proc. I.M.A. Conference on numerical methods for fluid dynamics II*, K. W. Morton, M. J. Baines, editors, p 635. Oxford University Press.

Leonard, A. (1980). Vortex methods for flow simulation. *Journal of Computational Physics*, 37 : 289.

Pullin, D. (1978). The large scale structure of unsteady self-similar rolled up vortex sheets. *Journal of Fluid Mechanics*, 88 : 401.

Raviart, P. (1986). Particle numerical models in fluid dynamics. In *Proc. I.M.A. Conference on numerical methods for fluid dynamics II*, K. W. Morton, M. J. Baines, editors, p 231. Oxford University Press.

Smith, P. and Stansby, P. (1987). Generalised discrete vortex method for cylinders without sharp edges. *A.I.A.A.Journal*, 25 : 199.

A streamwise upwind algorithm for the Euler and Navier–Stokes equations applied to transonic flows

P. M. Goorjian

NASA Ames Research Center, USA

1. Introduction

A new algorithm has been developed for the Euler and Navier–Stokes equations that uses upwinding based on the stream direction (Goorjian 1987a, 1987b). This algorithm is time accurate and can be used in codes for calculations of unsteady transonic flows over wings. Such codes can be used for flutter analysis of wings.

In this new algorithm, a coordinate system is used that has been rotated to align with the local streamwise direction. For differencing the convective terms in the streamwise direction, a new form of flux splitting is employed, in which the biasing depends on the local Mach number. In the plane perpendicular to the stream direction, the new flux splitting uses the condition of no flow in that local plane. (The formulae for the differencing in the rotated coordinate system are transformed to the original grid for the calculations.) By using a locally rotated coordinate system, the convective flux vector biasing depends on the total Mach number. Hence, the switching of the flux vector biasing occurs across shock waves and the proper domain of dependence is used in the supersonic regions. For comparison, many other upwind methods switch differencing based on Mach number components along coordinate lines. Such criteria allow downstream influences in supersonic regions and switching upstream of shock waves in multidimensional flows. The formulae for the convective flux vector differencing do not contain any user specified parameters. Hence, the amount of numerical dissipation is automatically determined. For viscous flows, calculations using the Navier–Stokes equations showed improved agreement with experimental results near the body, in a case of separated flow, when compared to calculations that use central differencing with fourth-order dissipation terms.

2. Flux splitting algorithm

The thin-layer, Reynolds-averaged, Navier–Stokes equations in generalized curvilinear coordinates are given by

$$\partial_\tau \hat{Q} + \partial_\xi \hat{E} + \partial_\eta \hat{F} + \partial_\zeta \hat{G} = Re^{-1}\partial_\zeta \hat{S} \qquad (2.1)$$

where

$$\hat{Q} = J^{-1}\begin{bmatrix} \rho \\ \rho u \\ \rho v \\ \rho w \\ e \end{bmatrix}, \hat{E} = J^{-1}\begin{bmatrix} \rho U \\ \rho u U + \xi_x p \\ \rho v U + \xi_y p \\ \rho w U + \xi_z p \\ U(e+p) \end{bmatrix}.$$

Here $\hat{Q}$ is the vector of conservative flow variables and $\hat{E}$ is the convective flux vector in the ξ direction. See (Goorjian 1987a, 1987b) for details. In this paper the flux splitting of $\hat{E}$ will be given. The variables $\hat{F}$ and $\hat{G}$ are split in a similar manner. The pressure is given by

$$p = (\gamma - 1)[e - (1/2)\rho(u^2 + v^2 + w^2)] \qquad (2.2)$$

and U is the contravariant velocity in the ξ direction.

In solving Eq.(2.1) at mesh point (i,j,k), the flux vector $\hat{E}$ at (i+1/2,j,k) is split into $\hat{E}^+$ and $\hat{E}^-$, where $\hat{E}^+$ is evaluated at (i,j,k) and $\hat{E}^-$ is evaluated at (i+1,j,k). Consider the case where $U \geq 0$; similar formulae hold for $U < 0$. First the split flux will be given for the mass flux component $\hat{E}_1 = J^{-1}\rho U$.

$$\hat{E}_1^+ = \frac{1}{2}J^{-1}\rho U[1 + (1 - s^+)(\tilde{U}/q)^2 + s^+(\rho\tilde{U}/\rho^* q^*)^2]$$
$$-s^+ J^{-1}(1 - 1/\overline{M}^2)(\rho - \rho^*)\tilde{B}_u(q + q^*)/2 \qquad (2.3)$$

$$\hat{E}_1^- = \frac{1}{2}J^{-1}\rho U[1 - (1 - s^-)(\tilde{U}/q)^2 - s^-(\rho\tilde{U}/\rho^* q^*)^2]$$
$$+s^- J^{-1}(1 - 1/\overline{M}^2)(\rho - \rho^*)\tilde{B}_u(q + q^*)/2 \qquad (2.4)$$

The split energy flux terms $\hat{E}_5^+$ and $\hat{E}_5^-$ are obtained by multiplying the respective split mass flux terms $\hat{E}_1^+$ and $\hat{E}_1^-$ by the local values of the total enthalpy $(e + p)/\rho$. The split momentum flux components in the x direction are obtained by replacing the mass flux terms $\hat{E}_1 = J^{-1}\rho U$ in Eqs.(2.3) and (2.4) by $\hat{E}_2 = J^{-1}(\rho u U + \xi_x p)$, which is the x component of the momentum flux from $\hat{E}$ and also replacing the factor $(q + q^*)/2$ by

$((q + q^*)/2)^2 (J\hat{Q}_2/\rho^* q^*)$, where $\hat{Q}_2 = J^{-1}\rho u$ is the second component of $\hat{Q}$. Similar substitutions yield the split momentum flux components $\hat{E}_3^+$, $\hat{E}_3^-$ and $\hat{E}_4^+$, $\hat{E}_4^-$ and in the y and z directions, respectively.

The quantities ρ^* and q^* in Eqs.(2.3) and (2.4) are local sonic values of the density ρ and speed q. The quantity $\overline{M}$ is an average Mach number between flow quantities and their sonic values. The switches s^+ and s^- switch the flux vector biasing at sonic values, where $M = 1$. The quantity $\tilde{U}$ is the physical component of the velocity corresponding to the contravariant velocity component U, where $(\tilde{U})^2 = U^2/(|\nabla\xi|)^2$. The variable $\tilde{B}_u$ is the maximum value of $\rho|U|/\rho^* q^*$ at the two mesh points used to split $\hat{E}$. For a more detailed explanation of this splitting see the papers by Goorjian (1987a, 1987b). The only changes in the formulae from Cartesian coordinates for curvilinear coordinates are to use contravariant velocity components such as U and their corresponding physical velocity components such as $\tilde{U}$.

3. Computed results

Unsteady flow

An unsteady calculation using the Euler equations was made for flow at $M = 0.85$ over an airfoil whose thickness varies in time. Fig. 1 shows the pressure coefficient plots for three times at which the shock wave is increasing in strength and moving downstream as a result of the initial thickening of the airfoil from zero thickness. Notice that the shock profiles at times $T = 11.5$ and $T = 18.25$ are sharply captured without any numerical oscillations. The shock profiles at those times contain one and zero mesh points, respectively. Fig. 2 shows Mach contour plots at $T = 18.25$. Notice that the Mach contours are smoothly varying in space. Also notice that the tight clustering of the contour lines, where there is a strong transonic shock. This clustering indicates the sharpness of the shock capture.

Wing C: Separated flow

Fig. 3 shows some features of the flow field's tip vortex. It shows traces of particles that were released near the tip. Fig. 4 shows a comparison of pressure profiles at the 90% span station, which is the only station for which there is experimantal data in the separated flow region, which is a small region near the tip. At the other four span stations for which there is experimental data, the flow is attached and the two computation methods are in close agreement. Notice in Fig. 4, that the new method, (denoted TNS-G), produces a shock location and a suction peak that are in closer agreement to the experimental results than the original method, (denoted by TNS). The original method in the TNS code used a coefficient, DIS =

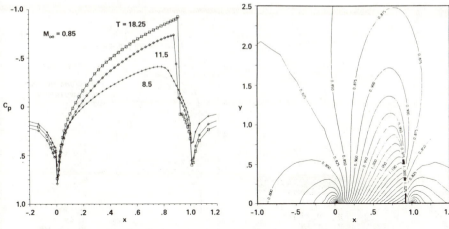

Fig. 1: Pressure coefficients for the formation and downstream propagation of the shock wave

Fig. 2: Mach number contours at $T = 18.25$

0.2, (a relatively low value), for the fourth-order dissipation. In the region that is influenced by the separated flow, which is from the 90% span station outward to the tip, the two methods show differences, not only in the shock wave location, but also along the entire upper and lower surfaces.

Fig. 5 shows Mach number contours at a height above the wing, where the Mach contours reached their highest values in the TNS calculation. Notice the higher contour levels reached by the TNS-G method. Also, notice that the TNS results show numerical oscillations in the contour levels at the downstream boundary of this block of the flow domain, whereas the TNS-G results do not. User specification of more dissipation in the TNS calculations to suppress these oscillations; e.g., DIS = 0.64, decreases the region of separated flow. By comparison, the amount of dissipation in the TNS-G results, which is automatically determined, suppresses the oscillations without either diminishing the separated region or diminishing the acceleration of the flow over the leading edge.

Fig. 6 shows a comparison of boundary-layer profiles at the span station, $\eta = 0.9362$, which is near the tip, and at the chord locations $x/c = 0.34$ and $x/c = 0.40$, which are in the shock wave. Notice at both locations that the new method has a fuller velocity profile, which is another indication that near the body, it is automatically adding less numerical viscosity than that used with central differencing.

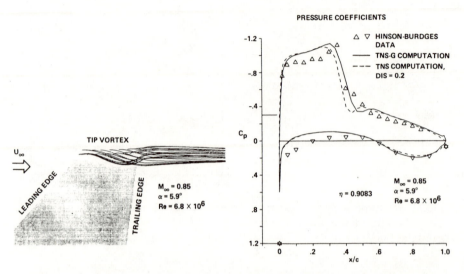

Fig. 3: Features of the flow field's
tip vortex for Wing C. Plane view
of particle traces near the tip

Fig. 4: Comparison of experimental
and computed pressure coefficients
for Wing C

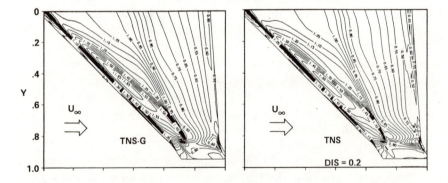

Fig. 5: Comparison of Mach contour levels at height above Wing C

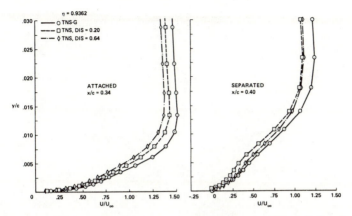

Fig. 6: Comparison of the boundary-layer profiles for Wing C for the two computational methods

NACA 0012 Wing: Normal shock flow

Next the shock capturing features of the two methods will be compared for flow in which there is a supersonic to subsonic shock wave; i.e., a normal shock wave. On the wing surface, the pressure profiles obtained by the two methods are similar. However, as the flow field is examined farther away from the wing, the physical viscosity diminishes and the shock capturing features of the two calculations become determined by the differencing of the convective terms.

For comparison, results are presented of pressure coefficient from a height of 0.07 above the wing in units of span length. Fig. 7 shows a comparison at two span stations. The TNS method used a value of DIS = 0.64, which is relatively large. Note the improvement obtained in the TNS-G results. The TNS-G results show no overshoots and a sharp capture of the reexpansion singularity. Also the shock is captured at the various span stations with either one or no points in the shock profile. For Euler calculations, these improvements are expected to occur at the wing surface.

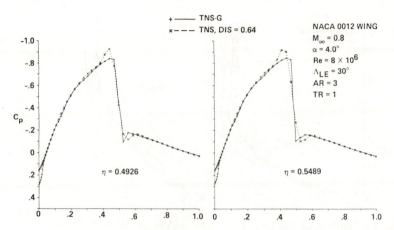

Fig. 7: Comparison of the shock capturing properties of the TNS-G and TNS. Pressure coefficients in block 2 (inviscid block), at height 0.07 (in span lengths)

References

Goorjian, P.M. (1987a). Algorithm developments for the Euler equations with calculations of transonic flows. AIAA Paper 87-0536. AIAA 25th Aerospace Sciences Meeting, Reno, NV, Jan 12-15, 1987.

Goorjian, P.M. (1987b). A new algorithm for the Navier-Stokes equations applied to transonic flows over wings. AIAA Paper 87-1121-CP. AIAA 8th Computational Fluid Dynamics Conference, Honolulu, Hawaii, June 9-11, 1987.

Computation of diffracting shock wave flows

R. Hillier

Department of Aeronautics

Imperial College, London

1. Introduction

This paper presents a numerical study of shock wave diffraction at a 90 degrees convex edge using a second-order Godunov-type scheme based upon a generalised Riemann solver (GRP). Results are presented for both a fully unsteady computation and also by writing the equations in a pseudo-stationary form which is then time marched to a 'steady' state. For this large diffraction angle the resultant flow is characterized by flow separation at the diffraction edge, and by the formation of a secondary shock wave system. These features are reproduced well by the computations.

Two-dimensional shock wave diffraction at an isolated convex edge has received little attention numerically, despite the fact that it forms a well defined problem for code validation as well as being a phenomenon of practical importance. Experimental studies have covered a wide range of diffraction angles and incident shock wave Mach number, presented for example by Skews (1967a, 1967b) and Bazhenova, Gvozdeva and Nettleton (1984). There are several different diffraction regimes, but the schematic in Fig. 1 shows the flow field typical of strong shock diffraction (that is the post-shock Mach number is supersonic). In particular, we see the transmitted incident wave (A), as yet unaffected by the reflected acoustic wave (B) propagating outwards from the diffraction edge, and the diffracting wave (C). The flow separates at the edge (for large enough angles), turning and accelerating through a certred rarefaction wave; this flow in turn is eventually decelerated through a secondary shock wave (S2) which is required to match the flow with that behind the appropriate segment of the diffracting shock. Further shock waves (e.g. S3 and S4) can be observed in some Schlieren visualisations.

The numerical method used here is based upon the one-dimensional second-order scheme developed by Ben-Artzi and Falcovitz (1983, 1984), implemented in two-dimensions by operator-splitting. The main feature of the method is the evaluation of fluxes at cell interfaces from the solution of the GRP. This assumes initial linear gradients of variables across cells and solves a GRP which includes discontinuities in both the variables and their derivatives at the interface. The method as used here is their E1 formulation, giving the simplest GRP consistent with second-order accuracy, and as such is very close to the MUSCL formulation (van Leer 1979). For the present computations the initial gradient across a cell is obtained using a second order central difference; a monotonicity constraint is applied to this if the cell interface values are extrema relative to the appropriate neighbouring cell average values, in which case the gradient is reduced until the extremum is removed. No other artificial dissipation is applied.

2. One-dimensional test cases

The basic method is illustrated here using a one-dimensional test case due to Harten(1983). Figs. 2a, 2b show the computation using the original (first order) Godunov scheme, and the current formulation with monotone-limited second-order spatial gradients. In Fig. 2c a contact surface steepener is applied similiar to that of Colella and Woodward(1984). This is invoked only in regions where pressure variations are weak relative to those in density (here we use $d\rho/dp > 10/a^2$ as the required threshold condition for a cell), in which case the density gradients are steepened further, if possible, up to the monotone limit. This clearly improves significantly the contact surface resolution. Fig. 2d finally shows the computation advanced with governing equations written in pseudo-stationary form, that is

$$\frac{\partial}{\partial t}(a_\infty Qt) + \frac{\partial}{\partial x^*}(F - a_\infty x^* Q) = 0 \tag{2.1}$$

where

$$Q = \begin{bmatrix} \rho \\ \rho u \\ \rho e \end{bmatrix}, F = \begin{bmatrix} \rho u \\ p + \rho u^2 \\ u(p + \rho e) \end{bmatrix}, x^* = \frac{x}{a_\infty t}$$

a_∞ is a reference speed of sound, and

$$e = \frac{p}{(\gamma - 1)\rho} + \frac{u^2}{2}. \tag{2.2}$$

This is advanced, for cell i, from time level n to $n+1$ by

$$a_\infty(t^{n+1} Q^{n+1} - t^n Q^n) + \frac{\Delta t}{\Delta x}(F_{i+\frac{1}{2}} - a_\infty x^*_{i+\frac{1}{2}} Q_{i+\frac{1}{2}}$$
$$-F_{i-\frac{1}{2}} + a_\infty x^*_{i-\frac{1}{2}} Q_{i-\frac{1}{2}})^{n+\frac{1}{2}} = 0 \tag{2.3}$$

where the flux values at the cell interfaces $(i \pm \frac{1}{2})$ are evaluated at the mid-time step $n + \frac{1}{2}$ using a simple modification to the GRP. The initial data are then time-marched to a steady state. In this special case, where the waves ultimately become stationary relative to the mesh, the waves are captured within one transition value, the best achievable in the absence of a sub-cell resolution technique.

3. Two-dimensional diffraction problem

The sequence in Fig. 3 shows density contours for the diffraction of a Mach 3 shock wave (with $\gamma = 1.403$) at a 90 degrees edge. The computation was started at time t=0 with an exact Rankine shock wave located at x=0. The computation was then advanced in two-dimensions using an operator-split formulation. This used exactly the same one-dimensional algorithm outlined earlier, with the contact steepener routine applied. The GRP needed a slight reformulation to handle the transverse component of velocity, essentially advecting it as a passive scalar. The contact steepener routine was applied to this transverse velocity, as well as to the density, to aid resolution of the slip surface. Figs. 3a to 3c are unsteady solutions of the flow field as it expands through a fixed mesh of square cells. The relative cell size is shown in each case, Fig. 3c corresponding to some 15,000 cells, and the axes are normalised in pseudo-stationary form (i.e. $x^* = x/(a_\infty t), y^* = y/(a_\infty t)$ where a_∞ is the initial ambient speed of sound.) Fig. 3d shows the computations with the equations themselves coded in the pseudo-stationary variables.

The best overall flow field resolution, not surprisingly, is given by the pseudo-stationary formulation, and salient data from the experiments of Skews are included on this figure. Fig. 3e also presents data by Carofano(1984), for computations of the same test problem using the TVD scheme of Harten, and the two methods appear to be in good agreement. Fig. 4a and 4b present, respectively, vorticity contours and velocity vectors for a restricted portion of the flow field. Again estimates from Skews' experiments of the separation angle and vortex position are included on these figures, showing satisfactory agreement with the computation. This modelling of unsteady separation by an Euler scheme is similar, in many respects, to the modelling of the leading edge separation problem for highly swept wings. In the present case inviscid modelling is satisfactory provided that the diffraction angle is large. In reality the diffraction angle range from zero degrees to about thirty degrees corresponds to the transition from no separation to separation (of the boundary layer on the diffraction surface) some distance downstream of the edge, essentially under the influence of the secondary shock wave. As the diffraction angle increases further then the separation position moves up to the diffraction edge.

4. Concluding remarks.

The computations simulate well the details of shock wave structure and flow separation for these large diffraction angles. Smaller diffraction angles provide viscous dominated flow separation. To this end a Navier-Stokes version of the code is under development, using the 'upwind' differencing of the Godunov scheme for the inviscid terms and central differencing for the viscous part.

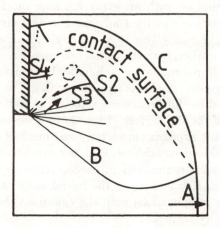

Fig.1: Schematic of shock diffraction

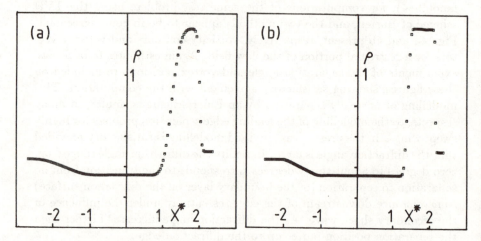

Fig.2: One–dimensional test problem. ($\rho_L = 0.445$, $\rho_R = 0.5$, $p_L = 3.554$, $p_R = 0.575$, $M_L = 0.3145$, $M_R = 0.0$, CFL = 0.95. L and R refer to initial left and right states)

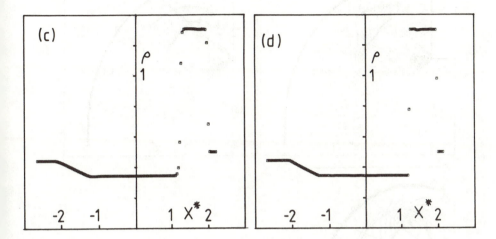

Fig. 2 (continued)

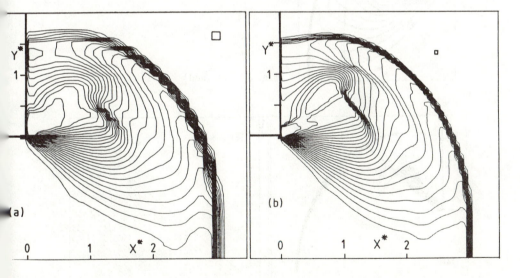

Fig. 3: Density contours (25 levels) for Mach 3 shock diffracted by 90⁰ edge. (a), (b), (c) show progressive growth of perturbed region up to 15,000 cells. Cell size is shown. (d) pseudo-stationary computations with 15,000 cells. Data from Skews: □ , diffracting shock; ——, initial angle of slipstream from diffraction edge; location of second shock, • and contact surface, ■ , on projection of slipstream; ◇ , vortex center. (e) Carofano computations

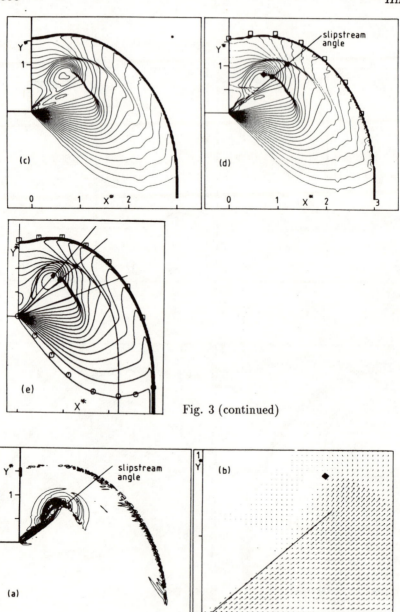

Fig. 3 (continued)

Fig. 4: (a) Vorticity contours (25 levels); (b) velocity vectors

References

Bazhenova, T. V., Gvozdeva, L.G. and Nettleton, M.A.(1984). Unsteady interactions of shock waves. *Progress in Aerospace Sciences*, 21: 249–331.

Ben-Artzi, M. and Falcovitz, J. (1984) A second–order Godunov–type scheme for compressible fluid dynamics. *Journal of Computational Physics*, 55: 1–32.

Carofano, G.C. (1984) Blast computations using Harten's total variation diminishing scheme. *Benet Weapons Laboratory Technical Report*, ARLCP–TR–84029.

Glaz, H.M., Colella, P., Glass, I.I., and Deschambault, R.L. (1985) A numerical study of oblique shock wave reflections with experimental comparisons. *Proceedings Royal Society of London (A)*, 117–140.

Harten, A. (1983) High resolution schemes for hyperbolic conservation laws. *Journal of Computational Physics*, 49:357–393.

Van Leer, B. (1979) Towards the unltimate conservative difference scheme (V), A second–order sequel to Godunov's method. *Journal of Computational Physics*, 32:101–136.

Skews, B.W. (1967a) The shape of a diffracting shock wave. *Journal of Fluid Mechanics*, 29:297–304.

Skews, B.W. (1967b) The perturbed region behind a diffracting shock wave. *Journal of Fluid Mechanics*, 29:705–720.

Multiple mesh simulation of turbulence

P. R. Voke

Dept. of Aeronautics, Queen Mary College, London

1. Introduction

The techniques of Direct Numerical Simulation (DNS) and Large-Eddy Simulation (LES) are popular for the investigation of turbulence at a fundamental level. Both methods are expensive in terms of computing time (Gavrilakis *et al.* 1986), if accurate results on fine meshes are to be obtained, since turbulence is intrinsically three-dimensional and time-dependent in nature.

To gather adequate statistics on a particular flow, a simulation must continue for some finite multiple of the underlying time scale T_L of the large-scale motions, typically for 5 to 10 T_L after settling from the arbitrary initial state which may itself require 3 to 4 T_L. If L is the length scale of the problem and u_τ is some turbulence-generating velocity scale, T_L is of order L/u_τ, and the simulation time required is EL/u_τ, where E is of order 10.

The size of the time step Δt, however, is determined by a Courant-Friedrich-Lewy (1928) limit involving the maximum flow velocity and the mesh spacing: for instance $\Delta t < 0.3\Delta x/U$. Typically $U \approx 20u_\tau$ and $L/\Delta x \approx 100$ for adequate resolution of the eddy motions. It is not difficult to see that a simulation on such a mesh requires $ET_L/\Delta t = (E/0.3)(L/\Delta x)(U/u_\tau)$ time steps: that is, on the order of 70,000 time steps.

Most of such a simulation is spent performing time advancement of a detailed flow field involving a range of eddy sizes (from L to Δx in scale), simply in order to reach a statistically distinct realisation of the flow. To reduce the scatter of the statistical quantities that are being computed, the large eddies have to be 'turned over' many times. The large amount of work required for the simulation arises from the small size of the time step compared with the correlation period of the large structures, which is essentially T_L.

2. The multiple mesh concept

Multiple Mesh Simulation (MMS) improves the efficiency of these types of simulations by working on a sequence of two or more nested meshes. The meshes are geometrically similar to those used in multi-grid algorithms

(Brandt 1977), though the motivation is different and the implementation and effects of the technique are quite distinct. MMS has been implemented as a staggered-mesh finite volume code; the principle is equally applicable to spectral or other varieties of finite-difference methods.

An initial flow field is generated on the finest mesh to be used, which may be as fine or finer than those typical of DNS and LES. The initial field satisfies continuity and the boundary conditions, but is otherwise arbitrary; it is usually constructed so as to mimic some mean velocity profile and to contain some artificial, pseudo-random 'turbulence' to kick off the simulation. Alternatively, a velocity field from another related simulation may be used as an initial state. From this a succession of velocity fields on coarser meshes (the multi-mesh levels) are generated by an injection method which preserves the continuity condition $div\mathbf{u} = 0$: in other words the injection process maintains mass conservation. Each mesh normally has half the number of mesh points of the next finer mesh, in each direction.

The difference between the velocity field on each mesh and that on the next coarser mesh is stored in the memory reserved for the finer mesh simulation, while the simulation proceeds on the coarser levels. This difference field also satisfies continuity (on its own mesh) and is added back in when the coarse velocity fields are later interpolated back to the fine mesh. All the processes involved preserve the discrete solenoidal condition. By retaining the difference fields on the finer meshes, the important fine scales of eddy motion are not destroyed while the coarse mesh part of the simulation proceeds; as a result, the simulation on the finer mesh proceeds more realistically after the interpolation process.

Each mesh is normally twice as coarse as the next finer mesh in each direction, and so has eight times fewer mesh points. Since the CFL stability limit is weaker on the coarser mesh, the time step is also automatically doubled on injection to the next coarser mesh (higher level), resulting in simulation that proceeds sixteen times faster in terms of physical time per cpu second (eight times faster per time step). Better vectorisation of the fine-mesh calculations reduces these ratios slightly in practice.

3. Subgrid and wall models

As the mesh is coarsened, the code automatically introduces and changes a subgrid model according to the resolution available on that mesh. For the present study, the resolution on each of the two meshes is such that the simulation is nearly full; that is, the resolution is not quite good enough to allow full simulation, but the effects of the subgrid-scale model are small, being just sufficient to keep the turbulence sustained and prevent a build-up of energy at high wave-numbers.

The model used on all the levels is similar to that of Moin and Kim (1982), which contains two eddy viscosities; (angle brackets denote a planar average parallel to the walls of the channel):

$$\tau_{ij} = \nu_1 \hat{s}_{ij} + \nu_2 \langle s_{ij} \rangle \tag{3.1}$$

$$\hat{s}_{ij} \equiv s_{ij} - \langle s_{ij} \rangle \tag{3.2}$$

$$\nu_1 = (c_1 L_1)^2 (\hat{s}_{ij} \hat{s}_{ij})^{1/2} \tag{3.3}$$

$$\nu_2 = (c_2 L_2)^2 (\langle s_{ij} \rangle \langle s_{ij} \rangle)^{1/2} \tag{3.4}$$

The length scales L_1 and L_2 are related to the mesh spacing, but decrease exponentially as the wall is approached. The parameters c_1 and c_2 were both fixed equal to 0.1 in the simulatons reported here. An attractive option for future work is to use the models suggested by Grotzbach (1986) to phase out the subgrid model as the resolution increases towards the point where the subgrid model is no longer reuqired and a full simulation is possible: this modification has not yet been incorporated.

The multi-mesh code implements log-law boundary conditions (wall models) as these become necessary at coarser resolutions. At the finest resolutions, when the point nearest the wall is closer than $z^+ = 5$, the stress in the wall-adjacent cell is assumed to be that given by the linear relationship $u^+ = z^+$. At the coarsest resolutions, when the nearest point is further than $z^+ = 30$, the stress is assumed to be given by the logarithmic law $u^+ = 5.5 + 2.5 \log(z^+)$. For intermediate resolutions, the Von Karman interpolation $u^+ = 5 + 5 \log(z^+/5)$ is used. This *adaptive boundary condition* has been used successfully in some of our multi-mesh simulations, though in the low Re runs reported here the boundary condition was always linear.

4. Simulation strategies

The simulation is performed for a period of order T_L on the coarsest mesh. The velocity fields are then interpolated onto the next finer mesh, where the difference fields stored there are added to them. The simulation is continued on the finer mesh for a period $T_L/16$, expending roughly the same amount of computer time as was used for the part of the simulation on the coarser mesh. the time step is half as long, so the number of time steps performed on the finer mesh is 1/8 the number on the coarser mesh, while the number of mesh points is greater by a factor of 8.

This process is repeated until a physically reasonable flow field is obtained on the finest mesh. The gathering of statistics is achieved using the same strategy: statistics are accumulated only on the finest mesh, but most of the simualtion is performed on the coarsest mesh. Immediately after accumulating the statistics from a flow realisation on the fine mesh, the flow field is injected onto successively coarser meshes, storing the difference velocity fields at each stage.

After simulating forward on the coarse mesh for a time $T \approx T_L$, so that a new flow realisation is obtained, the simulation is interpolated back onto finer and finer meshes, the difference fields that were left on those meshes always being added back in. In principle, it would be desirable to expend sufficient simulation time on each mesh to regenerate the eddies resolved on that mesh and to correct some of the errors introduced at the coarser level by the subgrid models used there. In practice, it is expedient to keep the computation time on each level roughly comparable, and our results have been obtained using this strategy.

The strategy used is that the same amount of cpu time should be expended on each of the levels; this is termed an *equal cost* strategy. It is also possible to perform simulations using a *double cost* strategy, which spends twice as much on each level as on the next coarser level. Note that computer resources are only affected by time spent, since the full memory allocation of the fine-mesh calculation is required at all stages of the simulation.

Physically, the fine mesh eddies have characteristic time scales T_S which are smaller than $\Delta x^{2/3} L^{1/3} / u_\tau$ (D C Leslie, private comunication, 1988; Leslie and Quarini 1979), and hence at most $(L/\Delta x)^{2/3}$ times the time scales $T_L = L/u_\tau$ of the large eddies. For a two-level simulation with $L/\Delta x = 32$ for the fine mesh, as in the simulation reported here, $T_S/T_L = 0.1$, while the actual time spent simulating on the fine mesh was $0.06 T_L$. The time spent on the fine mesh was therefore barely sufficient for the small eddies to adjust to the large eddies. A multi-mesh simulation in which the time spent on the finer meshes is determined solely from physical considerations such as these is said to be adopting a *physical strategy* as opposed to a *cost strategy*. For $L/\Delta x \approx 100$ and a two-level multi-mesh simulation, the strategies are equivalent, making this type of simulation very attractive.

5. Test results

The test simulation used a $32 \times 64 \times 32$ mesh for a low Reynolds number channel flow; this is a flow which is well understood, and represents a good test case. The Reynolds number of the results presented here is 3760, based on peak velocity and full channel width. In wall units the centre of the channel is at $h^+ = 100$. This is a transitional flow similar to those frequently used for testing simulation codes at relatively lwo cost; the performance of the mult-mesh code is generally close to that encountered previously in simulating such flows.

The simulation performed was a two-level multiple mesh simulation utilising meshes of 32 (streamwise) × 64 (spanwise) × 32 (between the walls) at the fine level and $16 \times 32 \times 16$ at the coarse level. The multi-mesh simulation spent 1750 time steps $(0.7h/u_\tau)$ on the coarse mesh, followed by 250 steps $(0.05h/u_\tau)$ on the fine mesh during each multi-mesh cycle. The computational box was $2\pi h \times \pi h \times 2h$ in size. This test is of particular significance since it is known that the simulation laminarises if performed solely on the coarse mesh, but sustains realistic transitional turbulence if performed solely on the fine mesh.

Results of the simulations are given in Figs. 1 to 5. The fine mesh results were obtained using an initial velocity field taken from the multiple mesh simulation, and allowed to evolve on the fine mesh for 6000 time steps $(1.2h/u_\tau)$. In Fig. 1, the coarse mesh result is obtained from a separate simulation on a $16 \times 32 \times 16$ mesh, after $6.4h/u_\tau$. This simulation laminarised, and so all the intensities were very close to zero, while the mean velocity was slowly approaching the parabolic profile. All the results shown are averaged over the planes parallel to the walls but are instantaneous so that some lack of symmetry, together with localised fluctuations, is to be expected.

The test shows that the multiple mesh simulation is capable of maintaining the turbulence, and of obtaining a mean velocity profile very close to that produced by the fine mesh simulation. The fact that the multi-mesh simulation succeeds at all is a striking success. The cost ratio between the multi-mesh simulation actually performed and an equivalent simulation performed wholly on the fine mesh is 7.5. The actual fine mesh results were obtained rather more cheaply by using the multi-mesh technique to condition the flow before switching to the fine mesh permanently. If nothing else, the multi-mesh technique is evidently a cheap way of conditioning simulations; that is, of turning pseudo-random initial velocities into something like real turbulence. At best, it allows simulations to be performed at a fraction of the cost of a full or large-eddy simulation of comparable accuracy.

The u' (streamwise) and u (cross-stream) intensities in the multi-mesh results show localised fluctuations that appear to be related to the positions of the coarse mesh planes. The spanwise (v) fluctuations are larger than they should be, but do not show mesh-related oscillations. The spatial oscillations in $\langle w^2 \rangle$ reveal that higher intensities occur on those planes that are explicitly represented on both the meshes, while on those planes only present in the fine mesh the intensity is lower. The effect is smaller closer to the walls. Attention is currently focussed on improving the interpolation procedure to eliminate such spurious effects.

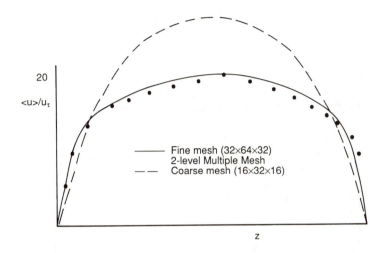

Fig. 1: Instantaneous averaged velocity profile $\langle u \rangle / u_\tau$

6. Conclusions

Test simulations show that MMS can produce results far better than those that would be obtained on the coarser mesh by a simple LES, and in some respects comparable in quality to those that would be obtained on the finer mesh. The quality of the results depends on the amount of simulation time spent on the finer mesh, and on the interpolation procedure. The computer time required is a small multiple of the time such a simulation, if it were possible at all, would take on the coarse mesh alone. Although MMS can never be as accurate as direct simulation on the finest meshes, the results produced are adequate for many purposes. The method allows great flexibility in allocating resources and in trading off computer time and accuracy. The savings in computer time, compared to direct simulations on the finest mesh, are considerable: the two-level multi-mesh simulation reported here achieved a saving ratio of 7.5 in cost.

In the future I expect to use information from the simulation on the fine mesh to change the subgrid scale models being used on the coarse meshes. Such *adaptive subgrid models* will maximise the use of the information available in the simulation. Multiple mesh siimulation is particularly attractive for high resolution simulations, though in cases where the

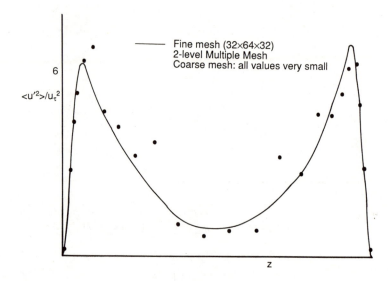

Fig. 2: Instantaneous averaged resolved intensity $\langle u'^2 \rangle / u_\tau^2$

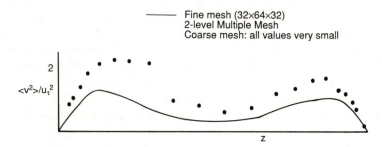

Fig. 3: Instantaneous averaged resolved intensity $\langle v^2 \rangle / u_\tau^2$

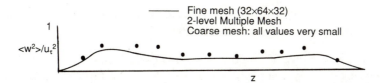

Fig. 4: Instantaneous averaged resolved intensity $\langle w^2 \rangle / u_\tau^2$

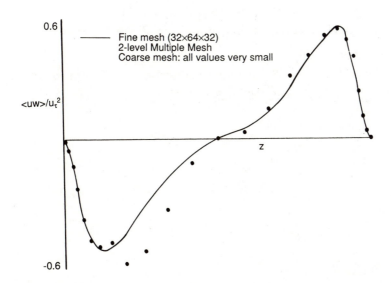

Fig. 5: Instantaneous averaged resolved Reynolds stress $\langle uw \rangle / u_\tau^2$

simulation on the finest mesh is direct, the phasing in of the subgrid model as the simulation moves onto the coarser meshes will need to be carefully controlled.

References

Brandt, A. (1977). Multi-level solutions to boundary value problems. *Mathematics of Computation*, 31:333–390.

Courant, R., Friedrichs, K., and Lewy, H. (1928). Uber die Partiellen Differenzengleichurgen der Mathematschen Physik. *Mathematische Annalen 100*, 1:541–543.

Gavrilakis, S., Tsai, H., Voke, P., et al. (1986). Large-eddy simulation of low Reynolds number channel flow by spectral and finite difference methods. In *Direct and large-eddy simulation of turbulence, (Proceedings of EUROMESH 199)*, Schumann, U. and Friedrich, R., editors, pages 105–188. Braunschweig, Vieweg.

Grotzbach, G. (1986). Direct simulation and large-eddy simulation of channel flows. In *Encyclopedia of fluid mechanics*, Cheremisinoff, N., editor. Gulf Publishing.

Leslie, D. and Quarini, G. (1979). Application of classical closures to the formulation of subgrid modelling procedures. *Journal of Fuid Mechanics*, 91:65.

Moin, P. and Kim, J. (1982). Numerical investigation of turbulent channel flow. *Journal of Fluid Mechanics*, 118:341.

Some experiences with grid generation on curved surfaces using variational and optimisation techniques.

C. R. Forsey and C. M. Billing.

Aircraft Research Association, Bedford.

1. Introduction.

Although grid generation remains one of the main pacing items in performing satisfactory calculations of flows about complex aerodynamic configurations, much progress has already been made. For example at ARA during the past few years a comprehensive grid generation capability has been developed (Weatherill et al. 1986). This capability is based on the concept of subdividing the flow domain into a number of sub-regions ('blocks') with locally generated grids in each block and suitable linking between them. Due to the success of this pioneering 'Multiblock' system, BAe have adopted the Multiblock concept as a framework for the majority of future CFD code development. However, despite the proven success of the existing Multiblock systems, there are still areas where further improvements are required.

One such area is the generation of grids on the actual surfaces of the configuation and on the other bounding surfaces of the flow. Surface grid generation is important in two respects. Firstly, it is the response of the flow to the precise shaping of the configuration surfaces that is usually of most interest to the user of the flow code. Secondly, the surface grids act as boundary conditions for the generation of the three-dimensional grids filling the flow domain, thus the quality of these grids is dependent on the quality of the underlying surface grids. For these two reasons it is necessary both to define the surface geometry with high accuracy and then generate grids with the highest possible quality on the surfaces so defined. Surface geometry definition is a well established area of CAD although the interface between CAD packages and CFD codes is not yet entirely satisfactory. The generation of high quality grids on surfaces is less well developed and is the subject of the work described herein.

In the current Multiblock systems both the surface and field grids are generated by solving the well known elliptic grid generation equations usually associated with Thompson et al. (1974). These grid generation equations consist of a set of coupled partial differential equations in which the dependent variables are the coordinates of the grid points. In three dimensions these coordinates are just the cartesian coordinates (x, y, z) of the grid points, as is also the case for planar two-dimensional surfaces. However, for three-dimensional curved surfaces some other coordinates, consistent with the surface definition, must be employed.

Now in order to accurately model the surface geometry, the technique adopted in the existing Multiblock systems is that due to Coons (1967). In this approach the surface is defined by a logically rectangular set of patches. The user defines the cartesian coordinates of each patch corner and then within each patch the cartesian coordinates are each expressed as a bicubic function of two surface parametric coordinates (u, w). These parametric coordinates are derived from the arc lengths along the two families of lines formed by joining patch corners and conventionally $0 \leq u \leq 1$, $0 \leq w \leq 1$ in each patch. Clearly it is possible to define global surface parametric coordinates (t_1, t_2) which correspond to (u, w) locally in each patch and thus form suitable coordinates for surface grid generation (see figure 1). However, if t_1, t_2 simply replace x, y in the planar two-dimensional grid generation equations, as is done in the existing Multiblock systems, then the grids generated by the solution of these equations will depend on the way in which the user defines the geometry. This is because the t_1, t_2 coordinates of each patch corner remain the same regardless of the physical cartesian coordinates of the corner assigned by the user. In the hands of a highly experienced user this can be an asset as it allows control of the local grid quality by altering the input geometric data. However, in general, this is an undesirable feature of the current Multiblock surface grid generators and a technique has been developed, as described below, to decouple the surface grid generation from the input geometry definition.

The technique developed to decouple the grid generation and the geometry definition could be applied to the Thompson grid generation equations. However, in an attempt to improve still further the surface grid quality, alternative grid generation techniques have been considered. One family of techniques which look particularly attractive are those based on a variational or optimisation approach. In such an approach various integral measures of grid quality are defined and a suitably weighted sum of the various measures minimised. Two methods of this type have been investigated: one based on the work of Brackbill and Saltzman (1982), the other on the work of Carcaillet et al. (1985). Both methods use similar integral measures but differ in their minimisation procedures. Most of our experience to date has been with the former method which will be described, firstly in its planar two-dimensional form including an effective multigrid solution procedure, followed by its extension to curved surfaces and including some recent enhancements. The latter method will be more

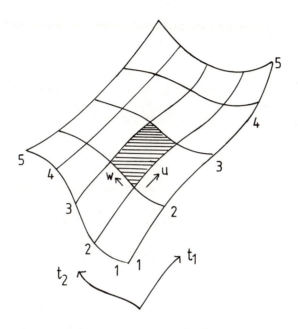

Fig. 1: Coons patches

briefly described again in both planar and curved surface forms.

2. Brackbill/Saltzman method

The Brackbill/Saltzman method is based on defining three integral measures of grid quality:

$$
\left.
\begin{array}{ll}
\text{Smoothness:} & I_S = \int\int(\nabla\xi^2 + \nabla\eta^2)\,dxdy \\[2ex]
\text{Orthogonality:} & I_O = \int\int(\nabla\xi.\nabla\eta)^2 J^3\,dxdy \\[2ex]
\text{Cell area control:} & I_A = \int\int wJ\,dxdy
\end{array}
\right\} \tag{2.1}
$$

where $w = w(x,y)$ is a cell area control function and J is the cell area. The coordinates x, y are the physical grid point coordinates and ξ, η the associated computational coordinates. If these three integrals are combined to give a cost function

$$
I = \lambda_S I_S + \lambda_O I_O + \lambda_A I_A
$$

then the grid obtained by minimising I will reflect the three properties in the relative proportions specified by $\lambda_S, \lambda_O, \lambda_A$; although in practise it is necessary to first scale λ_O, λ_A relative to λ_S to account for the different dimensions implied by each integral.

The minimum of I is obtained from the associated Euler-Lagrange equations (Weinstock 1952) where, as in the Thompson approach, the terms in the integrals are first inverted to produce integrals with respect to ξ, η rather than x, y before applying the Euler-Lagrange operators. Two partial differential equations result which have the form:

$$
\left.
\begin{aligned}
a_1 x_{\xi\xi} + a_2 x_{\xi\eta} + a_3 x_{\eta\eta} + c_1 y_{\xi\xi} + c_2 y_{\xi\eta} + c_3 y_{\eta\eta} &= R_1 \\
b_1 x_{\xi\xi} + b_2 x_{\xi\eta} + b_3 x_{\eta\eta} + a_1 y_{\xi\xi} + a_2 y_{\xi\eta} + a_3 y_{\eta\eta} &= R_2
\end{aligned}
\right\}
\qquad (2.2)
$$

where $a_1 - c_3$, R_1, R_2 are all functions of x_ξ, x_η, y_ξ, y_η and additionallly R_1, R_2 are functions of $\partial w / \partial x$, $\partial w / \partial y$. These equations are of similar form to the more common Thompson equations in that the grid coordinates are the dependent variables, however, they are more highly coupled as each contains second derivatives of both x and y as well as first derivatives. By taking suitable linear combinations of equations (2.2) and setting $\lambda_O = \lambda_A = 0$ the normal Thompson equations result.

To obtain the grid coordinates, the partial derivatives in equations (2.2) are replaced by central differences and the resulting discrete equations solved by a suitable iterative technique. For such highly coupled equations, however, normal successive line overrelaxation (SLOR) is inapplicable and we have used a block successive line overrelaxation (block SLOR) scheme in which the elements of the tri-diagonal matrices in the normal SLOR scheme are replaced by 2×2 matrices. Also, noting that suitable linear combinations of equations (2.2) with $\lambda_O = \lambda_A = 0$ will give the Thompson equations (i.e. set the off-diagonals of the 2×2 matrices to zero) which usually converge well, we have chosen to use the the same linear combinations of the equations (2.2) rather than equations (2.2) directly even when $\lambda_O \neq \lambda_A \neq 0$. However, even with this block SLOR technique convergence can be relatively poor, thus a full approximation storage (FAS) multigrid scheme has been developed which uses the block SLOR technique, applied in alternating directions, as the smoothing operator.

The FAS scheme used is essentially that described in Stuben and Linden (1986). Using the notation that h refers to fine grid and H to coarse grid and that RELAX denotes two sweeps of the block SLOR scheme (i.e. once in each direction), then with $u = (x, y)$ one multigrid cycle can be concisely defined as follows.

Solution of $L^h u^h = f^h$ is given by :

Smooth : $\overline{u}_h = RELAX(u^h_{old})$

Residual calculation : $r^h = f^h - L^h \overline{u}^h$

Restriction to coarse grid : $r^H = I^H_h r^h$; $\overline{u}^H = J^H_h \overline{u}^h$; $f^H = r^H + L^H \overline{u}^H$

Solve $L^H u^H = f^H$ at coarse grid : repeat multigrid cycle or just RELAX

Coarse grid correction : $v^H = u^H - \overline{u}^H$

Interpolate to fine grid : $v^h = K^h_H v^H$

Correct fine grid : $u^h = \overline{u}^h + v^h$

In the above multigrid cycle, I^H_h, J^H_h, are operators which restrict fine grid values to coarse grid and vice versa for K^h_H. In practice, following Stuben and Linden, for I^H_h a weighted average of surrounding values has been used, while for J^H_h the fine grid value at the corresponding coarse grid point ('injection') has been used. Likewise bilinear interpolation has been used for K^h_H. A single sweep of the RELAX operator is performed at each grid level per cycle ('V-cycle').

This multigrid scheme has proved very effective in most calculations to date, although with poor inital conditions it has been found necessary to perform about five RELAX sweeps at the finest level before introducing the multigrid cycles. Figure 2 shows the rate of the residual reduction as a function of cpu time for a typical 33×25 grid using 3 multigrid levels. In practice the scheme is essentially independent of the grid size. Actually figure 2 is with the same scheme applied to the curved surface form of equations (2.2) described later for which it is equally effective. The scheme has also been applied to the standard Thompson equations with SLOR smoothing and is, if anything, even more rapidly convergent (typically 10 orders of magnitude reduction in 20 cycles).

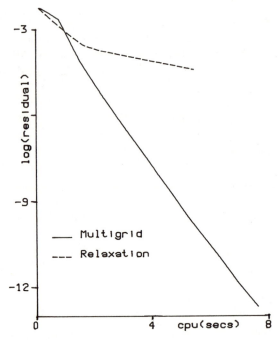

Fig. 2: Convergence

3.   Extension of Brackbill/Saltzman method to curved surfaces

In order to apply the Brackbill/Saltzman method to curved surfaces the
integral measures of grid quality, equations (2.1), must be reformulated
in terms of suitable surface parametric coordinates. Following Saltzman
(1986) orthogonal arc-length coordinates s_1, s_2 are specified locally at each
grid point. If these coordinates are also locally geodesic then they play
the same role on the surface that the cartesian coordinates x, y play on a
planar surface. Hence equations (2.1) remain the same except that x, y are
replaced by s_1, s_2 respectively. However, as discussed in the introduction,
surfaces are usually specified in terms of two surface parametric coordinates
t_1, t_2 which are neither orthogonal nor arc-length. Thus it is necessary to
locally transform from s_1, s_2 to t_1, t_2 coordinates. Saltzman does this
by first transforming equation (2.1) and then applying the Euler-Lagrange
operator. However, for simplicity we have preferred to first derive equations
(2.2) in terms of s_1, s_2 (i.e just replacing x, y in equations (2.2) by s_1, s_2
respectively) and then transform these equations. The resulting equations
have the same form as equations (2.2) except that the dependent variables
are t_1, t_2 rather than x, y and the coefficients $a_1 - c_3$, R_1, R_2 also contain
first and second derivatives of s_1, s_2 wrt t_1, t_2. As the equations have the
same form the same multigrid scheme may be used to solve them with only
minor modifications.

The derivatives of s_1, s_2 wrt t_1, t_2 may be derived from the surface definition as follows. Since $\underline{x} = (x, y, z)$ and $\underline{x} = \underline{x}(t_1, t_2) = \underline{x}(s_1, s_2)$ then from the chain rule for derivatives:

$$\frac{\partial \underline{x}}{\partial t_i} = \sum_{j=1}^{2} \frac{\partial \underline{x}}{\partial s_j} \frac{\partial s_j}{\partial t_i} \quad \text{for i=1,2} \tag{3.1}$$

Now since $\partial \underline{x}/\partial s_j = \underline{T}_j$ is a unit tangent vector in the s_j direction,

$$\underline{T}_j . \partial \underline{x}/\partial t_i = \partial s_j/\partial t_i \quad \text{for j=1,2 and i=1,2}$$

If the coordinates s_1 and t_1 are locally aligned then

$$\underline{T}_1 = \frac{\partial \underline{x}}{\partial s_1} = \frac{\partial \underline{x}}{\partial t_1} / \left| \frac{\partial \underline{x}}{\partial t_1} \right|, \quad \underline{T}_2 = \frac{\partial \underline{x}}{\partial s_2} = \underline{N} \times \underline{T}_1, \quad \underline{N} = \left(\frac{\partial \underline{x}}{\partial t_1} \times \frac{\partial \underline{x}}{\partial t_2} \right) / \left| \frac{\partial \underline{x}}{\partial t_1} \times \frac{\partial \underline{x}}{\partial t_2} \right|$$

By differentiating equation (3.1) the second derivatives of s_1, s_2 wrt t_1, t_2 may likewise be obtained.

The derivatives $\partial \underline{x}/\partial t_i$ can be calculated from the surface definition since $\underline{x} = \underline{x}(t_1, t_2)$. However, it should be noted that if the surface is defined using Coon's patches the calculation of these derivatives is computationally expensive and in addition must be redone at every grid point at every multigrid cycle (in fact in practice we have also found it necessary to recalculate them at every grid level during a multigrid cycle in order to obtain satisfactory convergence rates).

4. Results

The primary aim in investigating the curved surface form of the Brackbill/Saltzman equations was to decouple the surface grid generation from the geometry definition. To illustrate that this has been achieved, Figs. 3 and 4 show a surface defined by two quite different parameterisations. In each case, the two intersecting families of patch edges represent the same constant values of t_1 and t_2. For viewing convenience the surface shown is planar although similar results are obtained for a non-planar surface. Superimposed on each surface is a rectangular region within which a grid has been generated using the techniques currently employed in the Multiblock systems. For the boundary point distributions chosen both grids should be identical and of a simple cartesian form. Clearly this is not so, showing that the grids are dependent on the surface parameterisation. Figs. 5 and 6 on the otherhand show the same case when the grids are generated using the curved surface form of equations (2.2) and now the grids are identical and of the form expected.

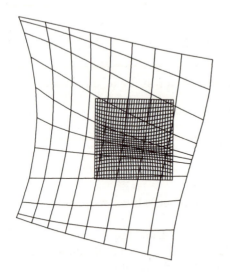

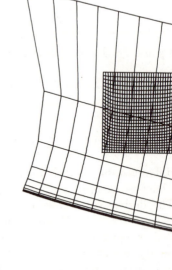

Fig. 3: Grid coupled   Fig. 4: Grid coupled

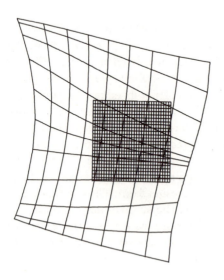

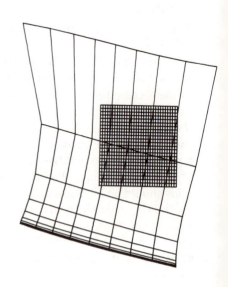

Fig. 5: Grid decoupled   Fig. 6: Grid decoupled

A secondary aim of our investigations has been to assess the additional controls available in the Brackbill/Saltzman method to improve grid quality. The controls are orthogonality and cell area control. To illustrate the orthogonality control, figure 7 shows a trapezoidal region superimposed on one of the surfaces described earlier. The grid in this region has been generated using the curved surface form of equations (2.2) with $\lambda_O = 25$. Clearly the grid is essentially orthogonal, and although not shown, an identical grid is obtained from the other surface parameterisation. Furthermore, convergence of the multigrid scheme is not significantly affected by the addition of orthogonality.

Tests using the cell area control were also successful in that the area could be controlled by appropriate changes to the control function w and without too large a reduction in convergence rates, although some degradation was found for $\lambda_A > 20$. However, the control was found to be less than satisfactory in that increasing the range of w did not produce similar variation in the cell areas. Instead the large areas increased while the small areas remained essentially constant. This is the reverse of the usual control requirements and thus the cell area control has been abandoned. Instead the grid control terms due to Thomas and Middlecoff (1980) have been introduced into equations (2.2). As with the other terms in the equations, for curved surfaces these terms are first formulated in s_1, s_2 and these transformed to t_1, t_2. This grid control technique has performed well to date and without significant convergence degradation.

5. Carcaillet *et al.* method

The Brackbill/Saltzman method has proved effective for curved surface grid generation, but it has two drawbacks. Firstly the algebra in applying the Euler–Lagrange operator to the integral methods is long and tedious, so that investigation of alternative grid quality measure is time consuming. Secondly, the coefficients in the resulting equations are computationally expensive, particularly when the additional curved surface transformation is included. Thus an alternative approach due to Carcaillet et al. has been briefly investigated. Again this scheme starts with integral measures of grid quality, however, the measures are formulated directly in physical coordinates. Refering to figure 8, at any grid point measures of local smoothness and orthogonality can be defined :

Smoothness: $SM_{ij} = k_1|\underline{r}_1|^2 + k_2|\underline{r}_2|^2 + k_3|\underline{r}_3|^2 + k_4|\underline{r}_4|^2$

Orthogonality: $ORT_{ij} = (\underline{r}_1.\underline{r}_2)^2 + (\underline{r}_2.\underline{r}_3)^2 + (\underline{r}_3.\underline{r}_4)^2 + (\underline{r}_4.\underline{r}_1)^2$

$$(5.1)$$

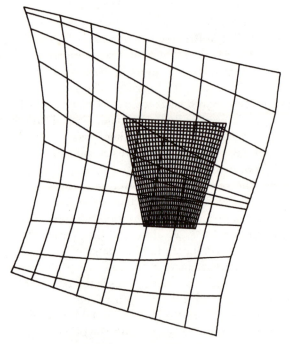

Fig. 7: Orthogonality

By summing over all grid points a cost function can be defined :

$$F = \sum_i \sum_j (\lambda_S SM_{ij} + \lambda_O ORT_{ij})$$

As in the Brackbill/Saltzman method, with suitable scaling on λ_O, minimisation of F produces the requisite grid.

The function F can be minimised using a standard conjugate gradient algorithm. Since such algorithms require only first derivatives of F wrt x_{ij}, y_{ij} which may be quickly and simply obtained by differentiating equations (5.1), alternative grid quality measures may be easily examined. Furthermore, the extension to curved surfaces is straightforward. Firstly, equations (5.1) are extended to three dimensions by including z in the definition of $\underline{r}_i$. Secondly, since x, y, z are all functions of t_1, t_2 on the given surface, F may be minimised directly wrt $t_{1_{ij}}$, $t_{2_{ij}}$. The derivatives of F wrt t_1, t_2 required by the conjugate gradient algorithm are easily obtained from the derivatives w.r.t x, y, z using the chain rule.

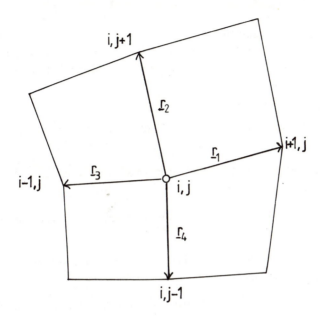

Fig. 8: Control area

This method has been tested using the integral measures defined by equations (5.1) applied to the same pair of surfaces used in testing the Brackbill/Saltzman method. Identical grids were obtained, however convergence was rather slow, requiring several hundred iterations to obtain the grids. This problem is thought to be due to incorrect scaling of the gradients. The method's originators show good convergence rates but note (without details) the importance of correct scaling. Further work on this aspect of the method is planned.

6. Concluding remarks

Two methods of grid generation based on optimisation have been investigated. In particular, modifications of these methods to generate grids on non-planar surfaces, decoupled from the surface geometry definiton, have been developed and tested and a fast multigrid solution algorithm has been successfully devised for one of the methods. Work is due to start shortly on incorporating this method into an existing Multiblock grid generation system for complex aircraft configurations.

Acknowledgements

This work was performed with the support of British Aerospace plc (Filton). The authors would like to thank Mr R H Doe of BAe (Filton) for devising the method of calculating certain transformation derivatives described in section 3 and to the BAe (Hatfield) Multiblock team for their encouragement.

References

Brackbill, J.U. and Saltzman, J.S. (1982). Adaptive zoning for singular problems in two dimensions. *J Comp Phys*, 46:342–368.

Carcaillet, R. et al. (1985). Optimization of three dimensional computational grids. AIAA 85-4087.

Coons, S.A. (1967). Surfaces for computer aided design of space forms. MIT MAC-TR-41.

Saltzman, J.S. (1986). Variational methods for generating meshes on surfaces in three dimensions. *J Comp Phys*, 63:1–19.

Stuben, K. and Linden, J. (1986). Multigrid methods: An overview with emphasis on grid generation processes. In Hauser, J. and Taylor, C., editors, *Numerical grid generation in computational fluid dynamics*. Pineridge Press, Swansea, UK.

Thomas, P.D. and Middlecoff, J.F. (1980). Direct control of the grid point distribution in meshes generated by elliptic equations. *AIAA Journal*, 18:652–656.

Thompson, J.F. et al. (1974). Automatic numerical generation of body-fitted curvilinear coordinate system for field containing any number of arbitrary two-dimensional bodies. *J Comp Phys*, 15:299–319.

Weatherill, N.P. et al. (1986). A discussion on a mesh generation technique applicable to complex geometries. AGARD-CP-412.

Weinstock, R. (1952). *Calculus of variations*. McGraw-Hill.

Adaptive orthogonal curvilinear coordinates

R. Arina

CNR-CSDF

Politecnico di Torino, Italy

1. Introduction

When a fluid flow is numerically calculated in a region of arbitrary shape, the theoretical basis for the simulation is a combination of numerical, fluid-mechanic and geometric analyses. The geometrical analysis, usually termed grid generation, consists in the representation of the shape of the domain of definition of the governing equations by a suitable choice of space coordinates.

One of the major problems which has been encountered in the context of grid generation lies in the difficulty of controlling the coordinate transformation, that is to obtain an appropriate functional distribution of the metric terms characterizing the curvilinear coordinate system, satisfying certain conditions of regularity, in order to preserve the smoothness of the solution as well as the formal accuracy of the numerical scheme. Moreover, in the case of adaptive grid techniques, the major problems that result from the grid movement are an excessive departure from orthogonality and a possible occurrence of folded grids.

In the present work, the conditions which must be respected by a mapping to represent an unfolded coordinate system have been defined. Moreover a direct control of the properties of the curvilinear system has been obtained by enforcing specific constraints on the metric tensor components in each point of the domain. Results are presented for the case of two-dimensional adaptive orthogonal coordinates.

2. Curvilinear coordinates

The principal problem posed by the theory of curvilinear coordinates in Euclidean spaces is the construction of the coordinate system $\{\xi^i\}$ corresponding to given metric tensor components, that is to calculate the transformation $x^i(\xi^j)$, $\{x^i\}$ being the cartesian frame.

If the mapping $x^i = f(\xi^j)$ defined on the domain $\mathcal{D}$ represents an allowable mapping, the function $f : \mathcal{D} \to D(\mathcal{D}, D \in R^n)$ must satisfy the following local conditions: (a) f must be continuously differentiable on $\mathcal{D}$, and (b) the Jacobian determinant $J(f) \neq 0$ for all the points of $\mathcal{D}$.

By the inverse function theorem it follows that at each point P$\in$ $\mathcal{D}$ the function is 1-1 and has a local inverse which is 1-1 on the image of a neighbourhood of P. Then there exists in a neighbourhood of each point of $\mathcal{D}$ a regular coordinate system without singularities or local foldings. The inverse function theorem is a local theorem, and it does not give the properties of the function on a global scale. Even if $J(f) \neq 0$ at all the points of $\mathcal{D}$, it does not follow that f is 1-1 on $\mathcal{D}$. It is necessary to add a third condition: (c) f is 1-1 on $\mathcal{D}$. Conditions (a) and (b) must be satisfied by a proper choice of the metric tensor components g_{ij}. The global condition (c) is respected by an appropriate specification of the physical domain D.

The problem of constructing curvilinear coordinates corresponding to given metric tensor components can be studied as proposed by Cartan (1946). The knowledge of the metric element $ds^2 = g_{ij}d\xi^i d\xi^j$ at an arbitrary point P of the space, enables us to construct a frame $\mathcal{F}$ with origin at this point, with characteristics specified by the components g_{ij}. The local reconstruction of the space consists in localizing with respect to this frame $\mathcal{F}$, the frame $\mathcal{F}'$ relative to a point Q contained in a neighbourhood of P. It is then necessary to find a set of equations expressing the characteristics of $\mathcal{F}'$ as a function of the known frame $\mathcal{F}$. Since $d\bar{r} = dx^i\bar{e}_i$, by the definition of the vectors $\bar{g}_i$, tangent to the axes of $\mathcal{F}$, we have

$$d\bar{r} = d\xi^i\bar{g}_i \quad i = 1, n \tag{2.1}$$

A second set of relations is obtained by the definition of covariant derivative and of connection coefficients Γ_{ij}^k (Schutz 1980)

$$d\bar{g}_i = \Gamma_{ij}^k d\xi^j \bar{g}_k \quad i, j, k = 1, n \tag{2.2}$$

If the functions $g_{ij}(\xi^k)$ are continuous, the symbols Γ_{ij}^k can be expressed as functions of the derivatives of g_{ij} (Schutz 1980). Equations (2.1, 2.2) solve completely the problem in a neigbourhood of P, and form the *mapping system*.

The conditions of integrability for equations (2.1, 2.2) require that, for symmetric symbols $\Gamma_{ij}^k = \Gamma_{ji}^k$, the curvature tensor $R_{ijk}^r = 0$. The functions g_{ij} must satisfy this flatness condition in order to be viewed as the components of the Euclidean metric. Moreover, assuming $\mathcal{D}$ be simply-connected, the functions $g_{ij} \in C^2$ and the determinant of the matrix g_{ij} $g \neq 0$ for all the points of $\mathcal{D}$ (since $J = \sqrt{g}$), it can be proved (Arina 1987) that locally the solution $x_i = f(\xi_j)$ of the mapping system (2.1, 2.2) represents the curvilinear coordinates specified by the functions g_{ij}, and satisfies the properties (a)and (b).

The mapping system (2.1, 2.2) could be used as it stands. However, it turns out to be useful to convert it into a set of second-order partial differential equations. Differentiating equations (2.1) with respect to ξ^j, by using equation (2.2) and inner multiplication by g_{ij}, we obtain the system (Arina 1987)

$$g^{ij}\frac{\partial^2 x^r}{\partial \xi^i \partial \xi^j} - g^{ij}\Gamma_{ij}^k \frac{\partial x^r}{\partial \xi^k} = 0 \qquad (2.3)$$

System (2.3), proposed by Thompson and Warsi (1983), who derived it by a different approach, is the most general set of equations which can be used for grid generation in 3-D Euclidean spaces.

The development of an appropriate grid generation technique for n-dimensional spaces must begin by specifying the n-independent metric components in order to represent a particular curvilinear coordinate system. Their substitution into system (2.3) leads to a specific mapping system which, with a suitable set of boundary conditions, defines the mapping problem whose solution represents the appropriate coordinate transformation. As has been shown previously, the coordinate transformation satisfies the properties of local regularity if the metric is *regular*: $g_{ij} \in C^2$ and $g \neq 0$ in $\mathcal{D}$. An explicit consideration of the flatness condition is not necessary. It is a requirement of integrability which is satisfied when the existence of the coordinate transformation $x^i(\xi^j)$ is proved.

3. Orthogonal coordinates

The considerations of the previous section are valid for general curvilinear coordinates in n-dimensional spaces. We will now apply these results to the case of 2-D orthogonal systems. The corresponding metric tensor components are

$$g_{22} = F^2(\xi, \eta)g_{11} \quad , \quad g_{12} = g_{21} = 0 \qquad (3.1)$$

where the *dilatation function* F must be non-zero and continuously differentiable on $\mathcal{D}$, in order to represent a regular metric. With the metric components (3.1), system (2.3) reduces to the elliptic self-adjoint system, for $\xi^1 = \xi$ and $\xi^2 = \eta$,

$$\frac{\partial}{\partial \xi}\left(F\frac{\partial x^i}{\partial \xi}\right) + \frac{\partial}{\partial \eta}\left(\frac{1}{F}\frac{\partial x^i}{\partial \eta}\right) = 0 \quad i = 1,2 \qquad (3.2)$$

The constraint of orthogonality $g_{12} = 0$ yields the relations, with $x^1 = x$ and $x^2 = y$,

$$F\frac{\partial x}{\partial \xi} = \frac{\partial y}{\partial \eta} \quad , \quad \frac{\partial x}{\partial \eta} = -F\frac{\partial y}{\partial \xi} \qquad (3.3)$$

It can be shown that equations (3.2) represent the condition of integrability of equations (3.3). For $F = 1$ we have the case of conformal mapping of the rectangular domain $\mathcal{D}$ onto the physical domain D. For these reasons equations (3.3) are termed *generalized* Cauchy-Riemann equations. The formulation of the boundary-value problem for the conformal case and the proof of the existence of a regular solution have been extensively presented in Arina (1986,1987).

A similar result can be achieved in the case of orthogonal coordinates. The functions $x(\xi, \eta)$ and $y(\xi, \eta)$ can be interpreted as the real and the imaginary parts of a complex-valued function $f(\zeta) = x(\zeta) + iy(\zeta)$, with $\zeta = \xi + i\eta$. Introducing the complex derivatives $f_\zeta = \frac{1}{2}(f_\xi - if_\eta)$ and $f_{\bar{\zeta}} = \frac{1}{2}(f_\xi + if_\eta)$, system (3.2) can be rewritten in the form

$$f_\zeta = \mu f_{\bar{\zeta}} \qquad (3.4)$$

with $\mu = \frac{1-F}{1+F}$. From complex analysis (Letho and Virtanen 1973) it follows that if μ is a continuous function in $\mathcal{D}$, such that $sup_{\zeta \in \mathcal{D}}|\mu| < 1$, then every solution of equation (3.4) is a regular quasiconformal mapping of $\mathcal{D}$ with complex dilatation $\mu(\zeta)$. The geometric interpretation of quasiconformal mappings states that locally infinitesimal circles are mapped on infinitesimal ellipses and we can conclude that the dilatation function represents the local axes-ratio of the image ellipse.

To formulate the boundary-value problem we must add suitable boundary conditions to the system (3.2). If the associated b.c's satisfy the generalized CR equations (3.3), the mapping $x^i = f(\xi^j)$ will represent a quasiconformal mapping of the domain $\mathcal{D}$ with quasiconformal continuation through the boundary $\partial \mathcal{D}$. In Arina (1986,1987) a boundary point relocation procedure enabling us to satisfy equations (3.3) and ensuring at the same time the shape correspondence of the given boundary ∂D with the image of $\partial \mathcal{D}$, is explained in detail.

From the theory of quasiconformal mappings it follows that the solution of this boundary-value problem is unique. Moreover it is possible to state a mapping theorem, analogous to the Riemann mapping theorem for conformal mappings, ensuring the existence of a quasiconformal transformation between a given domain D with conformal module $\mathcal{M} = m(D)$, and a rectangular domain $\mathcal{D}$ with conformal module $m(\mathcal{D})$ given by the relation $m(\mathcal{D}) = \frac{\mathcal{M}}{\mathcal{K}}$, $\mathcal{K}$ being the upper bound of the dilatation function F. As a consequence of the mapping theorem it is not possible to map a given quadrilateral D onto an 'a priori' specified rectangular region; the modulus $m(\mathcal{D})$ must be treated as an unknown parameter. For a rectangular region the conformal module coincides with the side ratio, and then it is possible to normalize the region $\mathcal{D}$ into the unit square $\tilde{\mathcal{D}}$. It is then necessary to introduce an additional relation which can be obtained from equations (3.3) (Arina 1986),

$$m^2 = \frac{\int_{\tilde{D}} F\sqrt{\tilde{g}_{11}/\tilde{g}_{22}}\,d\tilde{\xi}\,d\tilde{\eta}}{\int_{\tilde{D}} \frac{1}{F}\sqrt{\tilde{g}_{22}/\tilde{g}_{11}}\,d\tilde{\xi}\,d\tilde{\eta}} \tag{3.5}$$

Equation (3.5) can also be interpreted as the Euler-Lagrange equation with respect to m of the functional $J = \int_D (Fg_{11} + \frac{1}{F}g_{22})d\xi\,d\eta$, which is minimized with respect to x and y by the solution of equations (3.2).

The control of the coordinate lines is obtained by specifying the dilatation function $F(\xi, \eta)$. For conformal coordinates F is equal to one in the whole domain. More generally, if we consider the arc length s along a η-const coordinate line we have $\frac{\partial s}{\partial \xi} = \sqrt{g}_{11}$, and similarly along a ξ-const line, denoting by t the corresponding arc length, $\frac{\partial t}{\partial \eta} = \sqrt{g}_{22}$. Then by the definition of the dilatation function it follows that $F(\xi, \eta) = \frac{t_\eta}{s_\xi}$. From this result we can argue that general stretching functions will represent the distribution of the arc length along the curvilinear coordinate lines with respect to constant increments of the functions ξ and η. Fig.1 shows an orthogonal grid in a square domain, with concentrations of the grid points near the corners, obtained by specifying the functions s_ξ and t_η as exponential stretchings. It is worth noting that this grid cannot be interpreted as the combination of a conformal grid and simple one-dimensional stretchings.

The dilatation function can also be specified by an adaptive control, in order to match specific constraints depending upon a given scalar function representing an important variable of an associated pde problem. The continuous adaptation of the coordinates can be viewed as a sequence of successive transformations $(D \to D, \overline{D} \to D \ldots)$ (Warsi and Thompson 1984). It is possible to find a recursive relation for the dilatation function $\overline{F}$, as a function of the previous value F (Arina 1987), which reads

$$\overline{F}^2 = \frac{\xi_{\overline{\eta}}^2 + F^2\eta_{\overline{\eta}}^2}{\xi_{\overline{\xi}}^2 + F^2\eta_{\overline{\xi}}^2}$$

F being known, the adaptive control of $\overline{F}$ consists in the specification of the transformation $\overline{D} \to D : (\overline{\xi}\overline{\eta}) \to (\xi\eta)$ which must take into account the associated pde problem. This can be accomplished by an equidistribution law, as explained in (J. F.Thompson et al. 1985). In fig.2 is shown the grid fitting the gradients of a test function based on the hyperbolic tangent function and reproducing a bow-shock-like solution induced by a typycal re-entry configuration placed at incidence into a supersonic stream. Note the grid fitting in a 'one-dimensional' fashion of the discontinuity, aligning one family of coordinate lines along the shock, while the other one is crossing it orthogonally.

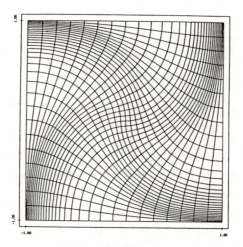

Fig. 1: Square domain (31 × 31)

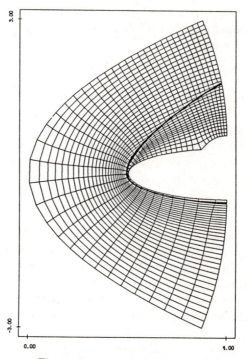

Fig. 2: Re-entry body (41 × 41)

4. Conclusions

In this work the problem of grid generation in Euclidean spaces has been analyzed. The meaning of regular metric has been described, and the expression for the most general mapping system proposed.

A grid generation method for adaptive orthogonal grids in 2-D domains has been devised. The existence and the uniqueness of the solution of the resulting elliptic self-adjoint system have been proved.

References

Arina, R. (1986). Orthogonal grids with adaptive control. 1st International Conference on Numerical Grid Generation in CFD. Landshut, FRG.

Arina, R. (1987). *Orthogonal Adaptive Grids and their Application to the Solution of the Euler Equations.* PhD thesis, Von Karman Institute, Bruxelles.

Cartan, E. (1946). *Leçon sur la géometriè des espaces de Riemann.* Gauthier-Villars, Paris, second edition.

Thompson, J. F. et al. (1985). *Numerical grid generation, foundations and applications.* North-Holland, New York.

Letho, O. and Virtanen, K. I. (1973). *Quasiconformal mappings in the plane.* Springer-Verlag, New York.

Schutz, B. F. (1980). *Geometrical methods of mathematical physics.* Cambridge University Press, Cambridge.

Thompson, J. F. and Warsi, Z. (1983). Three-dimensional grid generation from elliptic systems. AIAA Paper 83-1905.

Warsi, Z. and Thompson, J. F. (1984). The importance of dynamically-adaptive grids in the numerical solution of partial differential equations. Computational Fluid Dynamics, Lecture Series 1984-04, von Karman Institute.

An approximate equidistribution technique for unstructured grids

Peter K. Sweby

Department of Mathematics, University of Reading

1. Introduction

The technique of generating grid points by the equidistribution of a monitor function between them is well established in one dimension and can be extended in a number of ways to two dimensional structured grids. If, however, we wish to generate triangular unstructured grids which obey some equidistribution property then the process of generation is not so clear. In this paper we briefly overview equidistribution in one dimension and then propose an approach which yields an unstructured grid obeying an approximate equidistribution property.

Grids for numerical calculations are often generated based solely on the geometry of the computational region, e.g. body fitted grids (Thompson 1974). However, sometimes we may wish to take into account the behaviour of the function which we are trying to represent on the grid. For example, the Moving Finite Element (MFE) method (Wathen 1985) calculates updated nodal amplitudes and positions simultaneously using residual minimization based on the data representation at each timestep. It is therefore crucial that the initial data is represented as well as possible by the initial grid, since ultimately the whole solution process depends intimately on this representation. In one dimension the equidistribution of a monitor function based on the second derivative of the data has proved very successful in generating a suitable grid. However the extension of this technique to two dimensions is not straightforward. If we wished to confine our attention to structured quadrilateral grids then there are a variety of possible extensions of the one dimension technique which we could use (Sweby 1987; Anderson 1986), however if we want triangular grids, as for MFE, the extension of the procedure is not so clear.

In section 2 we briefly overview the one-dimensional process of equidistribution, including the choice of monitor function, then in section 3 we present an approximate equidistribution technique which can be used to generate unstructured grids.

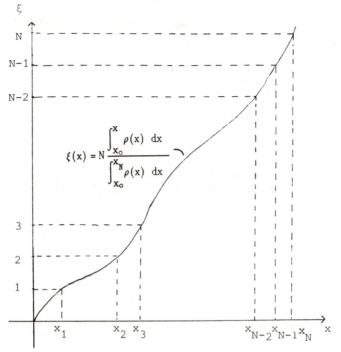

Fig. 1: The process of equidistribution

2. Equidistribution in one dimension

By equidistribution of a monitor function $\rho(x)$ we mean that we choose the grid points x_i to be such that

$$\int_{x_i}^{x_{i+1}} \rho(x)dx = \text{constant} \quad \forall i. \tag{2.1}$$

This can be achieved by defining a mapping

$$\xi(x) = N \frac{\int_{x_0}^{x} \rho(x)dx}{\int_{x_0}^{x_N} \rho(x)dx}, \tag{2.2}$$

where N is the total number of mesh points, and then solving

$$\xi(x_i) = i \quad \forall i. \tag{2.3}$$

The effect of this is illustrated graphically in Fig. 1. Carey and Dinh (1985) showed that the optimal monitor function $\rho(x)$ for minimizing the L_2 norm in the error in piecewise linear interpolation is given by

$$u_{xx}^{2/5} \tag{2.4}$$

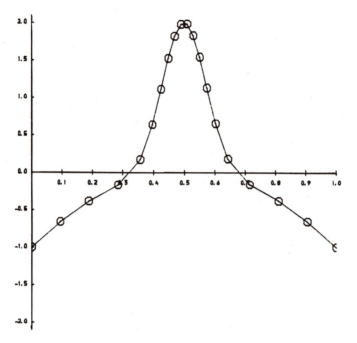

Fig. 2: A one-dimensional grid generated by (2.2)–(2.4)

where u is the data being represented (this is in fact a special case of their more general result). This is the representation we require for MFE and so we use this particular monitor function: Fig. 2 shows a typical grid thus obtained.

This approach can be extended to quadrilateral meshes in a number of ways (Sweby 1987), e.g. independent equidistribution in each coordinate direction, coupled equidistribution in both coordinate directions or equidistribution along grid lines (Anderson 1986). However for triangular meshes, since there is no unique line traversing the computational region along which to equidistribute, the extension of the procedure is not so clear. In the next section we propose one possible technique involving approximate equidistribution.

3. Approximate equidistribution in two dimensions

To extend the one-dimensional techniques of the last section to two-dimensional unstructured grids we adopt here the approach of approximate equidistribution, in the sense of requiring

$$\int_{\mathbf{r}_i}^{\mathbf{r}_j} \rho ds < \delta \qquad\qquad (3.1)$$

along each element edge $\mathbf{r}_j$—$\mathbf{r}_i$ where δ is some tolerance and the monitor function ρ is based on the directional second derivative along the edge, i.e.

$$\rho = (u_{xx} \cos^2 \theta + 2u_{xy} \cos \theta \sin \theta + u_{yy} \sin^2 \theta)^{2/5} \qquad (3.2)$$

where θ is the direction of the edge measured from the positive x–axis. The procedure is as follows:

1. Equidistribute nodes around the piecewise linear boundary of the region using one-dimensional equidistribution.

For each boundary node

2. Determine direction of contour passing through node from
$$\theta = \tan^{-1}\left(\frac{-u_x}{u_y}\right).$$

3. If contour leaves region
     delete node
   else

4. approximate contour by straight line and follow until
$$\int_{node}^{s} ds = \delta$$
   If still inside region
       place node
   else
       place node where line crosses boundary.

5. Triangulate (e.g. Delauney)

Fig. 3 demonstrates this procedure, and Fig. 4 shows a grid generated using it, together with contours and an isoparametric projection of the data representation on the grid.

The triangulation algorithm, however, will not take any account of the criterion used to generate the nodes and may therefore connect them in such a manner as to violate (3.1). Fig. 5 shows different data represented on a grid generated by the above technique, and as can be seen the quality is far from satisfactory. The solution to this problem is to regularise the grid after triangulation, i.e. consider the elements in pairs and if appropriate swap the mutual edge from one diagonal to the other in order to obtain the

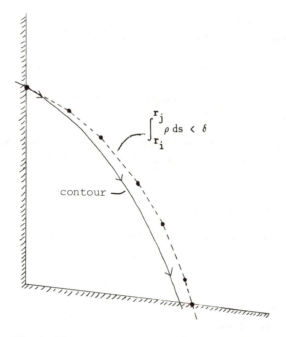

Fig. 3: The contouring process for node generation

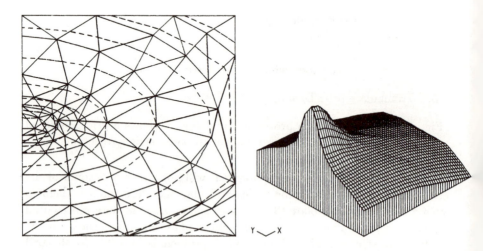

Fig. 4: The data representation on a grid generated by contouring

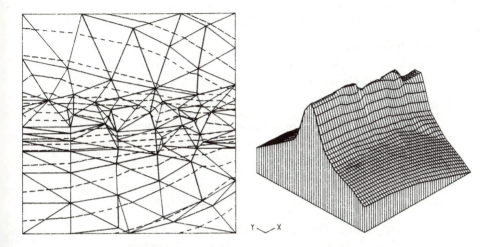

Fig. 5: A bad data representation due to incorrect connection

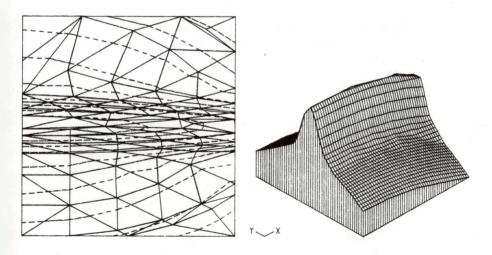

Fig. 6: The grid of Fig. 5 after regularisation

connection which minimises the integral in (3.1). This process is performed iteratively and Fig. 6 shows the improvement obtained.

An improvement to this generation/regularisation procedure would be to combine the triangulation with the node placement and work is currently being carried out on such an algorithm utilising a triangulation front.

Acknowledgements

This work forms part of the research programme of the Institute for Computational Fluid Dynamics at the Universities of Reading and Oxford.

References

D. A. Anderson (1986). Constructing adaptive grids with Poisson grid generators, *Proc. Numerical Grid Generation in Computational Fluid Dynamics*, J. Häuser & C. Taylor, editors, Pineridge Press.

G. F. Carey and H. T. Dinh (1985). Grading functions and mesh redistribution, *SIAM Journal on Numerical Analysis* 22:1028.

P. K. Sweby (1987). Data-dependent grids, *Reading University Numerical Analysis Report 7/87* and submitted to *Journal of Computational Physics*.

J. F. Thompson, F. C. Thames and C. W. Mastin (1974). Automatic numerical generation of body-fitted curvilinear coordinate systems for fields containing any number of arbitrary two-dimensional bodies, *Journal of Computational Physics* 15:299.

A. J. Wathen and M. J. Baines (1985). On the structure of the moving finite element equations, *IMA Journal of Numerical Analysis* 5:161.

Multiblock techniques for transonic flow computation about complex aircraft configurations

S. E. Allwright

British Aerospace, plc
Civil Aircraft Division, Hatfield

1. Introduction

Over the lifespan of an aircraft project, contrasting demands are made of theoretical aerodynamic flow prediction methods. Novel configurations must be investigated, numerous similar configuration geometries must be analysed systematically, and particularly complex configurations must be analysed in detail. The use of Multiblock techniques for grid generation and flow field calculation has allowed a unified computational approach to these contrasting demands.

Multiple blocks of curvilinear grid can be joined together to form the optimum grid structure for modelling the flow around each component geometry, and can be joined to cover the entire flow field without overlap or holes. A method is presented that through the use of interactive graphics enables the user to generate a schematic of the multiple-block grid structure required for a complex aircraft configuration, and also to associate labels to this schematic to achieve control of the resulting Multiblock grid. This description of the Multiblock grid is processed by a generalised interpreter that generates the surface grids for arbitary configurations. This capability is illustrated by the grid and flow field solutions presented for an executive jet with aft-fuselage mounted turbofan engine installation.

Conventional grid generation and flow solution techniques can be used over the major part of the flow field and it is only along certain block edges, where for example 3 or 5 blocks join in a non-cartesian structure, that alternative formulations must be considered. In the UK, Thompson (Thompson 1974) grid generation techniques and Jameson (Jameson 1981) Euler flow solution techniques have been developed for Multiblock at the Aircraft Research Association (ARA) at Bedford (Weatherill 1986) and British Aerospace (BAe) at Filton, exploiting the block decomposition to partition the computation for efficient processing on the Cray X/MP series supercomputers.

A number of issues must be addressed in completing the Multiblock flow field analysis for a new configuration type :-

- Geometry - definition of the configuration surfaces and the way in which these surfaces intersect.

- Topology - definition of a suitable Multiblock grid structure, and control of the mapping of this structure to the configuration surfaces.

- Surface grids - calculation of the mapped surface grids, for both the configuration and the outerboundary surfaces.

- Field grids - calculation of the 3.d field grid, covering the entire flow field.

- Flow solution - calculation of the flow field according to far-field, configuration and power-plant boundary conditions.

- Analysis - analysis of the flow field to validate the computational model, to identify flow structures, and to derive aerodynamic performance coefficients.

The field grid generation and flow solution stages are the most computationally expensive processes, however computing costs are reducing steadily each year. In that robust and efficient codes have been developed for these functions, attention at BAe Hatfield has focused on the issues of block topology specification and surface grid generation. These manpower intensive functions currently control the time-scales associated with applying Multiblock to a new configuration type. (Geometric definition is also man-power intensive and is addressed in computer aided design - CAD developments).

2.   Block structure definition

In devising the Multibock topology for a new configuration type, rough sketches of the block structure must be translated into a formal definition of the complete field grid topology (e.g. 30-40 parameters per block). A Graphical Multiblock Topology Generator (GMTG) has been developed to allow this "sketching" to be completed at a computer graphics terminal. The topology is represented by a 3.d wireframe schematic that is analysed automatically to derive the formal topology definition. Semi-automated block stacking operations facilitate generation of quasi-2.d block structures, while a block-by-block construction mode allows highly 3-dimensional block structures to be specified.

This graphical method has made it practicable to specify block structures for a wide range of configuration types, ranging from wing-body configurations with different wing-tip grid structures (50-70 blocks),through to highly integrated wing-pylon-propellor-nacelle configurations (884 blocks). In addition to the generation of specific topologies, the method has enabled investigation of a number of different block decomposition strategies.

Through the experience gained in block topology specification, it has been possible to devise automated block decomposition procedures, whereby on input of a simple block representation of the configuration, a full wireframe schematic of a suitable field grid block topology is generated (Fig. 1). The combination of this tool together with GMTG for graphic editing of the block structure, allows for the rapid specification of the block structure for a new configuration type.

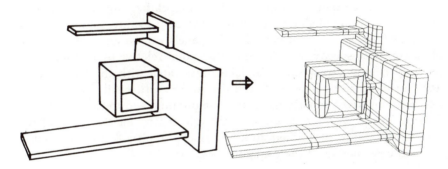

Fig. 1: Automatic block structure decomposition, for executive aircraft

3. Surface grid generation

The task of surface grid generation divides neatly into two parts, definition of boundary grid point positioning, and the generation of the surface grid according to these boundary points. Very close interaction between these two functions is required during development of a new Multiblock application, and this is provided within an interactive graphics environment.

3.1 Boundary point spacing

A number of authors refer to the use of interactive CAD tools for surface grid generation itself, whereby boundary points are positioned interactively along the edge lines of a geometric model and the intermediate surface grid is generated. While this technique was considered suitable for detailed grid control work - tailoring the grid for a specific geometry and for accommodating minor changes to that geometry, it was considered unsuitable for the main aerodynamic design function in which many hundreds of significantly different geometries must be analysed systematically.

To meet this requirement, a description of the grid point positioning and spacing throughout the block structure must be assembled, refering solely to generic features of the configuration. For example, instead of interactively fixing the absolute x,y,z position a grid point on the outer-boundary to be visually in line with the wing trailing edge point on a specific geometry, a position label is associated with that point referencing its position to the wing-trailing-edge as a feature of the configuration type. In this way, the description of the grid control can be applied to many different geometries maintaining consistent form and quality of grid.

The Graphical Multiblock Topology Generator allows the association of these generic grid control labels to block faces, edges and cornerpoints. Colour highlighting of the topology schematic is used to prompt for specification of the geometric components that a block face adjoins. Similarly, block edges are highlighted, and those that map to intersection lines or trailing edges can be identified. Finally, labels can be associated with a block corner to specify positioning of that point relative to another. Grid point spacing is also specified by labels associated with the block edges, non-dimensionalising by average cell size so that actual cell size adapts itself to the relative proportions of the configuration, and to changes in grid density. In this way, a description of the grid control for the whole of the Multiblock grid is established, in a way that is applicable to any number of specific geometries of the configuration type.

3.2 Surface grid generation

The surface grid generator comprises firstly, an interpreter that specifies the position of the boundary grid points according to the actual geometry and the grid control labels, and secondly the surface grid generation modules.

The primary grid generation method used is that due to Thompson (Thompson 1974) working in terms of the section/generator definition of the geometric surfaces, or an x-y type parameterisation for outerboundary surfaces:-

$$\alpha(x_{\zeta\zeta} + \phi x_\zeta) - \beta x_{\zeta\eta} + \gamma(x_{\eta\eta} + \psi x_\eta) = 0$$

$$\alpha(y_{\zeta\zeta} + \phi y_\zeta) - \beta y_{\zeta\eta} + \gamma(y_{\eta\eta} + \psi y_\eta) = 0$$

$$\alpha = x_\eta \cdot x_\eta + y_\eta \cdot y_\eta$$

$$\beta = x_\eta \cdot x_\zeta + y_\eta \cdot y_\zeta$$

$$\gamma = x_\zeta \cdot x_\zeta + y_\zeta \cdot y_\zeta.$$

The ϕ and ψ grid control terms are derived using the Thomas and Middle-coff formulation (Middlecoff 1979), both to propagate the boundary point stretchings into the grid, and to achieve orthogonality of the grid at the boundaries :-

$$\phi = (x_{\zeta\zeta} \cdot x_\zeta + y_{\zeta\zeta} \cdot y_\zeta)/\gamma + 2(x_\zeta \cdot x_\eta + y_\zeta \cdot y_\eta)/\alpha$$

$$\psi = (x_{\eta\eta} \cdot x_\eta + y_{\eta\eta} \cdot y_\eta)/\alpha - 2(x_\eta \cdot x_\zeta + y_\eta \cdot y_\zeta)/\gamma.$$

The formulation of the grid control and the recommendations for interpolating it through the field apply to a single block curvilinear grid structure. While analogous interpolation techniques can be applied in a multiple block environment, special consideration must be given to the formulation of grid control along grid lines approaching singular points, where the local grid structure is non-cartesian.

In the same way that the grid control is updated to achieve orthogonality at the fixed boundaries, so the grid control along lines approaching singular points can be updated to control the angle between the grid lines at the singular point itself (Fig. 2). If full mutual orthogonality is specified for the singular point then an equal angle between grid lines at the singular point will be achieved.

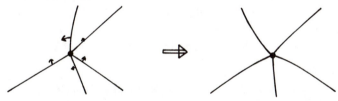

Fig. 2: Update of grid control at singularities

3.3 Grid initialisation

It is desirable to calculate an initial uncrossed grid as cheaply as possible to ensure that the final Thompson grid is itself un-crossed. An initial surface grid is calculated using the equation:-

$$\alpha(\underline{r}_{\zeta\zeta} + \underline{r}_{\eta\eta}) + (1 - \alpha)\underline{r}_{\zeta\zeta\eta\eta} = 0 \qquad \text{(e.g.} \qquad \alpha = 0.5).$$

The first Laplacean type term has solution characteristics analogous to an elastic membrane, and can result in grid crossover around highly curved boundaries. The second term has solution characteristics analogous to a plate in bending, so is more amenable to fitting around curved boundaries without crossover (Fig. 3).

4. Applications

The Graphical Multiblock Topology Generator and associated surface grid generator have been used within BAe on a number of aircraft projects. The geometric modelling and Multiblock field grid generation codes developed at ARA Bedford, and the Jameson Euler code developed at BAe Filton are used to complete the Multiblock flow field analysis for the configurations.

Wide-body civil transport aircraft with underwing turbofan engine installations are modelled using a 368-block 300,000-cell topology, that features C-grid structures around the wing and pylon, and around each section of the nacelle to form a C-O structure. An advanced twin-engined feederliner, for which the pylon-propellor-nacelle installation integrates closely with the wing, is modelled using an 884-block 500,000-cell topology. A propellor actuator disk model (Yu 1984) in the Multiblock Euler flow code is used to simulate both single and contra-rotation propellor slipstream effects. At BAe Warton, advanced fighter configurations have been studied, using a 319-block 250,000-cell topology to model wing-body-canard-fin configurations.

The block structure for modelling executive jets was specified automatically from a simple schematic (Fig. 1), and the resulting 1697-block 1,000,000-cell grid structure is illustrated in Fig. 4. The engine installation in fact lies forward of the wing trailing edge and there is a strong aerodynamic interference effect with the wing. This is illustrated by the contours of surface pressure in Fig. 5, showing the area of decelerated flow at the wing root trailing edge, caused by the proximity of the nacelle.

5. Conclusions

The use of graphical techniques in block structure specification has facilitated application of Multiblock grid generation and flow field analysis techniques to a number of new aircraft configuration types. In addition, the use of generic grid control labels has allowed systematic analysis of numerous aircraft geometries of each configuration type, with little additional user interaction. This experience is now steering development of automated block decomposition routines, further facilitating the exploitation of Multiblock.

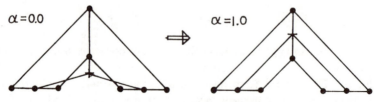

Fig. 3: Modified Laplacean method for grid initialisation

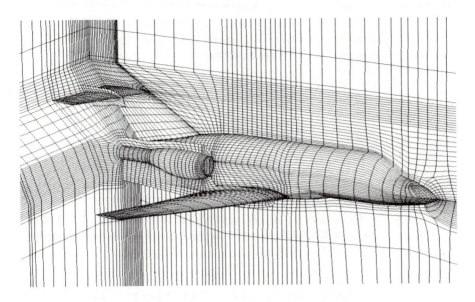

Fig. 4: Multiblock surface and field grids for executive jet

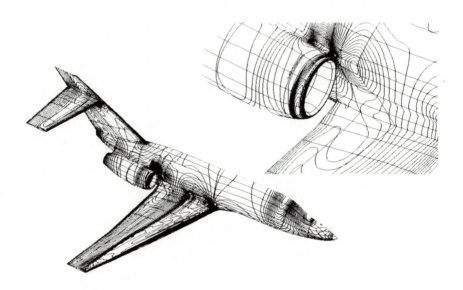

Fig. 5: Contours of surface pressure

References

Jameson, A., Schmidt, W. and Turkel, E. (1981). Numerical Solution of the Euler Equations by Finite Volume methods using Runge-Kutta time stepping schemes. AIAA Paper 81-1259.

Middlecoff, J. F. and Thomas, P. D. (1979). Direct Control of the Grid Point Distribution in Meshes Generated by Elliptic Equations. AIAA Paper 79-1462.

Thompson, J. F., Thames, J. P. and Mastin, C. W. (1974). Automatic Numerical Generation of Body-Fitted Curvilinear Coordinate System for Field Containing any Number of Arbitary Two-Dimensional Bodies. *J. Comput. Phys.*, 15:299.

Weatherill, N. P., Shaw, J. A., Forsey, C. R. and Rose, K. R. (1986). A Discussion on a Mesh Generation Technique Applicable to Complex Geometries. AGARD-CP-412.

Yu, N. H. and Chen, H. C. (1984). Flow Simulations for Nacelle-Propellor Configurations using Euler Equations. AIAA Paper 84-2143.

Cartesian grid methods for flow in irregular regions

R. J. LeVeque[1]

University of Washington

Seattle, Washington

1. Introduction

Although a variety of powerful mesh generation techniques are now available to create body-fitting grids, the generation of a good grid for complicated, multi-element structures (e.g. full airplane configurations in 3D) is still a very time-consuming and manpower-intensive task. This has lead to a resurgence of interest in the possibility of using Cartesian grids for fluid dynamical calculations in regions with complicated geometry. There has been recent work in this direction both for the full potential equation (e.g. Wedan and South 1983, Rubbert et al. 1986, Samant et al. 1987), and for the Euler equations, (Clarke et al. 1986, Falle and Giddings 1988).

This approach has several advantages:

- Mesh generation is straight forward and easily automated.

- On a uniform Cartesian grid, high resolution difference methods take a simple form and vectorize easily, leading to efficiencies of calculation.

- Grid refinement (needed to resolve nonsmooth regions of the boundaries as well as shocks and other flow features) is relatively simple, using for example the approach of Berger (1982) (see also Berger and Jameson 1985).

- Grid effects that sometimes arise from nonuniformities in the mesh are also eliminated (although possible difficulties remain at grid refinement interfaces).

---

[1]Supported in part by NSF Grants DMS-8601363 and DMS-8557319.

The primary difficulty with Cartesian grids is, of course, the boundary conditions. We wish to consider cell-centered finite volume methods for ease of guaranteeing conservation, but cells which are sliced off by the boundary may be orders of magnitude smaller than the regular cells away from the boundary. Figs. 3a and 4a show typical situations in 2D. With standard explicit finite volume methods, the time step restriction is proportional to the area of the smallest cell. This follows from the CFL condition, since typically the fluxes determined at each cell interface affect only the two neighbouring cells.

In this paper, I will describe a way to avoid this time step restriction by allowing information to propagate through more than one cell when necessary. Thus the CFL condition remains satisfied for larger time steps, and the time step can be chosen based on the regular Cartesian cells away from the boundary without restriction due to small cells at the boundary.

These boundary conditions are described here in 2D, but the same ideas would be applicable in 3D. The technique is briefly described in terms of wave propagation. More details may be found in (LeVeque 1987). These boundary conditions can also be rewritten to define fluxes at cell edges in cells near the boundary in such a way that the usual flux-differencing can still be applied to obtain updated cell values with the increased time step. This flux formulation has certain advantages. It allows the use of these boundary conditions together with any desired finite volume method (written in flux form) away from the boundary. Also, a joint project with Marsha Berger is currently under way to combine these boundary conditions with her Cartesian mesh refinement code, which requires fluxes for interpolating from one grid to another. This flux formulation will be described in a future paper.

2.   The wave propagation approach

The main idea of the wave propagation approach to defining a finite volume method is illustrated in Fig. 1 for a regular Cartesian grid. Roe's approximate Riemann solver (Roe 1981) for the Euler equations is used to solve the Riemann problem in the x-direction between cells (i, j) and $(i + 1, j)$. this gives a decomposition of $U_{i+1,j} - U_{ij}$ into three jumps, R_p $(p = 1, 2, 3)$,

$$U_{i+1,j} - U_{ij} = R_1 + R_2 + R_3 \qquad (2.1)$$

propagating with speeds $dx/dt = s_p$, where

$$s_1 = \bar{u} - \bar{c}, \quad s_2 = \bar{u}, \quad s_3 = \bar{u} + \bar{c}. \qquad (2.2)$$

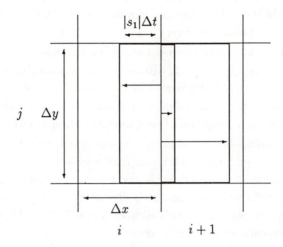

Fig. 1: Portion of a 2D Cartesian grid with three waves propagating from the interface between cells (i, j) and $(i + 1, j)$

Here $\bar{u}$ is an average fluid velocity and $\bar{c}$ an average sound speed. (There are actually four waves in the 2D Euler equations, but two of the waves always have the same speed $\bar{u}$ and can be combined.) These waves have the following effect on the neighbouring grid cells:

For each wave:

$$\text{if } s_p < 0 \text{ then} \qquad U_{ij} \leftarrow U_{ij} - \frac{s_p \Delta t \Delta y}{\Delta x \Delta y} R_p \qquad (2.3)$$

$$\text{if } s_p > 0 \text{ then} \qquad U_{i+1,j} \leftarrow U_{i+1,j} - \frac{s_p \Delta t \Delta y}{\Delta x \Delta y} R_p$$

Note that the area swept out by the wave in time Δt is

$$A_{\text{wave}} = |s_p| \Delta t \Delta y$$

while the full cell area for either cell (i, j) or $(i + 1, j)$ is

$$A_{\text{cell}} = \Delta x \Delta y.$$

The appropriate value of U is thus updated by the ratio $A_{\text{wave}}/A_{\text{cell}}$ times the jump $\text{sgn}(s_p)R_p$ carried by the wave.

The cell values are updated by each wave from each neighbouring Riemann problem in the x-direction and also by waves from Riemann problems solved in the y-direction in a similar manner (for example at the interface between cells (i, j) and $(i, j + 1)$).

An obvious stability restriction for the method, as described in (3), is that $A_{\text{wave}}/A_{\text{cell}} \leq 1$. However, it is possible to ease this stability restriction by allowing waves that propagate a distance $|s_p|\Delta t > \Delta x$ to affect more than one neighbouring cell. This is the idea behind the large time step generalization of Godunov's method described in one dimension in (LeVeque 1984, 1985).

In the work reported here the time step is restricted so that $|s_p|\Delta t < \Delta x$ (and $\Delta x = \Delta y$) and thus on the regular portion of the grid we have a standard finite volume method. By introducing piecewise linear approximations for the waves (with minmod slope limiters) this is converted into a high resolution (essentially second order accurate) method. See (LeVeque 1987) for details.

Near the boundary, however, the wave propagation viewpoint is used to deal with the small cells. Fig. 2a shows a wave originating from the Riemann problem between cells (i, j) and $(i + 1, j)$ that passes all the way through the small cell $(i + 1, j)$. The portion of this wave that lies beyond the boundary is then reflected across the boundary and back into the computational domain, as shown in Fig. 2b. The jump $\tilde{R}_p$ across this reflected wave is then used to update the values in each cell this wave overlaps, in this case $(i + 1, j)$ and $(i + 1, j - 1)$. As before, the update made to U in each cell is $\tilde{R}_p$ multiplied by the ratio of A_{wave} (the area of the cell intersected with the wave) divided by A_{cell} (the area of the cell). This ratio is always less than or equal to 1.

The jump $\tilde{R}_p$ across the reflected wave is equal to the jump across the incident wave R_p with the exception of the component of velocity normal to the boundary. This component of R_p is negated in $\tilde{R}_p$. This is the correct way to model the solid wall boundary condition and gives conservation (modulo our approximation of the boundary as piecewise linear).

In addition to waves coming from the cell interfaces, it is also necessary to solve a Riemann problem at the boundary face itself, in the direction normal to the boundary. This gives one additional wave propagating into the computational domain with speed $\bar{c}$, the local sound speed. The data for this Riemann problem consists of U from the interior boundary cell on one side, and $\tilde{U}$ on the other side, where $\tilde{U}$ equals U with the exception

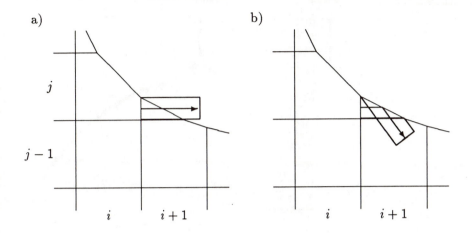

Fig. 2: (a) Wave propagating through a small boundary cell and out of computational domain. (b) Reflected wave actually used

of the component of normal velocity, which is again negated. This wave is propagated normal from the boundary and affects neighbouring cells in the same manner as described above for reflected waves.

In (LeVeque 1987) an additional wave splitting was also employed. Each wave R_p that arises from solving, for example, a Riemann problem in the x-direction can be further split into waves propagating oblique to the grid by solving a Riemann problem in the y-direction. In some situations this has stability advantages and may also give better resolution of 2D phenomena oblique to the Cartesian grid. However, in tests with stronger shocks (e.g. Mach 10) it has been found that this splitting can introduce some stability problems of its own. Moreover, it substantially increases the computational cost of the algorithm. More efficient splittings with better stability properties are currently being studied. In the calculations reported below, no secondary splitting has been used.

3. Numerical results

Fig. 3 shows density contours for a Mach 4.62 shock wave advancing up a 40° ramp, with a double Mach reflection. This calculation was performed on an underlying 200 × 200 Cartesian grid, with the ramp cutting through the grid at a 40° angle. A section of the grid near the corner is shown in

a)   b)

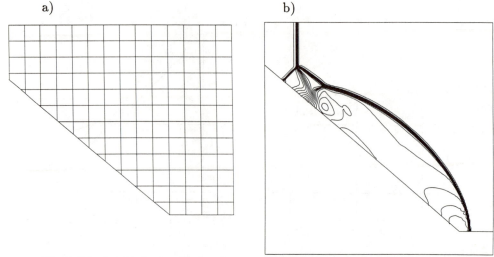

Fig. 3: Mach 4.62 shock reflection from a 40° ramp. (a) A portion of the Cartesian grid near the corner. (b) Density contours of the computed solution

Fig. 3a. The time step used gives a Courant number of about 0.5 relative to the regular cells (orders of magnitude larger relative to the smallest cells at the boundary). Ramp calculations of this nature are usually performed on a grid aligned with the ramp. For comparison purposes, one such calculation for this case may be found in Glaz et al. (1986).

Fig. 4 shows a Mach 2.81 shock wave after hitting a circular cylinder. Again a 200 × 200 grid is used, a portion of which is shown in Fig. 4a. For comparison see, for example, Yang and Lombard (1987), where the calculation is performed in polar coordinates.

As mentioned earlier, the method used here reduces to a very simple high resolution method on the regular portion of the grid. No attempt has been made yet to optimize this method. Techniques such a artificial compression of contact discontinuities have not been utilized. Some features are certainly smeared here, for example the slip lines within the double Mach reflection in the ramp calculation. Ultimately these boundary conditions will be used in conjunction with state-of-the-art finite volume methods on the regular portion of the grid and mesh refinement, which should yield superior results.

The main intent here is simply to demonstrate the feasibility of these boundary conditions for Cartesian grids in irregular regions. The accuracy obtained is comparable to what is produced by the same method in the

a)   b)

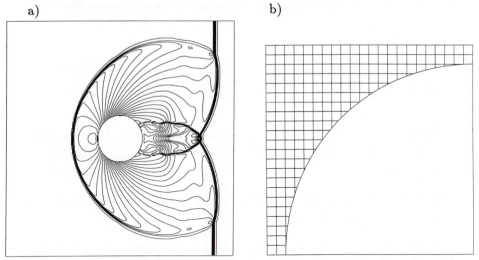

Fig. 4: Mach 2.81 shock passing a cylinder. (a) A portion of the Cartesian grid near the cylinder. (b) Density contours of the computed solution

absence of irregular boundaries (for example if the ramp is aligned with the grid). Most significantly, we obtain stable results using a time step based on the area of the regular grid cells.

References

Berger, M. (1982). Adaptive mesh refinement for hyperbolic partial differentail equations, Ph.D. dissertation, Computer Science Dept., Stanford University.

Berger, M., and Jameson, A. (1985). Automatic adaptive grid refinement for the Euler equations, *AIAA Journal* 23:561–568.

Clarke, D. K., Salas, M. D., and Hassan, H. A. (1986). Euler calculations for multielement airfoils using Cartesian grids, *AIAA Journal* 24:353–358.

Falle, S. and Giddings, J. (1988). An adaptive multigrid applied to supersonic blunt body flow. *These proceedings*.

Glaz, H. M., Colella, P., Glass, I. I., and Deschambault, R. L. (1986). A detailed numerical, graphical, and experimental study of oblique shock wave reflections, UTIAS Report No. 285, University of Toronto.

LeVeque, R. J. (1984). Convergence of a large time step generalization of Godunov's method for conservation laws, *Comm. Pure Appl. Math.* 37: 463–477.

LeVeque, R. J. (1985). A large time step generalization of Godunov's method for systems of conservation laws, *SIAM J. Num. Anal.*, 22:1051–1073.

LeVeque, R. J. (1987). High resolution finite volume methods on arbitrary grids via wave propagation, ICASE Report 87-68, to appear in *J. Comput. Phys.*

Roe, P. L. (1981). Approximate Riemann solvers, parameter vectors, and difference schemes, *J. Comput. Phys.* 43:357–372.

Rubbert, P. E., Bussoletti, J. E. et al. (1986). A new approach to the solution of boundary value problems involving complex configurations, in *Computational Mechanics — Advances and Trends*, AMD vol. 75, edited by A. K. Noor, 49–84.

Samant, S. S., Bussoletti, J. E. et al. (1987). TRANAIR: A computer code for transonic analysis of arbitrary configurations, AIAA Paper 87-0034.

Wedan, B., and South, J. C. (1983). A method for solving the transonic full-potential equation for general configurations, AIAA Paper 83-1889.

Yang, J. Y., and Lombard, C. K. (1987). Uniformly second order accurate ENO schemes for the Euler equations of gas dynamics, AIAA Paper 87-1166.

Numerical characteristic decomposition for compressible gas dynamics with general (convex) equations of state

P. Glaister

Dept. of Mathematics, University of Reading

1. Introduction

In the past decade much emphasis has been placed in the area of computational fluid dynamics on the accurate prediction of compressible flows governed by the Euler equations. One approach that has proved successful is to calculate approximate solutions of a set of linearized Riemann problems. A major contribution in this area was made by (Roe 1981). An alternative derivation of the resulting scheme was given by (Roe and Pike 1984). Roe's scheme, however, only applies to flows where the properties of the fluid are governed by the ideal equation of state. For many applications this can be a severe limitation.

In this paper we present an approximate linearized Riemann solver for the Euler equations with a general (convex) equation of state. The techniques involved in the construction of the scheme are similar to those used by (Roe and Pike 1984). We also show how the scheme can be extended to two and three dimensions using operator splitting and to generalized coordinates for use with body-fitted meshes. Finally, we show results for some standard test problems in one and two dimensions using ideal and non-ideal equations of state.

2. Equations of flow and state

The Euler equations for compressible flows in three dimensions are

$$\mathbf{w}_t + \mathbf{f}_x + \mathbf{g}_y + \mathbf{h}_z = \mathbf{0} \qquad (2.1)$$

where x, y, z are the Cartesian coordinates, t represents time,

$$\mathbf{w} = (\rho, \rho u, \rho v, \rho w, e)^T, \qquad (2.2)$$

$$\mathbf{f}(\mathbf{w}) = \left(\rho u, p + \rho u^2, \rho uv, \rho uw, u(e + p)\right)^T \qquad (2.3)$$

(with similar expressions for $\mathbf{g}(\mathbf{w})$ and $\mathbf{h}(\mathbf{w})$), and

$$e = \rho i + \frac{1}{2}(u^2 + v^2 + w^2). \qquad (2.4)$$

The quantities ρ, u, v, w, i, p and e represent the density, velocity in the three coordinate directions, specific internal energy, pressure and total energy respectively. After a transformation of coordinates from x, y, z to generalized coordinates ξ, η, ζ equation 2.1 becomes

$$(J\mathbf{w})_t + \mathbf{F}_\xi + \mathbf{G}_\eta + \mathbf{H}_\zeta = 0, \qquad (2.5)$$

where

$$J = \begin{vmatrix} x_\xi & y_\xi & z_\xi \\ x_\eta & y_\eta & z_\eta \\ x_\zeta & y_\zeta & z_\zeta \end{vmatrix} \qquad (2.6)$$

is the Jacobian of the grid transformation,

$$\mathbf{F} = \left(\rho U, J_\xi^x p + \rho u U, J_\xi^y p + \rho y U, J_\xi^z p + \rho z U, U(e + p) \right)^T, \qquad (2.7)$$

(with similar expressions for $\mathbf{G}$ and $\mathbf{H}$), J_ξ^x is the cofactor of x_ξ in J, ie

$$J_\xi^x = y_\eta z_\zeta - z_\eta y_\zeta, \qquad (2.8)$$

(with similar expressions for J_ξ^y, J_η^y), and

$$U = J_\xi^x u + J_\xi^y v + J_\xi^z w. \qquad (2.9)$$

In addition, we assume an equation of state for the fluid of the form

$$p = p(\rho, i). \qquad (2.10)$$

The use of generalized coordinates results in a scheme for use with body-fitted meshes, which is particularly useful for non-rectangular configurations.

3. Linearized Riemann problem

We propose solving equation 2.5 using operator splitting, i.e. we solve successively

$$(J\mathbf{w})_t + \mathbf{F}_\xi = 0 \qquad (3.1)$$
$$(J\mathbf{w})_t + \mathbf{G}_\eta = 0 \qquad (3.2)$$

and

$$(J\mathbf{w})_t + \mathbf{H}_\zeta = 0 \qquad (3.3)$$

along ξ, η, and ζ coordinate lines respectively. We consider a scheme for the solution of equation 3.1 only; the solution of equations 3.2 and 3.3 will follow in a similar way.

If the solution of equation 3.1 is sought along a ξ coordinate line given by $\eta = \eta_0$, $\zeta = \zeta_0$, constants, then we assume that the solution is known at a set of discrete mesh points $(\xi, \eta, \zeta, t) = (\xi_j, \eta_0, \zeta_0, t_n)$ at any time t_n. We follow the approach of (Godunov 1959) and assume that the approximate solution $\mathbf{w}_j^n$ to $\mathbf{w}$ at $(\xi_j, \eta_0, \zeta_0, t_n)$ can be considered as a set of piecewise constant states $\mathbf{w} = \mathbf{w}_j^n$ for $\xi \in (\xi_j - \frac{1}{2}\Delta\xi, \xi_j + \frac{1}{2}\Delta\xi)$ at time t_n where $\Delta\xi = \xi_j - \xi_{j-1}$ is a constant mesh spacing. A Riemann problem is now present at each interface $\xi_{j-\frac{1}{2}} = \frac{1}{2}(\xi_{j-1} + \xi_j)$ separating adjacent states $\mathbf{w}_{j-1}^n, \mathbf{w}_j^n$. We consider solving the linearized Riemann problem

$$(J\mathbf{w})_t + \tilde{A}(\mathbf{w}_{j-1}^n, \mathbf{w}_j^n)\mathbf{w}_\xi = 0 \tag{3.4}$$

where $\tilde{A}_{j-\frac{1}{2}} = \tilde{A}(\mathbf{w}_{j-1}^n, \mathbf{w}_j^n)$ is an approximation to the Jacobian matrix $A = \partial\mathbf{F}/\partial\mathbf{w}$ and is a constant matrix depending on the states either side of $\xi_{j-\frac{1}{2}}$. The matrix $\tilde{A}_{j-\frac{1}{2}}$ will be required to satisfy Property U as defined by (Roe 1981), namely

    (i) $\tilde{A}(\mathbf{w}_{j-1}^n, \mathbf{w}_j^n) \to A(\mathbf{w})$ as $\mathbf{w}_{j-1}^n \to \mathbf{w}_j^n \to \mathbf{w}$,

    (ii) $\tilde{A}_{j-\frac{1}{2}}\Delta\mathbf{w} = \Delta\mathbf{F}$

and

    (iii) $\tilde{A}_{j-\frac{1}{2}}$ has five linearly independent eigenvectors.

Roe shows that these conditions give good shock capturing properties when incorporated with an upwind scheme.

Once such a matrix has been constructed equation 3.4 can be solved approximately as

$$J_{j-\frac{1}{2}}\frac{(\mathbf{w}_k^{n+1} - \mathbf{w}_k^n)}{\Delta t} + \tilde{A}_{j-\frac{1}{2}}\frac{(\mathbf{w}_j^n - \mathbf{w}_{j-1}^n)}{\Delta\xi} = 0 \tag{3.5}$$

where k can be $j-1$ or j, $\Delta t = t_{n+1} - t_n$ is a constant time step and $J_{j-\frac{1}{2}}$ is a central difference approximation to the grid Jacobian at $(\xi_{j-\frac{1}{2}}, \eta_0, \zeta_0)$. To implement the algorithm we project $\Delta\mathbf{w}$ onto the eigenvectors $\tilde{\mathbf{e}}_i$ of $\tilde{A}_{j-\frac{1}{2}}$, i.e.

$$\Delta\mathbf{w} = \sum_{i=1}^{5} \tilde{\alpha}_i \tilde{\mathbf{e}}_i \tag{3.6}$$

so that

$$\Delta\mathbf{F} = \tilde{A}_{j-\frac{1}{2}}\Delta\mathbf{w} = \sum_{i=1}^{5} \tilde{\lambda}_i \tilde{\alpha}_i \tilde{\mathbf{e}}_i, \tag{3.7}$$

where $\tilde{\lambda}_i$ are the corresponding eigenvalues of $\tilde{A}_{j-\frac{1}{2}}$. Equation 3.5 can now be written as

$$J_{j-\frac{1}{2}}\frac{(\mathbf{w}_k^{n+1} - \mathbf{w}_k^n)}{\Delta t} + \frac{\sum_{i=1}^{5} \tilde{\lambda}_i \tilde{\alpha}_i \tilde{\mathbf{e}}_i}{\Delta\xi} = 0 \tag{3.8}$$

and this can be interpreted as a cell-based, first order upwind scheme as

$$\mathbf{w}_{j-1}^{n+1} = \mathbf{w}_{j-1}^n - \frac{\Delta t}{\Delta \xi} \tilde{\lambda}_i \tilde{\alpha}_i \tilde{\mathbf{e}}_i \qquad \text{if} \quad \tilde{\lambda}_i < 0$$

$$\text{or} \qquad\qquad i = 1, \dots, 5 \qquad (3.9)$$

$$\mathbf{w}_j^{n+1} = \mathbf{w}_j^n - \frac{\Delta t}{\Delta \xi} \tilde{\lambda}_i \tilde{\alpha}_i \tilde{\mathbf{e}}_i \qquad \text{if} \quad \tilde{\lambda}_i > 0.$$

$$(3.10)$$

Extensions of this scheme to be entropy satisfying, second order accurate and incorporate non-uniform meshes can easily be made, see (Glaister 1988).

4. Eigenvalues, eigenvectors and wavestrengths

In order to implement the algorithm of section 3 it is necessary to construct an approximate Jacobian matrix $\tilde{A}_{j-\frac{1}{2}}$ and determine its eigenvalues $\tilde{\lambda}_i$, eigenvectors $\tilde{\mathbf{e}}_i$, and the resulting wavestrengths $\tilde{\alpha}_i$ as given by equation 3.6. The construction is made by considering the continuous eigenvalues λ_i and continuous eigenvectors $\mathbf{e}_i$ of the Jacobian matrix A, and determining wave strengths α_i so that equations 3.6 and 3.7 are satisfied approximately. We then seek average eigenvalues $\tilde{\lambda}_i$ and average eigenvectors $\tilde{\mathbf{e}}_i$ such that equations 3.6 and 3.7 are satisfied exactly by retaining the form found for the α_i, for $\tilde{\alpha}_i$.

The matrix $\tilde{A}_{j-\frac{1}{2}}$ can then be found explicitly, but is not required as seen by equation 3.8. The required averages are

$$\tilde{\lambda}_i = \tilde{U} \pm \tilde{a}d, \tilde{U}, \tilde{U}, \tilde{U},$$

$$\tilde{\mathbf{e}}_{1,2} = \left(1, \tilde{u} \pm \frac{\tilde{a}X}{d}, \tilde{v} \pm \frac{\tilde{a}Y}{d}, \tilde{w} \pm \frac{\tilde{a}Z}{d}, \tilde{H} \pm \tilde{a}\tilde{U}\right)^T,$$

$$\tilde{\mathbf{e}}_3 = \left(1, \tilde{u}, \tilde{v}, \tilde{w}, \tilde{H} - \tilde{\rho}\tilde{a}^2/\tilde{p}_i\right)^T,$$

$$\tilde{\mathbf{e}}_4 = (0, -Y, X, 0, X\tilde{v} - Y\tilde{u})^T,$$

$$\tilde{\mathbf{e}}_5 = (0, -Z, 0, X, X\tilde{w} - Z\tilde{u})^T,$$

$$\tilde{\alpha}_{1,2} = \frac{1}{2\tilde{a}^2}\left(\Delta p \pm \tilde{\rho}\frac{\tilde{a}}{d}\Delta U\right),$$

$$\tilde{\alpha}_3 = \Delta\rho - \Delta p/\tilde{a}^2,$$

$$\tilde{\alpha}_4 = \tilde{\rho}\frac{\Delta v}{X} - \frac{\tilde{\rho}Y\Delta U}{d^2 X},$$

$$\tilde{\alpha}_5 = \tilde{\rho}\frac{\Delta w}{X} - \frac{\tilde{\rho}Z\Delta U}{d^2 X},$$

$$\tilde{U} = X\tilde{u} + Y\tilde{v} + Z\tilde{w},$$

$$\Delta\tilde{U} = \Delta(Xu + Yv + Zw) = X\Delta u + Y\Delta v + Z\Delta w,$$

$$\tilde{a}^2 = \tilde{p}\tilde{p}_i/\tilde{\rho}^2 + \tilde{p}_\rho,$$

$$\tilde{p} = \tilde{\rho}(\tilde{H} - \tilde{i} - \frac{1}{2}\tilde{u}^2 - \frac{1}{2}\tilde{v}^2 - \frac{1}{2}\tilde{w}^2),$$

$$\tilde{\rho} = \sqrt{\rho_{j-1}\rho_j},$$

$$\tilde{N} = \frac{\sqrt{\rho_{j-1}}N_{j-1} + \sqrt{\rho_j}N_j}{\sqrt{\rho_{j-1}} + \sqrt{\rho_j}}, N = u, v, w, i \text{ or } H,$$

$$d = \sqrt{X^2 + Y^2 + Z^2},$$

X, Y and Z denote constant approximations to $J_\xi^x, J_\xi^y, J_\xi^z$ in (ξ_{j-1}, ξ_j) and $\tilde{p}_\rho, \tilde{p}_i$ denote approximations to the first derivatives of the equation of state 2.10 that satisfy

$$\Delta p = \tilde{p}_\rho \Delta \rho + \tilde{p}_i \Delta i.$$

It is a straightforward matter to determine $\tilde{p}_\rho, \tilde{p}_i$ satisfying this constraint, (see (Glaister 1988)).

5. Test problems and numerical results

We show results for three test problems:

(a) One-dimensional shock reflection at a rigid wall. For details of this problem see for example (Glaister 1988). Two equations of state are chosen, (i) the stiffened gas equation of state $p = (\gamma - 1)\rho i + B(\rho/\rho_0 - 1)$ with the ratio of specific heat capacities γ taken to be $5/3, B = 1$ and $\rho_0 = 1$, (ii) a general equation of state for copper (see (Riley 1970)). Figures 1,2,3 and 4 show the results obtained for the density with 50 mesh points when the shock has move a distance of 0.3. Figures 1 and 2 are for the stiffened gas case with a pressure jump of 10 and 2, respectively. Figures 3 and 4 show similar results for the general equation of state for copper.

(b) Mach 3 flow past a forward facing step.
A description of this two-dimensional test problem can be found in (Emery 1968), or (Woodward and Colella 1984). We consider an equation of state for 'equilibrium air' due to Srinivasan, Tannehill and Weilmuenster, Simplified Curve fits for the thermodynamic properties of equilibrium air, Iowa State University Engineering Research Institute Project 1626, (1986). The density contours at $t = 4.0$ are displayed in figure 5 using a 120×40 mesh. The bow shock, Mach stem, rarefaction fan and reflections at the upper and lower walls are clearly visible.

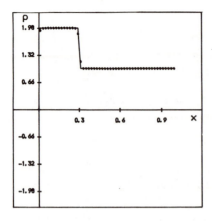

Figure 1

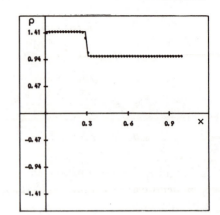

Figure 2

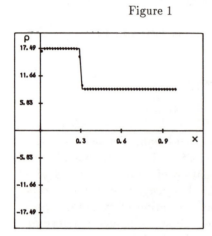

Figure 3

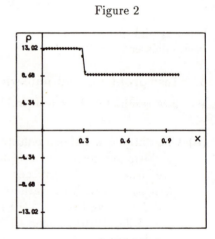

Figure 4

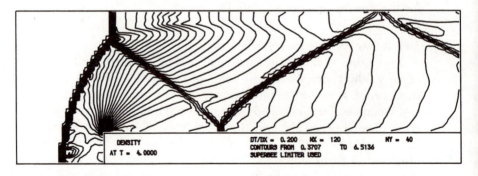

Figure 5

(c) Mach 8 flow past a cylinder.

This two-dimensional problem was defined at the GAMM workshop of 1986. The mesh consists of 33 concentric circles whose radii increase non-uniformly and 128 radial lines with a constant angle separating them. Figure 6 displays the isomach contours for a flow with the ideal equation of state $p = (\gamma - 1)\rho i$ where we have taken $\gamma = 1.4$.

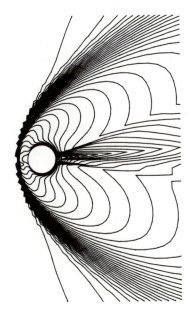

Figure 6

6. Conclusions

We have presented an approximate linearized Riemann solver capable of treating three-dimensional flows of an inviscid, compressible fluid with a general equation of state. The scheme incorporates operator splitting and applies to a generalized coordinate system for use with body-fitted meshes. The scheme has given satisfactory results for some standard test problems.

Acknowledgements

I acknowledge the financial support of AWRE Aldermaston.

References

Emery, A. (1968). An evaluation of several differencing methods for inviscid fluid flow problems. *Journal of Computational Physics*, 2 : 306–307.

Glaister, P. (1988). *Approximate Riemann solvers for systems of hyperbolic conservation laws*. Ph.D. thesis, University of Reading.

Godunov, S. (1959). A difference scheme for numerical computation of discontinuous solutions of equations of fluid dynamics. *Mat Sbornik*, 4 : 271–306.

Riley, T. (1970). *High-Velocity Impact Phenomena*. Academic Press, New York, 1st edition.

Roe, P. (1981). Approximate Riemann solvers, parameter vectors, and difference schemes. *Journal of Computational Physics*, 43 : 357–372.

Roe, P. and Pike, J. (1984). In *Computing Methods in Applied Sciences and Engineering VI*. North–Holland, Amsterdam, 1st edition.

Woodward, P. and Colella, P. (1984). The numerical simulation of two-dimensional fluid flow with strong shocks. *Journal of Computational Physics*, 54 : 115–173.

A hybrid scheme for the Euler equations using the Random Choice and Roe's methods

E. F. Toro and P. L. Roe

Cranfield Institute of Technology

1. Introduction

A combination of two methods for the Euler equations is presented. Roe's method is applied throughout the flow field except for computing cells which are transversed by strong discontinuities; these are updated by the Random Choice Method. The scheme works well in one space dimension. Smooth parts of the flow are accurately computed and discontinuities have zero width. For two dimensional problems the scheme can still provide zero-width contacts but some smearing of shocks is accepted.

We are interested in numerical methods for the Euler equations that are accurate in the smooth parts of the flow and resolve discontinuities with zero width. These requirements are somewhat contradictory for schemes that treat the entire flow field with a single method.

Traditional finite difference methods (e.g. Lax-Wendroff) provide accurate solutions for smooth flows but produce spurious oscillations near discontinuities or high gradients. High resolution methods (Roe 1981; Gottlieb et al. 1987), e.g. Roe's method, are capable of preserving accuracy in smooth parts of the flow while avoiding the over/under shoots. These methods smear discontinuities. Shocks are spread typically over 3/4 cells; contact discontinuities are spread typically over 6/7 grid cells. Shock waves have a natural compression mechanism that helps their sharper resolution by characteristic based methods (Roe 1986). There appears to be consensus that shock waves are no longer a serious problem for characteristic-based numerical methods. However, contact discontinuities remain a difficulty in these methods. They do not have the compressing mechanism (converging characteristics) that shock waves have. Numerically their spreading tends to increase with time. In applications such as Gas Dynamics coupled with combustion phenomena sharp resolution of contact discontinuities is as important (if not more) than that of shock waves, for they signal discontinuous changes of temperature, on which ignition criteria are based.

The Random Choice Method (Glimm 1955; Chorin 1976; Colella 1982) is a technique that has the ability to resolve discontinuities with zero width, but the smooth parts of the flow are not, due to the inherent randomness of the method, accurately represented. The idea behind the hybrid method is to combine a high-resolution type method with the Random Choice Method. The latter component is intended for the 'strong' discontinuities only. In (Toro and Roe 1987) we presented a hybrid scheme based on a Random Flux Method (SORF) presented in (Toro 1986). Success was somewhat limited by the still unresolved monotonicity aspects of SORF when applied to non-linear systems. In this paper we combine a well known high resolution method (Roe's method) with the Random Choice Method (RCM). The resulting ROE/RCM hybrid has proved considerably more successful than the previous one.

The rest of this paper is organised as follow: section 2 deals with the Euler equations and Riemann problems; in section 3 we briefly review the methods to be used. In section 4 we describe the hybrid method and show application to 1D and 2D problems. Conclusions are drawn in section 5.

2.   Euler equations and Riemann problems

The unsteady two-dimensional Euler equations can be written in conservation form as

$$U_t + F(U)_x + G(U)_y = 0 \qquad (2.1)$$

where $U = U(t, x, y)$, with t denoting time, x and y space. Subscripts denote partial differentiation. The vectors U, F and G are

$$U = \begin{bmatrix} \rho \\ \rho u \\ \rho v \\ E \end{bmatrix}, \ F = \begin{bmatrix} \rho u \\ \rho u^2 + p \\ \rho u v \\ u(E + p) \end{bmatrix}, \ G = \begin{bmatrix} \rho v \\ \rho u v \\ \rho v^2 + p \\ v(E + p) \end{bmatrix} \qquad (2.2)$$

where ρ is density, $V = (u, v)$ is velocity, p is pressure and E is total energy given by

$$E = \frac{1}{2}\rho(u^2 + v^2) + \rho e \qquad (2.3)$$

where e denotes specific internal energy.

The usual closure condition for (2.1)–(2.3) is an equation of state $e = e(p, \rho)$. Here we take the ideal gas case

$$e = \frac{p}{(\gamma - 1)\rho} \qquad (2.4)$$

where γ is the ratio of specific heats (constant).

System (2.1)–(2.4) is a set of non-linear partial differential equations of hyperbolic type. The corresponding initial value problem can only be solved analytically or in closed form for very special initial conditions. An important special initial value problem is the Riemann problem. For the unsteady one-dimensional case or the steady two-dimensional supersonic-flow case the Riemann problem can be solved exactly, although not in closed form. The initial data for the Riemann problem in one dimension consists of a pair of constant states joined by a discontinuity.

For the unsteady one-dimensional Euler equations the solution of the Riemann problem looks as depicted.

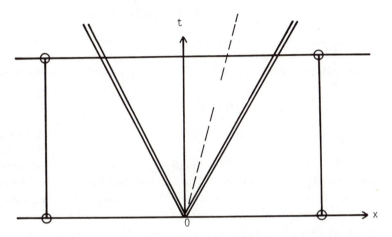

Fig. 1: Solution of the Riemann problem

Most modern numerical techniques for (2.1)–(2.4) use the Riemann problem locally. It is therefore important to find its solution in an efficient way. Difference-type methods can afford drastic approximations that will not affect the accuracy of the scheme. Finding the exact solution is more expensive although there have been significant advances in devising fast exact Riemann solvers (e.g. (Gottlieb et al. 1987) for ideal gases, (Toro 1987) for covolume equation of state).

The Random Choice Method uses the exact solution of the Riemann problem, although there are approximations that can be used with RCM.

Both component methods used in this paper utilise the Riemann problems. Roe's method (Roe 1981) uses an approximation designed to extract some of the information contained in the complete solution of the Riemann problem. RCM is used here with exact solution. These methods are described in more detail in the next section.

3. Review of methods used

The hybrid method presented in this paper is a combination of the high resolution method due to Roe (1981) and the Random Choice Method

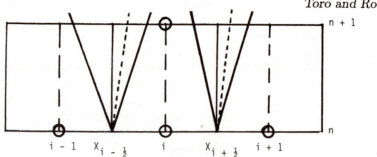

Fig. 2: Riemann problems affecting cell i

(Chorin 1976). The latter method is used only at cells transversed by large discontinuities. In this section we briefly review the methods used.

3.1   The Random Choice Method

This technique is based on an existence proof for non-linear hyperbolic systems due to Glimm (1955). As a computational technique it was first successfully implemented by Chorin (1976). Significant improvements were incorporated by Colella (1982). The method has been further refined to provide a successful numerical technique for unsteady one-dimensional, and steady two-dimensional (hyperbolic) problems (Shi and Gottlieb 1985).

RCM assumes piece-wise constant data. The corresponding sequence of Riemann problems are solved exactly. Fig. 2 illustrates the method implemented on a non-staggered grid.

The solution U_i^{n+1} at the new time level is defined as

$$U_i^{n+1} = \overline{U}_i(t_n + \Delta T, X_i + \theta_n \Delta X) \qquad (3.1)$$

where $\overline{U}_i$ is the solution of the two Riemann problems $RP(i-1, i)$ and $RP(i, i+1)$ at $t = t_n + \Delta T$, $X_{i-\frac{1}{2}} \leq X \leq X_{i+\frac{1}{2}}$ i.e. within cell i. Here θ_n is a member of a pseudo-random sequence θ_n of numbers in the interval $[0, 1]$. for best results θ_n must satisfy some desired statistical properties. Equidistribution in $[0, 1]$ is an important property. Colella (1982) introduced van der Corput sequences into RCM. They have been very successful in reducing randomness in the computed solution. Anderson (1987) has recently introduced a new sequence that has its merits.

A Courant condition must be imposed. The usual one for RCM is that which prevents all waves transversing a distance greater than $\frac{1}{2}\Delta X$ so that no wave interaction takes place in time ΔT. This is a little restrictive and can be extended.

Fig. 3 shows computed results for the shock tube problem

$$\left.\begin{array}{llllll} \rho_L & = & 1.0 & , & u_L & = & 0.0 & , & p_L & = & 1.0 \\ \rho_R & = & 0.125 & , & u_R & = & 0.0 & , & p_R & = & 0.1 \\ \gamma & = & 1.4 & , & \Delta X & = & 0.01 & , & \nu & = & \frac{1}{2} \end{array}\right\} \qquad (3.2)$$

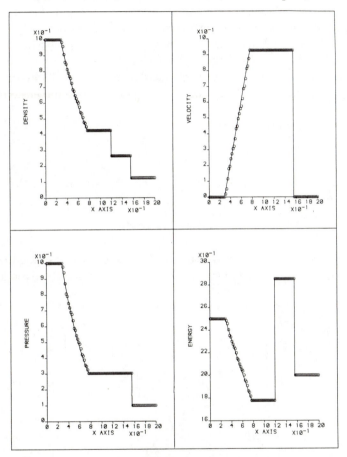

Fig.3: Sod's shock-tube test problem.
o : computed solution by RCM on fixed grid; — : exact solution

Note that discontinuities have zero width. Their positions at a given time, however, might not be the correct ones, but the errors are small. On the average, positions are correct. The negative aspect of the computed RCM solution is the randomness in the smooth parts of the flow. For this homogeneous problem (no sources) this is not a very serious problem, but when sources are present and are updated via time operator splitting, randomness is increased to intolerable levels. This is an important aspect behind the idea of hybridising.

3.2   Roe's Method

The original formulation of the method due to Roe can be found in (Roe 1981). For the two-dimensional problem (2.1) the method performs a linearisation by replacing the Jacobian matrices corresponding to the flux functions F and G by constant matrices $\tilde{A}$ and $\tilde{B}$ respectively. The construction of $\tilde{A}$ and $\tilde{B}$ assigns to them some desirable properties. Since the method applies to two and three -dimensional problems via space operator splitting (i.e. solving sequences of one-dimensional problems) it is sufficient to describe the method for a corresponding one-dimensional sweep (X-sweep, say).

One requires the property

$$\tilde{A}\Delta U = \Delta F \tag{3.3}$$

where $\Delta U = U_{i+1} - U_i$, $\Delta F = F_{i+1} - F_i$. ΔF in (3.3) can be uniquely expressed as

$$\Delta F = \sum_{k=1}^{4} \alpha_k \lambda_k \mathbf{e_k} \tag{3.4}$$

where $\mathbf{e_j}$ are the right eigenvectors of $\tilde{A}$, λ_j are the corresponding eigenvalues and α_j are the wave strengths. These are explicitly given as

$$\lambda_1 = \tilde{u} - \tilde{a}, \; \lambda_2 = \tilde{u}, \; \lambda_3 = \tilde{u}, \; \lambda_4 = \tilde{u} + \tilde{a} \tag{3.5}$$

$$\mathbf{e_1} = \begin{bmatrix} 1 \\ \tilde{u} - \tilde{a} \\ \tilde{v} \\ \tilde{H} - \tilde{u}\tilde{a} \end{bmatrix}, \mathbf{e_2} = \begin{bmatrix} 1 \\ \tilde{u} \\ \tilde{v} \\ \frac{1}{2}q^2 \end{bmatrix}, \mathbf{e_3} = \begin{bmatrix} 0 \\ 0 \\ 1 \\ \tilde{v} \end{bmatrix}, \mathbf{e_4} = \begin{bmatrix} 1 \\ \tilde{u} + \tilde{a} \\ \tilde{v} \\ \tilde{H} + \tilde{u}\tilde{a} \end{bmatrix} \tag{3.6}$$

$$\begin{aligned} \alpha_1 &= \frac{1}{2\tilde{a}^2}(\Delta p - \tilde{\rho}\tilde{a}\Delta u) \\ \alpha_2 &= \Delta\rho - \frac{\Delta p}{\tilde{a}^2} \\ \alpha_3 &= \rho\Delta v \\ \alpha_4 &= \frac{1}{2\tilde{a}^2}(\Delta p + \tilde{\rho}\tilde{a}\Delta u) \end{aligned} \tag{3.7}$$

where $q^2 = u^2 + v^2$ and $\tilde{a}$, $\tilde{\rho}$ etc., are the Roe average values. For details on the 1D basic scheme consult (Roe 1981).

The method advances the solution via the conservative formula

$$U_i^{u+1} = U_i^u - \frac{\Delta T}{\Delta X}\left[F_{i+\frac{1}{2}} - F_{i-\frac{1}{2}}\right] \tag{3.8}$$

which is obtained by integrating the conservation laws, expressed in integral form, around cell i (Fig.2).

The flux $F_{i+\frac{1}{2}}$ in Roe's scheme is given by

$$F_{i+\frac{1}{2}} = \frac{1}{2}(F_i + F_{i+1}) - \frac{1}{2}\sum_{k=1}^{4} \alpha_k |\lambda_k| \mathbf{e_k} \qquad (3.9)$$

The resulting scheme is first order accurate. Higher accuracy can be obtained by the addition of a correction term

$$S = \frac{1}{2}\sum_{k=1}^{4} \psi(r_k)(\sigma_k - \nu_k)\alpha_k \lambda_k \mathbf{e_k} \qquad (3.10)$$

Here $\nu_k = \frac{\Delta T \lambda_k}{\Delta X}$ is the Courant number associated with wave k, $\sigma_k = \text{sign}(\nu_k)$ and ψ is a flux limiter. ψ is a function of

$$r_i^k = \frac{\alpha_{i+1-\sigma}^k - \alpha_{i-\sigma}^k}{\alpha_{i+1}^k - \alpha_i^k} \qquad (3.11)$$

the ratio of the upwind difference to the local difference. The quantity α in (3.11) can, in principle, be any of the conserved or physical quantities; the choice $\alpha \equiv \rho$ works quite successfully.

A popular flux limiter is Superbee, defined as

$$\psi(X) = \begin{cases} 2 & , & 2 \leq X \\ X & , & 1 \leq X \leq 2 \\ 1 & , & \frac{1}{2} \leq X \leq 1 \\ 2X & , & 0 \leq X \leq \frac{1}{2} \\ 0 & , & X \leq 0 \end{cases} \qquad (3.12)$$

Fig. 4 shows computed results for problem (3.2). Note that the solution is accurate in the smooth parts (unlike RCM) but discontinuities are smeared, especially the contact discontinuities. The next section describes how these methods (ROE and SCM) can be put together to increase accuracy throughout the flow field.

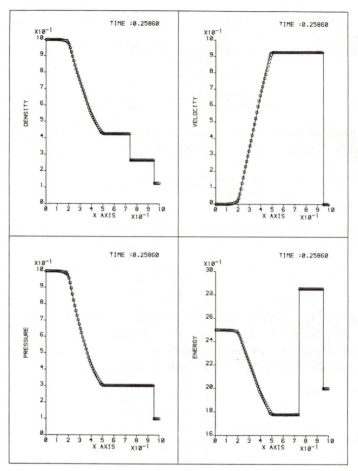

Fig.4: Sod's shock-tube test problem. Computed solution by Roe's method (with Superbee).

o : numerical solution; — : exact solution

4.   The hybrid method

The hybrid method consists of combining Roe's method with RCM. The basic algorithm uses Roe's method throughout and RCM is used whenever a cell is transversed by a strong wave. We distinguish between shocks and contacts as follows. If $p_s^*/p_i^n \geq p$ then a shock wave is transversing cell i and RCM must be used. Here p_s^* is the pressure between the acoustic waves in the solution of the Riemann problem. The subindex s here refers to the Riemann problems $RP(i-1,i)$ and $RP(i,i+1)$; p_i^n is the pressure in cell i at time level n (data). To detect contact discontinuities we use ρ_L^*/ρ_R^*, where ρ_L^* and ρ_R^* denote the values of density on either side (left and right) of the contact discontinuity $dx/dt = u^*$ in the solution of the Riemann problem. If $RP(i-1,i)$ has $|\rho_L^*/\rho_R^*| - 1 \geq D$ and $u^* \geq 0$ then RCM is used to update U_i^n. Similarly, if $RP(i, i+1)$ has $|\rho_L^*/\rho_R^* - 1| \geq D$ and $u^* \leq 0$ then RCM is used to update U_i^n.

The choice $P = 1.3$ and $D = 1.3$ works well, but there is a degree of arbitrariness in their selection. We remark that the wave strengths α_j of Roe's method are used to provide a first indication as to the strength of the waves present. If in doubt one then solves the Riemann problem exactly. This is done only a couple of times per time step.

Fig. 4 shows computed results using the ROE/RCM hybrid method for the one-dimensional shock-tube problem specified by (3.2). The solution has the expected properties, namely accurate in the smooth parts of the flow and zero-width discontinuities (shock and contact). The method is applied to the 2D Euler equations (2.1) via space operator splitting. Roe (Roe 1987) has recently demonstrated that such a procedure is incorrect for shock waves and shear waves but entropy waves come out clean. These findings will only manifest themselves in methods capable of resolving high frequencies (e.g. RCM). Fig. 5a shows the solution of problem (3.2) with data, an angle $\alpha = 45°$ to the $x - y$ computing grid, along the normal direction. The oscillations and overshoots are caused by the shock wave and these results are clearly unacceptable.

The hybrid ROE/RCM method will only treat the contact discontinuity via RCM. Roe's method will be applied elsewhere, including the shock. Fig. 5b shows computed results for the same problem as that of Fig. 5a. The contact discontinuity is of zero width as in 1D problems, but the shock wave has been compromised. However as pointed out previously, shock waves are very well resolved by high resolution methods. The procedure extends to 3D problems.

5. Conclusions

A hybrid ROE/RCM method for the 1D and 2D Euler equations has been presented. For 1D problems the method is accurate in smooth parts of the flow and resolves shock waves and contact discontinuities with zero width. For 2D (and 3D) problems the method can still resolve contact discontinuities with zero width but other waves must be smeared. For multidimensional problems the present method can be viewed as a procedure for sharpening contacts to infinite resolution.

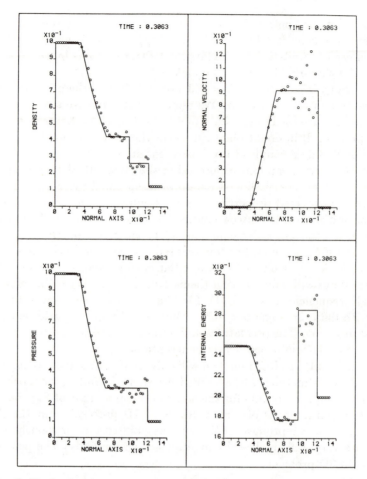

Fig.5a: Sod's shock-tube test problem in 2-D (angle = 45°)
Random Choice Method.

o : computed solution; — : exact solution

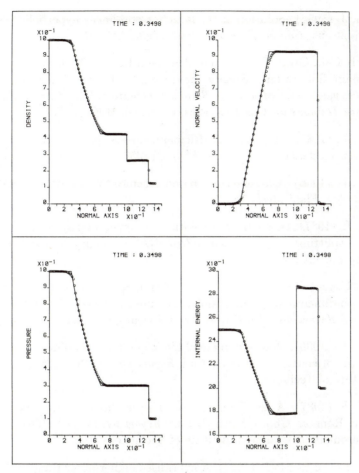

Fig.5b: Sod's shock-tube test problem in 2-D (angle = 45°)
Roe/RCM hybrid method with Superbee.
o : computed solution; — : exact solution

References

Anderson, D. S. and Gottlieb J. J. (1987). On random numbers for the Random Choice Method, pre-print of UTIAS Report 1987, *paper presented at the Second International Meeting on Random Choice Methods on Gas Dynamics*, Cranfield, U.K. 1987.

Chorin, A. (1976). Random choice solution of hyperbolic systems, *J. Comput. Phys.* 22:517.

Colella, P. (1982). Glimm's method for gas dynamics *SIAM J. Sci. Stat. Comput.*, Vol. 3, 1:76

Glimm, J. (1955). Solution in the large for non-linear hyperbolic systems of equations. *Comm. Pure and Appl. Math.* 697–715.

Gottlieb J. J., Groth C. P. T. and Anderson D. S. (1987). Assessment of Riemann Solvers for unsteady one-dimensional inviscid flows of perfect gasses, *paper presented at the Second International Meeting on Random Choice Methods on Gas Dynamics*, Cranfield, U.K. 1987.

Roe, P. L. (1981). Approximate Riemann solvers, parameter vectors and difference schemes, *J. Comput. Phys.* 27:357.

Roe, P. L. (1986). Characteristic-based schemes for the Euler equations, *Ann. Rev. Fluid Mech.* 18:337-65.

Roe, P. L. (1987). Discontinuous solutions to hyperbolic systems under operator splitting, *ICASE Report 87/64*, NASA Langley Research Center, Hampton, VA 23665, 1987.

Shi, Z. C. and Gottlieb J. J. (1985). Random Choice Method solutions for two-dimensional planar and axisymmetric steady supersonic flow, *UTIAS Report no. 297*, University of Toronto, Canada.

Toro E. F. (1986). A new numerical technique for quasi-linear hyperbolic systems of conservation laws, *CoA Report 86/20*, Dec. 1986, Cranfield Institute of Technology, U.K.

Toro E. F. (1987). A fast Riemann solver with constant covolume applied to the Random Choice Method, *CoA Report 87/19*, Oct. 1987, College of Aeronautics, Cranfield Institute of Technology, U.K.

Toro E. F. and Roe P. L. (1987). A hybridised higher-order Random Choice Method for quasi-linear hyperbolic systems, *Proc. 16th International Symposium on shock tubes and waves*, Aachen July 1987, W. Germany, p.701. H. Gronig (Editor).

A variational finite element formulation for three-dimensional incompressible flows

P. Ward, R. Desai, and W. Kebede

SDRC Engineering Services

Hitchin, England

A. Ecer

Purdue University

Indianapolis, USA

1. Introduction

Applications of the finite element method to the solution of compressible and incompressible flows have previously been presented by two of the authors (Ecer and Akay 1982; Ecer et al. 1983). A variational principle was obtained by adding physically appropriate constraints to the Lagrangian density of Hamilton's principle (Serrin 1959; Seliger and Whitham 1968) using Lagrangian multipliers. This led to a transformation of the velocity vector in terms of potential-like variables and the Lagrange multipliers, referred to as Clebsch transformation (Clebsch 1859). Numerical results have been presented using this approach (Ecer and Akay 1982; Ecer et al. 1983; Ecer et al. 1984)

In this paper a new variational form particularly suited to three-dimensional iso-thermal incompressible flow is presented. A unique feature of the approach is that it reduces to potential and Euler flows as special cases of the Navier–Stokes equations.

2. Theory

The development of the variational principle starts with the general form of Hamilton's principle in the following form:

$$\delta \int_{\Omega_0} \int L d\alpha dt = 0 \qquad (2.1)$$

where L is the Lagrangian density including the kinetic and potential energies of material particles occupying the Eulerian space Ω_0: $\underline{x}(\alpha, t)$ is the position vector of a particle at time t which can be defined by a Lagrangian coordinate vector $\underline{\alpha} = (\alpha_1, \alpha_2, \alpha_3)$. The variational statement of (2.1) implies that the equations of motion of the fluid particles occupying Ω_0 can be obtained by applying the variational operator on the integral of the Lagrangian. In order to apply this to a viscous incompressible fluid it is important to include all of the components of energy and work.

The Lagrangian density may be expressed in terms of the kinetic and potential energies of the fluid particles as

$$L = \frac{1}{2}\rho \underline{u} \cdot \underline{u} + \rho H \qquad (2.2)$$

where ρ is the density, $\underline{u}$ is the velocity vector, $H = \frac{1}{2}\rho\underline{u}\cdot\underline{u}+p$, is the total head and p is pressure. At this point all of the variables in the Lagrangian are functions of the material coordinates $\underline{\alpha}$. In the case of velocities $\underline{u}$ one can operate in an Eulerian system. However, for the other variables additional equations are required to describe their distribution in Eulerian space $\Omega(x,t)$. This can be achieved by adding each of these equations as constraints on the Lagrangian using a set of Lagrange multipliers.

The incompressibility condition is imposed through the conservation of mass equation

$$\underline{\nabla} \cdot (\rho\underline{u}) = 0. \qquad (2.3)$$

The transport of the total head, H is defined as follows (Serrin 1959):

$$\underline{\nabla} \cdot \rho(\underline{u}H - \mu\underline{u} \times \underline{\omega}) + \mu\underline{\omega} \cdot \underline{\omega} = 0 \qquad (2.4)$$

where

$$\underline{\omega} = \underline{\nabla} \times \underline{u} \qquad (2.5)$$

is the vorticity vector. The final conservation equation is for the material coordinates

$$\underline{\nabla} \cdot (\rho\underline{u}\alpha) = 0. \qquad (2.6)$$

The desciption is completed by considering the boundary conditions for each of the equations on boundaries S. These are

$$\rho\underline{u} \cdot \underline{n} = a \qquad (2.7)$$

$$\underline{n} \cdot (\rho\underline{u}H - \mu\underline{u} \times \underline{\omega}) = g \qquad (2.8)$$

$$\underline{n} \times \underline{u} = \underline{b} \qquad (2.9)$$

$$\rho\underline{u} \cdot \underline{n}\alpha = c \qquad (2.10)$$

for equations (2.3), (2.4), (2.5) and (2.6) respectively. In the above equations, a, g, $\underline{b}$ and c are the known values on the boundaries. It should be noted that the total head, H, of a fluid particle at any point in the domain is a function of a) the material coordinate of the particle, α, which defines where the particle enterd the flow domain and the path it has followed and b) the viscous dissipation. The relationship between the total head, H, and the material coordinate α, may be described by decomposing the total head into two parts as follows:

$$H = H_0(\alpha) + H' \tag{2.11}$$

where H_0 is the total head of the particle when it entered the flowfield which is a function of the material coordinates. H' represents the change in the total head due to viscous dissipation and is a new variable independent of the material coordinates. This new variable can be used in the variational principle instead of H which is not independent of α.

Including all of these considerations leads to the following constrained functional.

$$
\begin{aligned}
\pi = & \int_\Omega \left\{ \tfrac{1}{2}\underline{u}\cdot\underline{u} + \phi(\underline{\nabla}\cdot\rho\underline{u}) + \rho H_0 + \rho H' + \eta\underline{\nabla}\cdot(\rho\underline{u}\alpha) \right. \\
& + \gamma\left[\underline{\nabla}\cdot\rho(\underline{u}[H' + H_0] - \mu\underline{u}\times\underline{\omega}) + \mu\underline{\omega}\cdot\underline{\omega}\right] \\
& + \left. \underline{\beta}\cdot(\underline{\nabla}\times\underline{u} - \underline{\omega})\right\}d\Omega \\
& - \int_{\Gamma_\phi} \phi(\rho\underline{u}\cdot\underline{n} - a)d\Gamma - \int_{\Gamma_\beta} \underline{\beta}\cdot(\underline{n}\times\underline{u} - \underline{b})d\Gamma \\
& - \int_{\Gamma_\gamma} \gamma\underline{n}(\rho\underline{u}[H' + H_0] - \mu\underline{u}\times\underline{\omega} - g)d\Gamma - \int_{\Gamma_\eta} \eta(\rho\underline{u}\cdot\underline{n}\alpha - c)d\Gamma.
\end{aligned}
\tag{2.12}
$$

3. Determination of velocity: Clebsch transformation

Taking variations of the functional in (2.12) with respect to velocity yields the following:

$$\underline{u} = \underline{\nabla}\phi + \alpha\underline{\nabla}\eta + H'\underline{\nabla}\gamma - \underline{\omega}\times\underline{\nabla}\gamma - \frac{1}{\rho}\underline{\nabla}\times\underline{\beta}. \tag{3.1}$$

Substituting (3.1) into the constrained variational functional (2.12) yields the following

$$
\begin{aligned}
\pi = & \int_\Omega \left\{ -\tfrac{1}{2}\rho\underline{u}\cdot\underline{u} + \rho H_0 + \rho H' + \mu\gamma\underline{\omega}\cdot\underline{\omega} - \underline{\beta}\cdot\underline{\omega} \right\} \\
& + \int_{\Gamma_\phi} \phi a d\Gamma + \int_{\Gamma_\beta} \underline{\beta}d\Gamma + \int_{\Gamma_\eta} \eta c + \int_{\Gamma_\gamma} \gamma g.
\end{aligned}
\tag{3.2}
$$

4.   Derivation of governing equations

Considering variations of equations (3.2) with respect to the independent variables ϕ, α, η, H' and $\underline{\beta}$ the following equations describing the fluid motion may be obtained.

$$\delta\phi \quad : \quad \underline{\nabla} \cdot (\rho\underline{u}) = 0 \tag{4.1}$$

$$\delta\eta \quad : \quad \underline{\nabla} \cdot (\rho\underline{u}\alpha) = 0 \tag{4.2}$$

$$\delta\alpha \quad : \quad \underline{\nabla} \cdot (\rho\underline{u}\eta) = \rho\frac{\partial H_0}{\partial\alpha} \tag{4.3}$$

$$\delta\gamma \quad : \quad \underline{\nabla} \cdot (\rho\underline{u}H' - \mu\underline{u} \times \underline{\omega}) + \mu\underline{\omega} \cdot \underline{\omega} \tag{4.4}$$

$$\delta H' \quad : \quad \underline{\nabla} \cdot (\rho\underline{u}\gamma) = \rho \tag{4.5}$$

$$\delta\beta \quad : \quad \underline{\nabla} \times \underline{u} = \underline{\omega} \tag{4.6}$$

$$\delta\underline{\omega} \quad : \quad \underline{\beta} = \mu(\underline{\nabla} \times \gamma\underline{u} + \gamma\underline{\omega}) \tag{4.7}$$

It can be seen that the above variations produce the transport equations which were specified as constraints to the variational functional together with the boundary conditions of equations (2.7)–(2.10). In addition a set of equations is produced for the Lagrange multipliers η, γ and $\underline{\beta}$ as specified by equations (4.2), (4.5), and (4.6). No Neumann boundary conditions are produced for these equations. The conservation of mass equation (4.1) is used to determine the Lagrange multiplier ϕ by substituting the Clebsch transformation of (3.1).

In the case of inviscid flows, $\mu = 0$ and hence $H' = 0$. This leads to the following Clebsch transformation

$$\underline{u} = \underline{\nabla}\phi + \alpha\underline{\nabla}\eta \tag{4.8}$$

which has previously ben presented (Ecer and Akay 1982) for Euler flows. For irrotational, inviscid flows the Clebsch transformations simply reduces to

$$u = \underline{\nabla}\phi. \tag{4.9}$$

5.   Finite element approximation

Finite element approximation of the governing equations after a weak variational formulation results in the following system

$$\underline{K}_{\phi\phi}\underline{\phi} \;=\; \underline{f}_\phi \tag{5.1}$$

$$\underline{K}_{\alpha\eta}\underline{\eta} \;=\; \underline{f}_\alpha \tag{5.2}$$

$$\underline{K}_{\eta\alpha}\underline{\alpha} \;=\; \underline{f}_\gamma \tag{5.3}$$

$$\underline{K}_{\gamma H'}\underline{H}' = \underline{f}_{\gamma} \tag{5.4}$$

$$\underline{K}_{H'\gamma}\underline{\gamma} = \underline{f}_{H'} \tag{5.5}$$

with element matrices

$$\underline{K}^e_{\phi\phi} = \int_{\Omega_e} \rho \underline{\nabla} N \cdot \underline{N} d\Omega \tag{5.6}$$

$$\underline{f}^e_{\phi} = -\int_{\Omega_e} \underline{\nabla} N \cdot (\rho\alpha\underline{\nabla}\eta + \rho H'\underline{\nabla}\gamma - \mu\underline{\omega} \times \underline{\nabla}\gamma - \underline{\nabla} \times \underline{\beta})d\Omega \tag{5.7}$$

$$\underline{K}_{\alpha\eta} = \underline{K}_{\eta\alpha} = \underline{K}_{\gamma H'} = \underline{K}_{H'\gamma} = \int_{\Omega_e} (\rho\underline{u} \cdot \underline{\nabla} N)(\rho\underline{u} \cdot \underline{\nabla} N)d\Omega \tag{5.8}$$

$$\underline{f}^e_{\alpha} = \int_{\Omega_e} \rho\underline{u} \cdot \underline{\nabla} N \rho \frac{\partial H_0}{\partial \alpha} d\Omega \tag{5.9}$$

$$\underline{f}^e_{\eta} = 0 \tag{5.10}$$

$$\underline{f}^e_{\gamma} = \int_{\Omega_e} \rho^2 \underline{u} \cdot \underline{\nabla} N d\Omega \tag{5.11}$$

$$\underline{f}^e_{H'} = \int_{\Omega_e} \rho\underline{u} \cdot \underline{\nabla} N \{\underline{\nabla} \cdot (\mu\underline{u} \times \underline{\omega}) - \mu\underline{\omega} \cdot \underline{\omega}\}d\Omega. \tag{5.12}$$

6. Developing flow in a square straight duct

Developing laminar flow in a square straight duct is a popular and well documented test case for finite difference methods (Ponagare and Lakshminarayana. 1985). Due to symmetry only a quarter of the duct needs to be modelled. The mesh in the cross-section is a 6x6 regular grid. The mesh in the flow direction was divided into 50 equal steps corresponding to a distance of 8.333 duct heights. The total model therefore is comprised of 1800 elements. The Reynold's Number, based on inlet velocity and duct height, was taken as 150.

A uniform velocity $u = 1.0$ is specified at the inlet. The follwing values are specified for the Clebsch variables at the inlet; $\gamma = 0$, $H' = 0$. In this particular example as the inlet is iso-energetic it is not necessary to solve for the variables α and η. The velocities u, v, and w on the solid wall are specified equal to zero to enforce the no-slip condition. The symmetry conditions are required for the $\underline{\beta}$ variables. The ϕ variable is specified to be zero at one internal node. The downstream velocity boundary condition is satisfied iteratively. A uniform irrotational starting value is specified. This uniform value is chosen to satisfy conservation of mass. In the present example an iterative velocity of $u = 1$ is specified downstream.

The governing non-linear equations are solved iteratively. The starting solution is taken from an irrotational flow. The velocity boundary conditions for this inital solution are determined by computing a uniform velocity of the same mass as the fixed inlet velocity. In this example this lead to an inital velocity field of $u = 1.0$ and $v = w = 0.0$. The stability of the non-linear iteration is controlled by the specification of relaxation factors for the primary variables. In this example a relaxation factor of 0.05 was specified for all variables .

Contours of the developed axial velocity profile are show in Fig. 1. Even on a relatively crude mesh good accuracy is obtained. The development of the axial velocity in the streamwise directions is illustrated in Fig. 2.

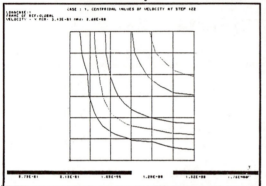

Fig. 1: Outlet Axial Velocity   Fig. 2: Development of Axial Velocity

References

Akay, H. U. and Ecer, A. (1982). Application of a Finite Element Algorithm for the Solution of Steady Transonic Euler Equations. AIAA Paper 82-0970. AIAA/ASME 3rd Joint Thermophysics, Fluids, Plasma and Heat Transfer Conference, St. Louis, Missouri, 7-11 June, 1982. Also published in *AIAA Journal*, volume 21, No. 11, 1518 (1983).

Clebsch, A. (1859). Über eine Allgemeine Transformation d. Hydrodynamichen. *J. Reine Agnew. Math*, 56.

Ecer, A. and Akay, H. U. (1982). A Finite Element Formulation of Euler Equations for the Solution of Steady Transonic Flows. AIAA Paper 82-0062. AIAA 20th Aerospace Sciences Meeting, Orlando, Florida. January 11-14, 1982. Also published in *AIAA Journal*, volume 21, No. 3, 343 (1983).

Ecer, A., Akay, H. U., and Senen, B. (1984). Solution of Three-Dimensonal Inviscid Rotational Flows in a Curved Duct. AIAA Paper 84-0032. AIAA 22nd Aerospace Sciences Meeting, Reno, Nevada. January 11-12, (1984).

Ecer, A., Rout, R. K., and Ward, P. (1983). Investigation of Solution of Navier–Stokes Equations using a Variational Formulation. *International Journal for Numerical Methods in Fluids*, 3:23–31.

Ponagare, M. and Lakshminarayana, B. (1985) A Space-Marching Method for Incompressible Navier–Stokes Equations. AIAA Paper 85-0170. AIAA 23rd Aerospace Sciences Meeting, January 14-17, Reno, Nevada.

Seliger, R. L. and Whitham, G. B. (1968). Variational Principles in Continuum Mechanics. *Proceedings of Royal Society A.*, 305:1.

Serrin, J. (1959). *Mathematical Principles of Classical Mechanics*, volume VIII/I of *Handbuch der Physic.* Springer-Verlag, Berlin.

A comparison of multigrid methods for the incompressible Navier–Stokes equations

S. Sivaloganathan, G.J. Shaw, T.M. Shah, and D.F. Mayers

ICFD, Oxford University Computing Laboratory, Oxford

1. Introduction

Sivaloganathan and Shaw (1988b) introduced a multigrid pressure correction algorithm (MGPC) and in Shaw and Sivaloganathan (1988a) presented a local mode analysis of the underlying smoothing procedure - the SIMPLE algorithm. Their theoretical analysis and practical results demonstrated that the smoothing procedure damped high frequency error components sufficiently to enable the construction of an efficient multigrid method. In a series of papers Vanka (1986), Gaskell and Wright (1986) and others have presented a Block Implicit Multigrid algorithm (BLIMM) using a symmetric coupled Gauss - Seidel algorithm (SCGS) as the smoothing procedure and have demonstrated its efficacy on a number of standard test problems. However, to date, no Fourier analysis of the procedure has been presented.

The purpose of this paper is twofold : firstly to compare the theoretical predictions of a local mode analysis of the two smoothing procedures and secondly to compare the practical performance of the two methods on a standard test problem.

2. Governing equations and discretisation

The equations expressing conservation of mass and momentum in two dimensions for an ideal, incompressible, Newtonian fluid are given by:

$$\frac{\partial \rho u^2}{\partial x} + \frac{\partial \rho uv}{\partial y} = -\frac{\partial p}{\partial x} + \frac{\partial}{\partial x}\left(2\mu\frac{\partial u}{\partial x}\right) + \frac{\partial}{\partial y}\left[\mu(\frac{\partial u}{\partial y} + \frac{\partial v}{\partial x})\right] \qquad (2.1)$$

$$\frac{\partial \rho v^2}{\partial y} + \frac{\partial \rho uv}{\partial x} = -\frac{\partial p}{\partial y} + \frac{\partial}{\partial y}\left(2\mu\frac{\partial u}{\partial y}\right) + \frac{\partial}{\partial x}\left[\mu(\frac{\partial u}{\partial y} + \frac{\partial v}{\partial x})\right] \qquad (2.2)$$

$$\frac{\partial \rho u}{\partial x} + \frac{\partial \rho v}{\partial y} = 0. \qquad (2.3)$$

The discretised equations are formed using a MAC-type staggered grid (see Sivaloganathan and Shaw (1988b)). There are several methods of devising finite difference approximations but to ensure that the scheme is conservative, a finite volume approach is adopted. Due to the staggering of the mesh, three different types of control volume are required for the two momentum and continuity equations in the interior, with straightforward modifications near the boundaries. A detailed description is given in Sivaloganathan and Shaw (1988b). The finite volume equations are then derived in the standard manner by integrating equations (2.1) - (2.3) over there respective control volumes assuming a linear variation between nodes for the dependent variables but constant fluxes over each control volume surface. The source terms are assumed constant over each control volume. Hybrid differencing is used whereby artificial viscosity is added sufficient to keep the coefficients of the discretised equations positive. This procedure is discussed in some detail in Sivaloganathan and Shaw (1988b) and simply results in a change from central to donor cell/upwind differencing of the convection term plus neglect of diffusion in the direction considered whenever the appropriate cell Reynolds number is greater than two.

The multigrid components of grid coarsening, restriction, prolongation and smoothing operators are summarised below. Further details can be found in Sivaloganathan and Shaw (1988b) for MGPC and the corresponding details for BLIMM can be found in Shah (1987).

3. Multigrid components

3.1 Grid coarsening

Both multigrid methods have been programmed using *continuity control volume lumping* as presented in Sivaloganathan and Shaw (1988b). In short, we ensure that each coarse grid continuity control volume is composed of four fine grid continuity control volumes. This ensures that the discrete compatibility condition is enforced on all grids, which is crucial to obtaining optimal multigrid convergence rates.

3.2 Restriction and prolongation

For MGPC and BLIMM, the restricted coarse grid velocities are defined to be the mean of their two nearest neighbouring fine grid velocities. Coarse grid pressures are defined to be the mean of the four neighbouring fine grid pressures. Prolongation operators are derived in all cases using bilinear interpolation.

3.3 Smoothing procedures

MGPC uses the SIMPLE pressure correction method as the smoothing procedure. This has been described in detail in Sivaloganathan and Shaw

(1988b) and we shall not dwell on it here. BLIMM uses SCGS as the underlying smoothing procedure (see Gaskell et al. (1986) , Shah (1987), Vanka (1986)). Briefly, the finite difference equations are solved simultaneously by a point Gauss-Seidel procedure. The grid is scanned in a predetermined manner , and for each continuity control volume, the momentum equations corresponding to the velocities on all four sides of the control volume and the continuity equation are solved in a coupled fashion. The equations are linearised about old values and a set of five linear equations for the corrections is solved by inverting a bordered matrix. A more comprehensive account of the algorithmic details of SCGS can be found in Shah (1987) and Vanka (1986) . Thus in essence the methods differ only in their choice of smoothers.

A smoothing analysis for SIMPLE has already been presented in Shaw and Sivaloganathan (1988a). The analysis of SCGS proceeds in basically the same manner. The method is treated as a two stage process where each velocity is updated twice and the pressure once.

4. Fourier analysis of smoothing procedures

In this section a local mode analysis is used to examine SIMPLE and SCGS as smoothers. Since the reduction of high frequency error components is essentially a local process, the analysis of this reduction need not take account of distant boundaries. Consider an arbitrary local section of the mesh with the u, v velocities staggered along the x and y directions respectively and the pressure p located at nodes. Assume that at the start of the iteration, the errors in u, v, p are given by :

$$
\begin{bmatrix} e^u \\ e^v \\ e^p \end{bmatrix} = \sum_\theta \begin{bmatrix} e^u_\theta \\ e^v_\theta \\ e^p_\theta \end{bmatrix} \tag{4.1}
$$

and that $\theta = (\theta_1, \theta_2)$ components of the errors are defined by:

$$
\begin{bmatrix} e^u_\theta \\ e^v_\theta \\ e^p_\theta \end{bmatrix} = \begin{bmatrix} \alpha^u_\theta \\ \alpha^v_\theta \\ \alpha^p_\theta \end{bmatrix} \exp(i\theta.x/h) \tag{4.2}
$$

where $\theta.x/h = (\theta_1 x + \theta_2 y)/h$. After the first stage of SIMPLE (momentum relaxations), the error amplitudes become:

$$
\begin{bmatrix} \dot{e}^u_\theta \\ \dot{e}^v_\theta \\ \dot{e}^p_\theta \end{bmatrix} = \begin{bmatrix} \dot{\alpha}^u_\theta \\ \dot{\alpha}^v_\theta \\ \dot{\alpha}^p_\theta \end{bmatrix} \exp(i\theta.x/h). \tag{4.3}
$$

After the pressure correction stage:

$$
\left[
\begin{array}{c}
\ddot{e}^u_\theta \\
\ddot{e}^v_\theta \\
\ddot{e}^p_\theta
\end{array}
\right]
=
\left[
\begin{array}{c}
\ddot{\alpha}^u_\theta \\
\ddot{\alpha}^v_\theta \\
\ddot{\alpha}^p_\theta
\end{array}
\right]
\exp{(i\theta.x/h)}.
\tag{4.4}
$$

Thus we calculate the amplification matrix A which is defined by:

$$
\left[
\begin{array}{c}
\ddot{\alpha}^u_\theta \\
\ddot{\alpha}^v_\theta \\
\ddot{\alpha}^p_\theta
\end{array}
\right]
= A
\left[
\begin{array}{c}
\alpha^u_\theta \\
\alpha^v_\theta \\
\alpha^p_\theta
\end{array}
\right]
= A_2 A_1
\left[
\begin{array}{c}
\alpha^u_\theta \\
\alpha^v_\theta \\
\alpha^p_\theta
\end{array}
\right]
\tag{4.5}
$$

where A_1, A_2 are the amplification matrices after momentum relaxation and pressure correction respectively. The smoothing factor is then given by:

$$
\bar{\mu} = \sup_{\theta \, \epsilon \, \mathcal{H}} \; [\rho(A)]
\tag{4.6}
$$

where $\mathcal{H} = [-\pi, \pi]^2/[-.5\pi, .5\pi]^2$ is the set of high frequencies. In the case of convection dominated flows we use:

$$
\bar{\mu} = \sup_{\substack{|u_0|,|v_0| < 1 \\ \theta \, \epsilon \, \mathcal{H}}} \; [\rho(A)]
\tag{4.7}
$$

where u_0, v_0 are frozen velocities used to linearise the problem and are constrained in order to maintain the relevant Reynolds number. We shall not go into the technical detail of the analysis for the SIMPLE pressure correction algorithm as this has been presented in Shaw et al. (1988a) . The analysis of SCGS proceeds along the same lines. We assume at the start of the smoothing process, that the errors in u, v, p are given by equations 4.1 and 4.2. Then, during the smoothing process, the singly corrected and fully corrected errors are defined by:

$$
\left[
\begin{array}{c}
\dot{e}^u_\theta \\
\dot{e}^v_\theta
\end{array}
\right]
=
\left[
\begin{array}{c}
\dot{\alpha}^u_\theta \\
\dot{\alpha}^v_\theta
\end{array}
\right]
\exp{(i\theta.x/h)}
\tag{4.8}
$$

and

$$
\begin{bmatrix} \ddot{e}_\theta^u \\ \ddot{e}_\theta^v \\ \ddot{e}_\theta^p \end{bmatrix} = \begin{bmatrix} \ddot{\alpha}_\theta^u \\ \ddot{\alpha}_\theta^v \\ \ddot{\alpha}_\theta^p \end{bmatrix} \exp\left(i\theta.x/h\right) \tag{4.9}
$$

where $\dot{e}, \ddot{e}$ denote singly and fully corrected values. Notice that u and v are each corrected twice, but the p only once. After some technical manipulation it can be shown that (see Shah (1987)):

$$
\begin{bmatrix} \dot{e}_\theta^u \\ \dot{e}_\theta^v \\ \ddot{e}_\theta^u \\ \ddot{e}_\theta^v \\ \ddot{e}_\theta^p \end{bmatrix} = \begin{bmatrix} M_1 \\ M_2 \end{bmatrix} \begin{bmatrix} e_\theta^u \\ e_\theta^v \\ e_\theta^p \end{bmatrix} \tag{4.10}
$$

where M_1 is a two by three complex matrix and M_2 is a three by three complex matrix. Thus the smoothing procedure can be written:

$$
\begin{bmatrix} \ddot{e}_\theta^u \\ \ddot{e}_\theta^v \\ \ddot{e}_\theta^p \end{bmatrix} = M_2 \begin{bmatrix} e_\theta^u \\ e_\theta^v \\ e_\theta^p \end{bmatrix}. \tag{4.11}
$$

5. Comparison of theoretical and practical smoothing rates

To obtain practical smoothing rates, both multigrid methods were applied to the driven cavity problem (for problem specification and details see Shah (1987), Shaw et al. (1988b)). A more detailed account of the theoretical smoothing capabilities of each of the methods can be found in Shah (1987) and Sivaloganathan and Shaw (1988a); and a presentation of the numerical solutions obtained and a comparison with well-established results for this test problem are given in Shah (1987) and Shaw et al. (1988b). Local mode analysis is clearly a good predictor of practically obtainable smoothing rates as illustrated in Figs. 1 and 2. These figures also show the consistently better smoothing rates obtained using BLIMM as opposed to MGPC. Tables 1 and 2 confirm that the coupled strategy does result in a saving in CPU time over MGPC. However, at the higher Reynolds numbers (5000, 10000) and the finest mesh level (66 x 66), the two strategies take roughly the same CPU time on a DEC MicroVAX. The number of multigrid iterations taken to solve to the same accuracy demonstrates the fact that each BLIMM iteration is significantly more expensive than an MGPC iteration.

Table 1: Number of multigrid iterations to reduce residual norm by 10^{-4}.
(Figures in parentheses are CPU times on a DEC Microvax)

MGPC

Finest Grid	2	3	4	5	6
Re	6x6	10x10	18x18	34x34	66x66
1	6	7	8	7	6
($\omega = .5$)	(2s)	(7s)	(32s)	(1m 53s)	(6m 18s)
100	11	10	10	13	16
($\omega = .35$)	(3s)	(11s)	(41s)	(3m 33s)	(17m 15s)
400	12	12	14	16	17
($\omega = .4$)	(4s)	(13s)	(56s)	(4m 14s)	(18m 5s)
1000	17	16	19	22	24
($\omega = .25$)	(5s)	(16s)	(1m 15s)	(5m 45s)	(25m)
5000	24	28	34	35	35
($\omega = .15$)	(7s)	(32s)	(2m 37s)	(11m 45s)	(36m 1s)
10000	24	27	32	42	42
($\omega = .15$)	(7s)	(31s)	(2m 28s)	(12m 49s)	(43m 14s)

Table 2: Number of multigrid iterations to reduce residual norm by 10^{-4}.
(Figures in parentheses are CPU times on a DEC Microvax)

BLIMM

Finest Grid	2	3	4	5	6
Re	6x6	10x10	18x18	34x34	66x66
1	4	5	5	5	4
($\omega = .7$)	(1s)	(6s)	(24s)	(1m 20s)	(6m 7s)
100	7	5	6	6	4
($\omega = .7$)	(2s)	(6s)	(26s)	(1m 36s)	(6m 7s)
400	6	7	9	9	7
($\omega = .6$)	(1s)	(7s)	(36s)	(3m)	(10m 42s)
1000	8	8	9	10	8
($\omega = .5$)	(2s)	(9s)	(36s)	(3m 10s)	(17m 31s)
5000	8	9	11	18	18
($\omega = .5$)	(2s)	(10s)	(44s)	(5m 20s)	(32m 18s)
10000	10	11	14	22	22
($\omega = .5$)	(3s)	(12s)	(57s)	(6m 35s)	(45m)

Sivaloganathan et al.

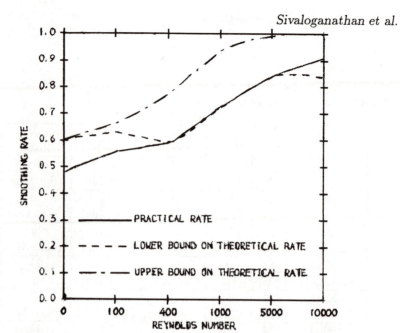

Fig. 1: Theoretical and practical smoothing rates for SIMPLE

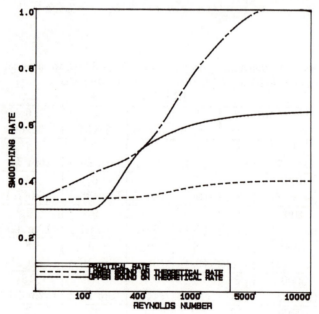

Fig. 2: Theoretical and practical smoothing rates for SCGS

6. Conclusion

A theoretical and practial comparison of the smoothing capabilities of SIM-PLE and SCGS has been presented. The theoretical analysis has shown that h-independent convergence rates are attainable by multigrid methods for the incompressible NSE, and such convergence rates have been achieved in practise by MGPC and BLIMM. In both cases, the analysis has been compared with the practical behaviour of the method and found to be an accurate predictor of convergence rates (see figures 1 and 2). The overall conclusion, given the limited numerical comparison carried out to date, must be that the coupled BLIMM strategy is a more efficient and faster procedure than MGPC; but whether this will carry over to higher Reynolds numbers and finer grids is perhaps debatable. However, it must be stated that BLIMM is less sensitive to relaxation parameters used in the smoothing procedure and has consistently better smoothing rates than MGPC. Against this must be balanced the ease with which MGPC can be incorporated into existing industrial codes, a great many of which are based on pressure correction iterative procedures.

Acknowledgements

The authors are grateful to Rolls Royce , Derby plc. and the Science and Engineering Research Council for financial support during the course of this research and to María D. Cantero for all her help.

References

Gaskell, P. and Wright, N. (1986). Multigrids applied to an efficient fully coupled solution technique for recirculating flow problems. *IMA Conference on Simulation and Optimisation of Large Systems.*

Shah, M. (1987). Analysis of multigrid methods. MSc. dissertation, University of Oxford.

Shaw, G. and Sivaloganathan, S. (1988). On the smoothing properties of the simple pressure correction algorithm. *International Journal for Numerical Methods in Fluids*, 8 : 441–462

Sivaloganathan, S. and Shaw, G. (1988). A multigrid method for recirculating flows. *International Journal for Numerical Methods in Fluids*, 8 : 417–440.

Vanka, S. (1986). Block implicit multigrid solution of the Navier–Stokes equations in primitive variables. *Journal of Computational Physics*, 65 : 138–158.

Multigrid calculations of jet flows

S. A. E. G. Falle and M. J. Wilson

School of Applied Mathematical Studies

The University of Leeds

1. Introduction

This paper describes numerical calculations of two types of steady jet flow,
and axisymmetric underexpanded jet and a wall jet. Both of these flows
are primarily supersonic, containing complicated patterns of shock and
rarefaction waves, although there are also significant embedded subsonic
regions.

Although our approach was to model these jets by using the inviscid
Euler equations, i.e. we ignored the effects of turbulence and combustion,
our results are of practical interest since such flows are good approximations to those found downstream of jet and rocket exhausts in a variety of
situations (e.g. VTOL aircraft, spacecraft exhaust and attitude thrusters).

1.1 The jet flow patterns
1.1.1 Underexpanded jets

The first type of jet flow is set up when a supersonic gas jet emerges from a
nozzle and exhausts into a reservoir containing gas at a different pressure.
For inviscid flow there are three dimensionless parameters in the problem,
the adiabatic index γ of the gas, the exit Mach number of the jet M_e and
the exit pressure ratio $P_r =$ jet pressure/ambient pressure. If $P_r < 1$ then
the jet is said to be overexpanded, whereas if $P_r > 1$ the jet is said to be
underexpanded.

Here we present results for an axisymmetric underexpanded jet ($P_r =$
4.1, $M_e = 1.0$, $\gamma = 1.4$). Fig. 1 shows the numerical flow pattern for such
a jet. Note the characteristic pattern of shock and rarefaction waves that
are present in the flow, and in particular incident shock, Mach disc and
reflected shock. The flow contains a subsonic region downstream of the
Mach disc and so, unlike a wholly supersonic flow, cannot be computed by
marching in the streamwise direction.

1.1.2 Wall jets

The second example of a jet we shall discuss is that found when a supersonic
jet impacts on a flat plate. Fig. 2 shows the numerical flow pattern we have

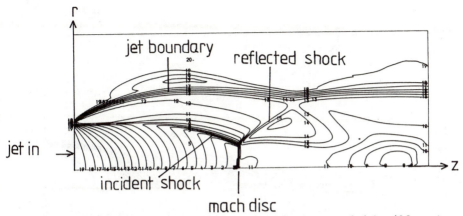

Fig. 1: Logarithmic contours of density for an underexpanded jet ($M_e = 1$, $P_r = 4.1$, $\gamma = 1.4$). Maximum density = 1.02, minimum density = 0.064. This calculation was carried out on a computing grid with 256 × 96 cells in the axial and radial directions respectively

calculated for such a jet (for a fuller description of walljets see (Lamont and Hunt 1980) and the references therein).

In the present case, the wall jet is set up when an axisymmetric supersonic jet ($M_e = 2.41$, $P_r = 1$, $\gamma = 1.4$) impinges normlly on a flat plate. There is obviously a shock across the jet just above the surface of the plate, behind which is a region of high pressure gas. Post-shock, gas is accelerated sideways by a rarefaction wave and forms a layer of gas moving supersonically along the surface of the plate. It is this layer which is termed a wall jet. Note that the gas in the layer is, in general, not pressure matched to its surroundings, so that there is a repeating pattern of shock and rarefaction waves, similar to those found in the supersonic jets of section 1.1.1.

2.   The numerical method

2.1   The iterative scheme

The jet flows we are considering are approximated as stationary solutions to the Euler Equations. They contain complicated wave patters within them so that our method for modelling these flows was to use a shock capturing technique designed for unsteady gas dynamcis: a first-order, explicit, Godunov finite-volume scheme. This code was used by starting from a suitable set of initial conditions, imposing appropriate boundary conditions and then iterating until the desired stationary numerical solution was reached. In other words a steady flow was achieved by a time evolution of the gas towards the steady state. Fig. 1 shows the flow for the underexpanded jet and Fig. 2 that for the wall jet.

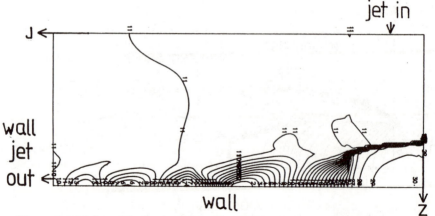

Fig. 2: 30 logarithmic contours of pressure for an axisymmetric wall jet ($M_e = 2.41$, $P_r = 1$, $\gamma = 1.4$) impinging normally upon a flat plate. Maximum pressure = 7.8, minimum pressure = 0.31. This calculations was carried out on a computing grid with 64 × 160 cells respectively in the axial and radial directions

2.2 Multigridding

Because of the complex nature of the flow, it is necessary to use high resolution in order for the numerical scheme to capture all the features of the flow pattern. However, this meant that the amount of work required to obtain converged solution was very large. We therefore employed a multigrid to accelerate the convergence. The scheme we employed was a FAS multigrid (Brandt 1977), the details of which will be described elsewhere. Here we confine ourselves to those aspects of the method that are novel or which require some explanation such as the definition of the residuals, the adaptive multigrid stategy and the criterion for creating a new fine grid.

2.2.1 Residuals

The time-dependent Euler Equations are of the form

$$\frac{\partial \mathbf{u}}{\partial t} + L(\mathbf{u}) = 0, \tag{2.1}$$

where $\mathbf{u}$ is the state vector for the system, and $L(u)$ is a non-liniear differential operator. The discrete version of these equations, which we aim to solve numerically, are of the form

$$\frac{\partial \mathbf{u}_n}{\partial t} + L_n(\mathbf{u}_n) = 0, \tag{2.2}$$

where $\mathbf{u}_n$ is a grid function (the discrete approximation to $\mathbf{u}$) defined on a grid G_n, and $L_n()$ is a discrete approximation to $L()$ on G_n. Now, we seek a grid function $\mathbf{u}_n^\infty$ which satisfies

$$L_n(\mathbf{u}_n^\infty) = 0, \tag{2.3}$$

i.e. the steady solution to the discrete equations. Therefore, if $\mathbf{u}_n^k$ is the current approximation to $\mathbf{u}_n^\infty$, then

$$\frac{\partial \mathbf{u}_n^k}{\partial t} = -L_n(\mathbf{u}_n^k) \equiv \mathbf{R}_n^k \tag{2.4}$$

can be identified as the residual. In other words, the residual is the (discretised) time rate of change of the current approximation to the steady state.

2.2.2 The adaptive strategy

Since the approach of the numerical solution to the steady state is controlled by the dissipation of sound waves passing across the computing grid, it is not possible to predict beforehand the most efficient way of cycling between grids during the multigrid solution process. However, we have used a technique that detects which wavelength regime is is making the dominant contribution to the error and then reduces its amplitude.

The size of the residual on a particular grid G_m, as defined above, is mainly determined by soundwaves of amplitude e_m that have a wavelength $\simeq h_m$ (the grid spacing on G_m). So we have approximately

$$\mathbf{R}_m^k = \frac{\partial \mathbf{u}_m}{\partial t} \propto e_m/h_m. \tag{2.5}$$

For an N-level multigrid ($G_1 = $ coarsest grid, $\dots$, $G_N = $ finest grid), a multigrid cycle is as follows:

(a) Restrict the solution from G_N to each of the coarser grids ($G_{N-1}, G_{N-2}, \dots, G_2, G_1$).

(b) Perform a single iteration on each grid and calculate the quantity $\| R_m \| h_m$.

(c) Note that the grid with the largest value of $\| R_m \| h_m$ (say G_m) is, in some sense, the grid which is most efficient at destroying the dominant frequency component of the error. Therefore, iterate on G_m until the residual has been reduced by an appropriate fraction.

(d) Prolong the correction onto G_{m+1}, and iterate on G_{m+1} until the solution on that grid is smooth. Then prolong the correction onto G_{m+2} and smooth etc.

(e) Repeat until G_N is reached. This constitutes one multigrid cycle. Our multigrid strategy could thus be described as an adaptive V-cycle.

2.2.3 New fine grid

Into the multigrid process we have incorporated the following procedure for creating a new finest grid. Say G_N is the current finest grid at some point in the solution process. We interpolate the solution onto a new fine grid G_{N+1} (with twice the resolution) and continue the solution process as a $N + 1$ level multigrid when

$$\| \mathbf{R}_n \| < \frac{1 - \frac{1}{2^p}}{2^d - 1} \| \tau_N^{N+1} \|, \tag{2.6}$$

whee p is the order of the iterative shcem, d is the dimensionality of the problem, τ_N^{N+1} the relative truncation error of the solution G_N with respect to G_{N-1}. This criterion is a rough guide as to when working on G_{N+1} is more efficient than working on G_N at approaching the true solution (Brandt 1982).

3. Results

3.1   Numerical flow pattern

The numerical flow pattersn calculated using our multigridding technique are shown in Figs. 1 and 2. These contour plots are indistinguishable from those calculated on a single grid with a resolution equal to that of the finest level of the multigrid. We used a 5 level multigrid for both the underexpanded jet and the wall jet. The CPU time required was about 3600 s and 600 s respectively on an Amdahl 580.

3.1.1   Convergence properties

In order to compare the convergence properties of the multigrid with a solution obtained by iterating on a single grid, a bench-mark solution was calculated. This was essentially the steady solution on the finest grid, or as close to it as we could practically manage. The bench-mark was calculated by using a combination of multigridding and single grid iterations (so as not to favour any one solution process in the error comparison). Fig. 3 shows the variation of the error (defined as $\|$ current solution - benchmark solution $\|$) with work for both the multigrid and the single grid solutions in the case of the underexpanded jet. Fig. 4 shows the variation of error with work in the case of the wall jet. A work unit is defined as the computational work expended in one iteration on level 1.

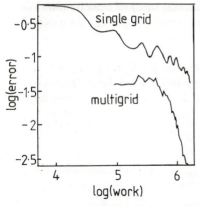

Fig. 3: The variation of the $\log_{10}$(error) with $\log_{10}$(work) for multigrid and single grid calculations of the underexpanded jet shown in Fig. 1. The multigrid had five levels, with the resolution of the finest level equal to that of the single grid calculation (256×96)

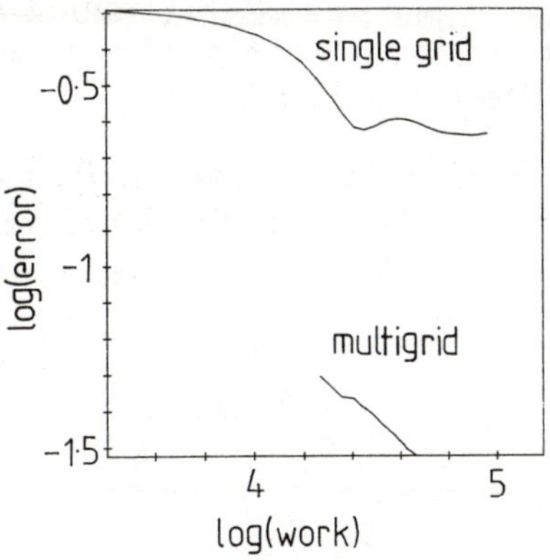

Fig. 4: The variation of $\log_{10}$(error) with $\log_{10}$(work) for multigrid and single grid calculations of the wall jet shown in Fig. 1. The multigrid had five levels, with the resolution of the finest level equal to that of the single grid calculation (64×160)

4. Conclusion

The multigrid is much more efficient at reducing the error than the single grid iteration. However, it is clear that we have not achieved h independent convergence. This is hardly surprising since the numerical solution is still changing significanlty with resolution between levels 4 and 5. One would therefore not expect the efficiency of the multigrid to reach its asymptotic value. Nevertheless, at this resolution the multigrid solver reduces the computational cost by about a factor of ten.

Finally the multigrid and single grid calculations give the same steady solution for these problems and both agree extremely well with experimental results (Love et al. 1959; Lamont and Hunt 1980), at least for the first shock cell.

References

Brandt, A. (1977). Multi-level adaptive solutions to boundary value problems. *Mathematics of Computation*, 31:333–390.

Brandt, A. (1982). Guide to Multigrid Development. In *Multigrid Methods*, Hackbusch, W. and Trottenberg, U., editors, pages 220–312. Springer-Verlag.

Lamont, P. J. and Hunt, B. L. (1980). The impingement of underexpanded axisymmetric jets on perpendicular and inclined flat plates. *Journal of Fluid Mechanics*, 100:471–511.

Love, E. S., Grigsby, C. E., Lee, L. P., and Woodling, M. J. (1959). Experimental and theoretical Studies of Axisymmetric Free Jets. Technical Report R-6, NASA.

The accurate approximation and economic solution of steady-state convection dominated flows

P. H. Gaskell, A. C. K. Lau, and N. G. Wright

The University of Leeds, U.K.

1. Introduction

The increased reliance on computers as a means of predicting complex fluid flow phenomena is placing ever greater demands on theoreticians to produce (a) more accurate discrete approximations to the governing equations of motion and (b) economically viable solutions to large systems of algebraic equations. In this paper we report some recent progress in these two key areas.

Our approach has been to promote the use of high-order approximations for the nonlinear convective transport terms present in such systems and to adopt a multigrid algorithm as a convergence accelerator. Although the results presented are for an ideal test problem they are nevertheless encouraging from the point of view of extending the method to examine recirculating flows of a turbulent or chemically reactive nature.

2. Numerical formulation

2.1 Discretisation

In line with common engineering practice a primitive variable formulation is used for the governing equations which are written in a discrete form using a staggered grid arrangement (Harlow and Welch 1965) and control volume formulation. Gaskell and Lau's high-order Curvature Compensated Convective Transport (CCCT) approximation is used to model the nonlinear convection terms. Fortunately, for the test problem considered here it was not found necessary to implement the CCCT scheme fully, just the generic form for the approximation to the value of some transport variable ϕ at the bounding faces of a control volume. So, for a control volume centered at the node $(i-1, j)$, on a two-dimensional uniform grid (with $u_{i-\frac{1}{2},j} > 0$), ϕ at the right hand face is given by :

$$\phi_{i-\frac{1}{2},j} = \left(\frac{3}{4} + 2\alpha\right)\phi_{i-1,j} + \left(\frac{3}{8} - \alpha\right)\phi_{i,j} - \left(\frac{1}{8} + \alpha\right)\phi_{i-2,j} \qquad (2.1)$$

which has a leading truncation error term of $O(\alpha h^2)$, where h is the mesh spacing and α is a parameter, $\alpha = 0$ representing maximum accuracy. Other well known approximations to convection can be generated by selecting the appropriate α (Gaskell and Lau 1988). The diffusion terms are approximated in the standard way using central differencing.

2.2 Solution strategy

2.2.1 Smoothing technique (SCGS)

Segregated or decoupling procedures have proved the most popular approach for solving discrete approximations to the Navier-Stokes equations: SIMPLE is perhaps the most well known of these and has been shown to possess good smoothing properties (Shaw and Sivaloganathan 1988). However, the obvious drawback of such an approach is that it weakens the coupling between the velocity and pressure fields (Raithby and Schnieder 1979). Consequently we have chosen to implement the Symmetric Coupled Gauss Siedel (SCGS) technique (Vanka 1986) in our computations. It is a simple unsegregated approach offering the combined advantage of a low operational count and minimal storage requirement.

For two-dimensional flows the SCGS smoother requires four velocities and the pressure to be updated simultaneously by inverting a 5×5 matrix for a series of contiguous control volumes spanning the solution domain. This matrix is doubly bordered, diagonal and sparse, and can be solved 'exactly' using a form of LU-decomposition. During iteration processes it is necessary to relax updated values as a consequence of the nonlinearity of the algebraic system of equations and the use of old values for **u** and p when the matrix coefficients and residuals are evaluated. Accordingly, the updated values of the velocity and pressure are multiplied by an appropriate relaxation factor once they have been determined (Gaskell and Wright 1988).

The standard method of sweeping through the grid (proposed by Vanka) is to solve for each control volume, first in the direction i-increasing and then in the direction j-increasing — consequently, each velocity is updated twice. It was observed that this ensured the stability lacking with a single update. However, we have taken this a stage further, opting to perform four sweeps per iteration — (i) i-increasing then j-increasing, (ii) i-decreasing then j-decreasing, (iii) j-increasing then i-increasing and finally (iv) j-decreasing then i-decreasing. This approach was found to be more efficient as a method of smoothing errors.

2.2.2 Multigrid method

Brandt (1981) has shown that the method of multigridding can be used to great effect to significantly reduce solution times. The method entails the use of a hierarchy of grids, of different mesh sizes, to solve the fine grid problem. The problem is set up on coarser grids which are used to calculate corrections to the fine grid solution. Each of these grids is very effective at eradicating a particular wavelength component of the error.

The problem considered in the next section is highly nonlinear and is solved by employing a Full Approximation Storage (FAS) algorithm, the finer points of which are well documented elsewhere (Brandt 1977) and therefore we confine our attention to those features of our approach that are different and warrant additional explanation.

Several strategies have been proposed and implemented for cycling between grids in order to smooth errors efficiently. However, the one employed is geared to home in on the grid with the largest residual, starting the smoothing cycle there; this approach has been used to great effect by Falle and Wilson (1988). In performing one FAS multigrid cycle the following steps are taken :

- A restricted solution to the fine grid problem is set up on each of the grids.

- The residuals of the solution of the restricted problem on each grid is calculated. The level with the highest residual is selected for smoothing.

- The solution is smoothed on this grid until the error has been reduced by a factor γ. This factor can be varied to give a fast rate of convergence. The final and initial solutions are used to calculate a correction.

- This correction is prolonged onto the next finer grid and added to the current solution there which is then smoothed until the error has been reduced by a factor γ.

- The last two steps are repeated until the solution on the finest grid has been corrected and smoothed.

The value of γ should be chosen such that the majority of smoothing is carried out on the coarse grids. It was found that by giving γ a value of 0.001 on the coarsest grid, thus providing an accurate solution to the problem there, led to improved convergence when compared with results reported in an earlier study (Gaskell, Lau and Wright 1988): γ was assigned the value 0.5 on the remaining grid levels.

2.3 Boundary conditions, interpolation, convergence criterion

In transfering information between grids linear interpolation is used for the restriction of velocities and bilinear interpolation for the restriction of scalar variables. The boundaries are not restricted here as they are assigned a fixed value. Prolongation is also carried out by linear and bilinear interpolation.

The grid configuration employed requires that special care be taken when prolonging the solution at nodes adjacent to the boundary. If the boundary values alone are used then prolongation can make the method unstable for high Reynolds number flows. It is better to use a zero derivative condition for these points setting them equal to their value on the first coarse grid line. However, this in itself is not as accurate as one might hope and so, after each prolongation, all boundary adjacent nodal points are updated after the correction has been applied and before the current fine grid iteration. It has been observed that such measures can reduce cpu time by a factor ∼30% and ensure stability. In the present study the first order upwind approximation is applied at the boundaries of the solution domain and the computational grids quoted in the text refer to internal nodes only.

The results presented in the next section are for converged solutions such that the residual norm $\| r \|$ given by :

$$\| r \| \;=\; \frac{\left(\sum_{i,j} \left((r_{\mathbf{u}})_{i,j}^2 + (r_\rho)_{i,j}^2 \right) \right)^{1/2}}{NV \times iN \times jN} \qquad (2.2)$$

is constrained, on the finest mesh, to be less than 10^{-5}. Here $r_{\mathbf{u}} = (r_u, r_v)$ are the residuals for the velocity components, r_ρ is the mass residual, NV is the total number of variables being solved for and iN and jN represent the number of nodes in the i and j directions, respectively.

3. Test problem and results

Laminar flow in a lid driven cavity represents an ideal test problem for evaluating the relative merits of the numerical procedure described above. Results are presented for flow at Reynolds numbers of 100 and 1000.

Table 1 lists the total cpu time and number of fine grid work units (FGWU's) required to obtain solutions with seven grid levels for the flows under consideration. The latter is a measure of the total work done on all grids expressed in terms of the work required for one fine grid iteration.

GRID	Re = 100			Re = 1000		
	SCGS	SCGS (mg)		SCGS	SCGS (mg)	
	cpu	cpu	FGWU	cpu	cpu	FGWU
4	0.05	0.10	4.00	0.15	0.21	17.00
8	0.03	0.47	9.99	2.34	2.05	54.25
16	2.03	1.50	8.69	15.45	5.93	39.06
32	16.44	4.32	6.36	73.26	19.11	31.14
64	200.16	15.37	5.73	710.44	50.41	20.13
128	(2436.98)	58.30	5.44	(6889.50)	157.80	15.52
256	(29670.73)	229.23	5.36	(66811.06)	557.56	13.70

Table 1. Comparison of solution times (cpu seconds - Amdahl 5860) obtained with both ordinary and multigridded versions of SCGS, for flows at Reynolds numbers of 100 and 1000. The results in brackets have been projected on the basis of the power relationship (3.1).

It is clear from this table that the cpu time is considerably less in the multigrid case. On the finest grid the cpu times for the multigrid solution are around 1% of those one might expect to obtain with standard SCGS — even on the coarsest grids the savings are significant. These results can be examined further to reveal some additional information.

In the multigrid case it can be seen that cpu times increase, between grids, by a factor less than four — this is borne out by inspecting the FGWU columns. One FGWU on a grid of size h should ideally be equivalent to four FGWU's on a grid of size $2h$, and so if the power law relationship :

$$cpu \propto N^\beta \qquad (3.1)$$

is obeyed, where N is the total number of grid points on that level, the FGWU counts on each level will be identical. In fact it can be seen that they decrease and therefore the relationship is more than satisfied. Satisfaction of the relationship (3.1) is equivalent to realising h-independent convergence.

Fig. 1 shows contour plots of the vorticity and streamfunction obtained on 256 × 256 grid for flow at a Reynolds number of 1000. They are in excellent agreement with the predictions of other authors, see for example Ghia, Ghia and Shin (1982).

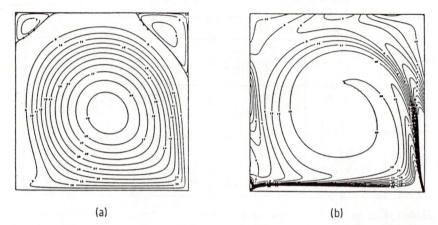

(a)   (b)

Fig. 1: (a) Streamfunction and (b) vorticity contours for flow in a two-dimensional lid driven cavity at Reynolds number 1000, obtained on a 256 × 256 grid (bottom boundary moving from left to right)

4.   Conclusions

An efficient SCGS/FAS multigrid technique has been presented which produces very accurate solutions, on fine grids, to the problem of flow in a lid driven cavity. A high order approximation has been employed to model the convection terms in the governing equations of motion and results show that the multigrid strategy achieves h-independent convergence. Work is presently underway to further develop this approach for the investigation of more complex flow situations.

References

Brandt, A. (1977). Multi-level adaptive solutions to boundary value problems. *Mathematics of Computation*, 31 : 333–390.

Brandt, A. (1982). Guide to multigrid development. In *Multigrid methods*, Hackbusch, W. and Trottenberg, U., editors, pages 220–312. Springer-Verlag.

Falle, S. A. E. G. and Wilson, M. J. (1988). Multigrid calculation of jet flow. In *ICFD Conference on Numerical Methods for Fluid Dynamics*. IMA Conference Series, Oxford University Press.

Gaskell, P. H. and Lau, A. K. C. (1988). Curvature compensated convective transport: SMART a new boundedness preserving transport algorithm. *International Journal for Numerical Methods in Fluids*. In press.

Gaskell, P. H., Lau, A. K. C. and Wright, N. G. (1988). Comparison of two solution strategies for use with higher order discretisation schemes in fluid flow simulation. *International Journal for Numerical Methods in Fluids*, 8:617–641

Gaskell, P. H., and Wright, N. G. (1988). Multigrids applied to a solution technique for recirculating fluid flow problems. In *Simulation and Optimisation of Large Systems*, Osiadacz, A., editor, pages 51–65. IMA Conference Series, Clarendon Press, Oxford.

Ghia, U., Ghia, K. N. and Shin, C. T. (1982). High Re solutions for incompressible flow using the Navier-Stokes equations and a multigrid method. *Journal of Computational Physics*, 43:387–411.

Harlow, F. H. and Welch, J. E. (1965). Numerical investigation of time-dependant viscous incompressible flow of fluid with a free surface. *Physics of Fluids*, 8:2182–2189.

Raithby, G. D. and Schneider, G. E. (1979). Numerical solution of problems in incompressible fluid flow : treatment of the pressure-velocity coupling. *Numerical Heat Transfer*, 2:417–440.

Shaw, G. J. and Sivaloganathan, S. (1988). On the smoothing properties of the SIMPLE pressure-correction algorithm. *International Journal for Numerical Methods in Fluids*, 8:441–461.

Vanka, S. P. (1986). Block-implicit multigrid solution of Navier–Stokes equations in primitive variables. *Journal of Computational Physics*, 65:138–158.

A 3D finite element code for industrial applications

J. P. Chabard and O. Daubert

Electricité de France

Chatou, France

1. Introduction

At EDF (the French Company for Electricity), the "Direction des Etudes et Recherches" is in charge of the numerical simulation of various kinds of flows. One domain of great importance is the computation of non isothermal isovolume flows because it is involved in the thermal-hydraulic studies of nuclear vessels. In most cases the geometries of these components are very complicated. Thus a 3D finite element code named N3S has been under development for some years. The use of this numerical method allows the treatment of complex shapes of solid boundaries and the refinement of the mesh where it is necessary.

This code has to be an industrial tool; hence we are looking for robustness, simplicity, efficiency, accuracy and reliability of the numerical algorithm. Of course, all these qualities have not yet been achieved, but they are underlying the choices made in our formulation.

In the first part of this paper, we describe the basic equations solved. Then more details are given concerning the numerical method implemented in the code and an industrial application will be presented.

2. Basic equations

The N3S code solves the Reynolds averaged turbulent Navier–Stokes equations for an incompressible Newtonian fluid. In a closed domain Ω whose boundary is Γ, these equations are:

$$\frac{\partial \mathbf{U}}{\partial t} + \mathbf{U}.\nabla \mathbf{U} = \frac{-1}{\rho}\nabla P + \nabla.\{(v+v_t)\nabla \mathbf{U}\} + \frac{\Delta \rho}{\rho}\mathbf{g} \qquad (2.1)$$

$$\nabla.\mathbf{U} = 0 \qquad (2.2)$$

where $\mathbf{U}$ denotes the velocity, P the pressure, ρ the density, $\mathbf{g}$ the gravity and v_t the eddy viscosity. These equations can be coupled to the energy equation:

$$\frac{\partial T}{\partial t} + \mathbf{U}.\nabla T = \nabla.\{(\lambda + \frac{v_t}{\sigma_T})\nabla T\} + \Phi \qquad (2.3)$$

where T denotes temperatute and Φ a source term.

A standard $k - \varepsilon$ model is used for turbulence modelling. Even if sometimes this model does not fit the physics well, it is very useful for the industrial applications because it has been fully validated and it is fully predictive when the standard set of constants is used (Launder and Spalding 1974). In this model, the eddy viscosity is related to the turbulent kinetic energy k and to its dissipation rate ε:

$$v_t = C_\mu \frac{k^2}{\varepsilon}. \qquad (2.4)$$

Two advection-diffusion equations govern the evolution of the turbulent quantities:

$$\frac{\partial k}{\partial t} + \mathbf{U}.\nabla k = \nabla.\{(v + \frac{v_t}{\sigma_k})\nabla k\} + \mathbf{P} + \mathbf{G} - \varepsilon \qquad (2.5)$$

$$\frac{\partial \varepsilon}{\partial t} + \mathbf{U}.\nabla \varepsilon = \nabla.\{(v + \frac{v_t}{\sigma_\varepsilon})\nabla \varepsilon\} + \frac{\varepsilon}{k}(C_{\varepsilon 1}(\mathbf{P} + (1 - C_{\varepsilon 3})\mathbf{G}) - C_{\varepsilon 2}\varepsilon).$$

Here $\mathbf{P}$ is a source term due to the shear stress and $\mathbf{G}$ is the production term induced by the temperature gradients. The standard set of constants is (Launder and Spalding 1974):
$C_{\varepsilon 1} = 1.44$, $C_{\varepsilon 2} = 1.92$, $C_{\varepsilon 3} = 0$, $C_\mu = 0.99$, $\sigma_k = 1.0$, $\sigma_\varepsilon = 1.3$, $\sigma_T = 1.3$,
$C_{\varepsilon 3} = 0$ if the thermal stratification is stable and $C_{\varepsilon 3} = 1$, if not.

Several kinds of boundary conditions are available in the code. The most useful are specified velocity, specified total normal stress, symmetry or wall conditions. These last boundary conditions, have to be emphasised because they are very important for the prediction of turbulent flows.

The natural boundary condition at the wall (vanishing velocity) cannot be applied for industrial simulations because of the very small size of the viscous sublayer compared to the size of the mesh. Therefore wall functions are used in order to take into account the effect of the wall on the flow. In the code we assume a logarithmic velocity profile near the wall:

$$\frac{\mathbf{U}.\tau}{U_*} = \frac{1}{K}\ln(\frac{yU_*}{v}) + C, \qquad (2.6)$$

where τ denotes a unit vector tangent to the wall. In fact the boundary of the computational domain is supposed to be at a distance y from the wall, and the logarithmic law gives the friction velocity U_* when the velocity is known (from the previous time step). Thus for the current time step, the boundary conditions are impermeability (essential boundary condition for on the normal component of velocity) and friction stress (natural boundary condition for the tangential velocities) conditions:

$$\mathbf{U.n} = 0 \tag{2.7}$$

$$(v + v_t)\frac{\partial}{\partial n}(\mathbf{U}.\tau) = -U_*^2, \tag{2.8}$$

where $\mathbf{n}$ is the outer normal. The boundary conditions for k and ε are deduced from the hypothesis of local equilibrium between production $\mathbf{P}$ and dissipation ε near the wall which gives:

$$k_W = \frac{U_*^2}{\sqrt{C_\mu}} \text{ and } \varepsilon_W = \frac{U_*^3}{K_y}. \tag{2.9}$$

3.   Time discretisation

The time discretisation is based on the fractional step method (Temam 1977). An approximation of the total time derivative is used in order to write the diffusion equations as:

$$\frac{C^{n+1} - \bar{C}}{\Delta t} - \frac{\partial}{\partial x_i}\{K_C\frac{\partial C^{n+1}}{\partial x_i}\} = \Sigma_C, \tag{3.1}$$

where C stands for T, k or ε and $\bar{C}$ is the result of pure advection:

$$\frac{\bar{C} - C^n}{\Delta t} + U_i^n\frac{\partial \bar{C}}{\partial x_i} = S_c. \tag{3.2}$$

In the same manner we obtain for the Navier–Stokes equations:

$$\frac{U_i^{n+1} - \bar{U}_i}{\Delta t} - \frac{\partial}{\partial x_i}\{(v + v_t^n)\frac{\partial U_i^{n+1}}{\partial x_i}\} + \frac{1}{\rho}\frac{\partial P^{n+1}}{\partial x_i} = \frac{\Delta\rho}{\rho}g_i \tag{3.3}$$

$$\frac{\partial U_i^{n+1}}{\partial x_i} = 0. \tag{3.4}$$

The main advantage of this method which is first order accurate in time is to keep a simple physical meaning for each fractional step. This leads us to choose for each of them a numerical method well suited to the single-type operator involved in each part.

4. Space discretisation

A standard Galerkin finite element method is used in the code. In the computational domain Ω a Taylor–Hood discretisation is used with continuous pressure from one element to another. Either parabolic (P2) or linear on sub-elements (IsoP2) approximation is used for the velocity and a linear approximation (P1) is used for the pressure. These choices satisfy the inf-sup (L.B.B) condition which insures the existence and uniqueness of the solution of the discrete Stokes problem (Hood and Taylor 1973).

5. Convection step

This stage is solved by a characteristics method. The advective field is supposed to be known (from the previous time step). So we have to solve equation (3.2) where C stands for the components of $\mathbf{U}$, the kinetic energy k and the dissipation rate ε, and S represents the source terms of the k and ε equations. The characteristics method is described in Benqué (1982) and Pironneau (1988) . This method has been shown to be unconditionally stable when the source terms vanish but when source terms are integrated along the characteristics, a semi-implicit treatment is necessary in order to improve the stability of the scheme. Moreover the quadratic interpolation of k and ε can generate overshoots and undershoots which give non-physical negative values of these quantities. When the quadratic interpolation is not monotone, a linear one is used on sub-elements.

6. Diffusion step

A classical finite element discretisation of the diffusion equation is used. It leads to a symmetrical linear system. In order to handle the very big matrices associated with a 3D industrial problem, a compact storage (non vanishing terms only) has been chosen. Thus the associated linear system is solved by a conjugate gradient method preconditioned by an incomplete Choleski factorisation of the matrix. Then the computational time and the memory requirement become nearly linear when dealing with 10,000 to 50,000 nodes.

7. Stokes problem

Let us consider the classical Stokes problem. When essential conditions are taken into account in the definition of the subspace V of $H_0^1(\Omega)$ and natural boundary conditions are treated in the right-hand side, the weak form of the Stokes problem can be written as follows: find $\mathbf{U}$ in V and P in $M = L^2(\Omega)/\mathbf{R}$ such that:

$$a(\mathbf{U}, \mathbf{N}) + b(\mathbf{N}, P) = l(\mathbf{N}) \; \forall \mathbf{N} \in V$$
$$b(\mathbf{U}, Q) = 0 \; \forall Q \in M, \qquad (7.1)$$

The classical choice for the variational test functions $\mathbf{N}$ is along the axis of the cartesian frame of reference. This choice leads to a natural uncoupling of the velocity components. However, in order to treat the impermeability boundary condition at the wall as an essential condition the test functions along the wall are in the tangential plane and this leads to coupling terms. Let U denote the velocities associated with the inner nodes and W those associated with the wall, then after space discretisation, the Stokes problem can be written in the following matrix form:

$$\begin{pmatrix} A_{UU} & A_{UW}^t & B_U^t \\ A_{UW} & A_{WW} & B_W^t \\ B_U & B_W & 0 \end{pmatrix} \begin{pmatrix} U \\ W \\ P \end{pmatrix} = \begin{pmatrix} S_U \\ S_W \\ 0 \end{pmatrix} \tag{7.2}$$

where the uncoupling of the velocity components is preserved only for the matrix A_{UU}:

$$A_{UU} = \begin{pmatrix} A_{11} & 0 & 0 \\ 0 & A_{22} & 0 \\ 0 & 0 & A_{33} \end{pmatrix} \tag{7.3}$$

The symmetry of the matrix is obtained by a correct application of the friction stress. This problem is solved by an iterative method (Uzawa algorithm) which can be interpreted as a gradient algorithm working on the equivalent pressure problem:

$$\begin{aligned} \begin{pmatrix} B_U & B_W \end{pmatrix} & \begin{pmatrix} A_{UU} & A_{UW}^t \\ A_{UW} & A_{WW} \end{pmatrix}^{-1} \begin{pmatrix} B_U^t \\ B_W^t \end{pmatrix} P \\ &= \begin{pmatrix} B_U & B_W \end{pmatrix} \begin{pmatrix} A_{UU} & A_{UW}^t \\ A_{UW} & A_{WW} \end{pmatrix}^{-1} S. \end{aligned} \tag{7.4}$$

This algorithm can be described as follows:

Step 0: Initialise P_m;

Step 1: Compute the velocity components $(U, W)_m$ by solving a coupled system;

Step 2: Compute the residual R_m of the pressure problem, which is in fact the divergence of the velocity field $(U, W)_m$;

Step 3: Compute G_m satisfying $C_P G_m = R_m$ where C_P is a preconditioning matrix;

Step 4: Update P_m as $P_{m+1} = P_m - \rho G_m$;

Step 5: Go to step 1 until convergence ($\|R_m\| \leq epsilon$).

In fact we use in the N3S code a preconditioned steepest descent conjugate gradient version of this algorithm which is fast when it is preconditioned (Cahouet and Chabard 1978). From numerical experiment we have deduced that the best preconditioner is:

$$\begin{pmatrix} B_U & B_W \end{pmatrix} \begin{pmatrix} Diag(A_{UU}) & 0 \\ 0 & Diag(A_{WW}) \end{pmatrix}^{-1} \begin{pmatrix} B_U^t \\ B_W^t \end{pmatrix}. \tag{7.5}$$

8. Industrial application

The code has been validated on a wide range of simple test cases (Chabard et al. 1987) and we want to present now the computation of the flow in the cold plenum of a 1500 MW fast breeder reactor (Chabard and Daubert 1988). The domain is sketched in Fig.1: the fluid (sodium) goes out from the window of the heat exchanger and it is sucked by the pump. Because of symmetry considerations, we have represented only one eighth of the total geometry. The purpose of this computation is on the first hand to study the regularity of the flow in the plenum and especially at the sucking by the pump. On the second hand it is necessary to know the order of magnitude of the friction velocities on the internal wall in order to apply thermal boundary conditions for the hot plenum flow simulation. A mesh of 13936 P1-isoP2 tetrahedra and 20108 velocity nodes (cf. Fig. 2) has been generated using the sold modeler GEOMOD and the automatic mesh generator SUPERTAB from the commercial software Ideas from SDRC. The boundary conditions are:

- specified values at the inlet,

- logarithmic law at the walls,

- symmetry in the medidian planes,

- homogeneous Neuman boundary conditions at the outlet.

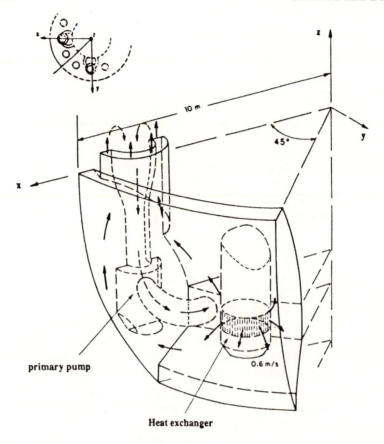

Fig. 1: Sketch of one-eighth of the cold plenum of a fast breeder reactor

Using the flow rate at the entrance, the height of the inlet window and the molecular viscosity of the sodium, we obtain a Reynolds number of $1.5 \, 10^6$. The $k - \varepsilon$ turbulence model has been used for this computation. The steady state has been obtained after 78 time steps of $0.5 s$. The velocity field and the contours of turbulent kinetic energy are given in Fig. 3 and Fig. 4. The computational time for each time step is about $100 s$ (i.e., $5 s$ for 1000 nodes), and the total time for this simulation has been about 2.5 hours on a CRAY X-MP 2-16. The memory requirement is 4 million words. The flow compares qualitatively well with experimental results. Further comparisons are underway.

9.   Conclusion

With this turbulent industrial application of the N3S code, it has been shown that realistic flow simulations are now possible using finite elements.

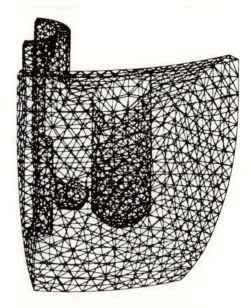

Fig. 2: View of the finite element mesh on the boundary of the cold plenum

This code will therefore be widely used for industrial design computations at EDF but it is also available for customers. Moreover, work is still under way to make N3S a more general purpose code. So we plan to use it for flow simulation in the components of turbomachinery or heat exchangers. To reach this goal we are implementing periodic boundary conditions, head losses and porosity. A next step will be the treatment of variable density.

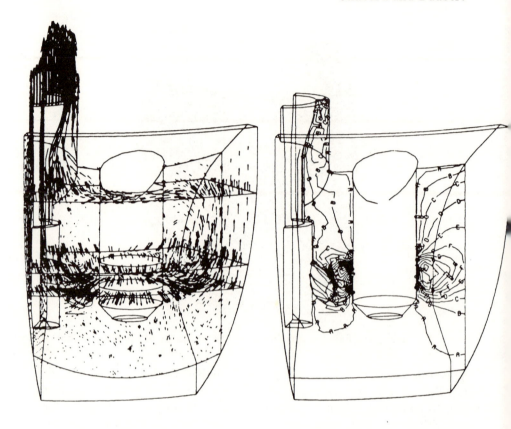

Fig.3: Velocity field at the steady
state

Fig.4: Contours of turbulent kinetic
energy at the steady state

References

Benqué, J., Ibler, B., Keramsi, A., and Labadie, G. (1982). A new finite
element method for Navier–Stokes equations coupled with a temperature
equation. In *4th International Symposium on Finite Element in Flow
Problems*, pages 295–301. North Hollands.

Cahouet, J. and Chabard, J. P. (1978). Some fast 3D finite element solvers
for generalised Stokes problems. *International Journal For Numerical
Methods in Fluids*, to appear in 1988.

Chabard, J. P. and Daubert, O. (1988). N3S: Bilan des calculs tridimen-
sionnels effectués en 1987. EDF Report, HE-41/87.29.

Chabard, J. P., Daubert, O., Grégoire, J. P., and Hemmerich, P. (1987). A
finite element code for the efficient computation of turbulent industrial

flows. *Numerical Methods in Laminar and Turbulent flow*, 5:672–683. Pineridge Press.

Hood, P. and Taylor, C. (1973). A numerical solution of Navier–Stokes equations using the finite element technique. *Computers and Fluids*, 1:73–100.

Launder and Spalding (1974). The numerical computation of turbulent flows. *Computing Methods in Applied Mechanical Engineering*, Vol. 3.

Pironneau, O. (1988). *Méthodes d'éléments finis pour les fluides*. Masson.

Temam, R. (1977). *Theory and numerical analysis of the Navier–Stokes equations*. North Hollands.

The behaviour of Flux Difference Splitting schemes near slowly moving shock waves

Thomas W. Roberts

FFA, the Aeronautical Research Institute of Sweden

1. Introduction

A comparison of Godunov's (1959), Roe's (1981), and Osher's (1982) up-wind schemes has been made for a flow consisting of a slowly moving shock wave. "Slowly moving" means that the ratio of the shock speed to the maximum wave speed in the domain is $\ll 1$. The relevance of this flow condition is twofold: first, unsteady flows can have regions of nearly steady waves, and the behaviour near these shocks can have a global influence on the quality of the results (Woodward 1984); second, these algorithms are being widely used to compute steady flows in a time asymptotic fashion, and the asymptotic convergence rate is highly dependent on the slow motion of the shocks to their equilibrium positions.

2. Model problem

The flux difference splitting schemes mentioned above were compared for three sets of equations: the inviscid Burgers equation,

$$u_t + \left(\frac{1}{2}u^2\right)_x = 0; \qquad (2.1)$$

the Euler equations for an isothermal gas,

$$
\begin{aligned}
\rho_t + (\rho u)_x &= 0, \\
(\rho u)_t + \left(\rho \left(u^2 + a^2\right)\right)_x &= 0,
\end{aligned}
\qquad (2.2)
$$

where a is the speed of sound, a constant; and the full Euler equations for a nonisentropic ideal gas,

$$
\begin{aligned}
\rho_t + (\rho u)_x &= 0, \\
(\rho u)_t + \left(\rho u^2 + p\right)_x &= 0, \\
(\rho E)_t + ((\rho E + p)u)_x &= 0,
\end{aligned}
\qquad (2.3)
$$

where $E = u^2/2 + p/((\gamma - 1)\rho)$. Each equation was solved for a flow consisting of a single shock wave propagating slowly to the left. The initial

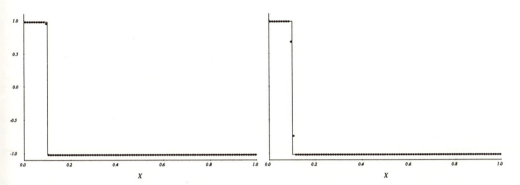

Fig. 1: Inviscid Burgers equation: Roe's/Godunov's scheme (left), Engquist-Osher scheme (right)

conditions for each case were two constant states separated by the exact shock jump at the midpoint of a domain of 100 cells. The conditions were chosen to give a shock speed such that the shock traversed a single cell in approximately 50 time steps at a Courant number of 0.95. All the results below are shown after approximately 2000 time steps. For equations (2.2) and (2.3), the shock jumps were chosen to give about a one order of magnitude pressure rise across the shock (shock Mach number $M_s \sim 3$).

Note that for this model problem, Godunov's and Roe's schemes are identical in the scalar case, Eq. (2.1). For systems of equations, these two schemes remain virtually identical. Indeed, the differences in the results using Godunov's and Roe's schemes for equations (2.2) and (2.3) are insignificant for the above model problem. In the results shown in section 3 solutions obtained using Godunov's scheme are not shown; what is said for Roe's scheme applies equally to Godunov's scheme.

3. Results

In Fig. 1, the results obtained for Roe's (Godunov's) scheme and Osher's (Engquist-Osher) scheme are displayed. (In Fig. 1 and all subsequent figures, Roe's scheme is shown on the left and Osher's scheme on the right.)

Each scheme performs for this equation as expected, with monotone shock profiles. The Engquist-Osher scheme has two-zone shocks compared to Roe's single zone, as should be the case.

To examine the results for equation (2.2), the Riemann invariants ($\log \rho \pm u/a$) associated with the $u \pm a$ characteristic fields respectively, are shown

$$\log \rho - u/a$$

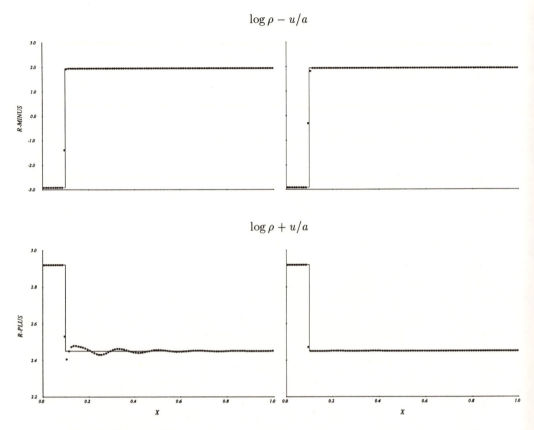

Fig. 2: Isothermal Euler equations, Roe's and Osher's schemes: $\log \rho - u/a$ (top), $\log \rho + u/a$ (bottom)

in Fig. 2. The shock belongs to the $u - a$ characteristic family. This characteristic family behaves qualitatively like the solution to Burgers equation, and it is seen that the $(\log \rho - u/a)$ Riemann invariant is monotone for both Roe's and Osher's schemes, as in the scalar case. However, the $(\log \rho + u/a)$ Riemann invariant shows a very smooth, long wavelength error behind the shock for Roe's scheme. Because the error is so smooth it is very persistent, as it can only be slowly damped by the dissipation in the scheme, even though the solution is only first order accurate. The error with Osher's scheme, on the other hand, is barely visible.

The behaviour observed for the isothermal Euler equations (2.2) is also seen for the full Euler equations (2.3). In this case we no longer have Riemann invariants, since the integrability conditions along characteristics do not hold for more than a two equations system. However, we may use Roe's linearization to compute the strengths of the wave at the cell interfaces and sum these wave strengths through the domain. The results are displayed in this format in Fig. 3. Again, both schemes are monotone in the characteristic family associated with the shock, but Roe's scheme generates significant noise in both downstream running waves. The level of noise in Osher's scheme is again quite small.

4. Discussion

The results of section 3 show that for Roe's scheme there is a significant error behind slowly moving shocks which occurs when solving nonlinear systems of equations. Scalar equations are well behaved. Although close examination of Osher's scheme shows a post-shock error as well, it is not nearly as pronounced.

It must be emphasized that the error observed using Roe's scheme is not peculiar to that method. Godunov's scheme gives results virtually indistinguishable from Roe's scheme for equations (2.2) and (2.3). One could expect this, since the differences at a shock are very small with both methods. One would also expect that any approximate Riemann solver that resolves a Rankine-Hugoniot jump exactly will exhibit this behaviour.

Given these results, an understanding of the source of the error has been sought. Following a suggestion by P. L. Roe (private communication), the error can be explained in terms of the discrete shock structure. The requirement that no noise be generated behind the shock is equivalent to requiring that the internal zones of the discrete shock must move along the shock curve (the Hugoniot curve for the full Euler equations) passing through the post-shock state. In the scalar case, this condition is trivially satisfied, since any state between the pre- and post-shock states lies on the appropriate shock curve. For nonlinear systems, this is not true, and it can be shown that this requirement implies that no unsteady shock can be represented with only one internal zone (Roberts 1988). In a sense, Roe's and Godunov's schemes have a shock transition layer that is too narrow. On the other hand, Osher's use of a differentiable numerical flux function and his choice of integration path results in a shock that more nearly satisfies this requirement than does either Roe's or Godunov's scheme, and consequently has a "thicker" shock.

This is best illustrated by considering the isothermal case, equation (2.2), and examining the shock structure in state space. These results are shown in Fig. 4 for the solutions shown in Fig. 2. The symbols show the evolution of a zone as it is crossed by the shock; the solid line is the shock curve. It is seen that Roe's scheme gives a shock transition that deviates more from the shock curve as the post-shock state is approached. Osher's scheme yields a smooth transition that stays very close to the required

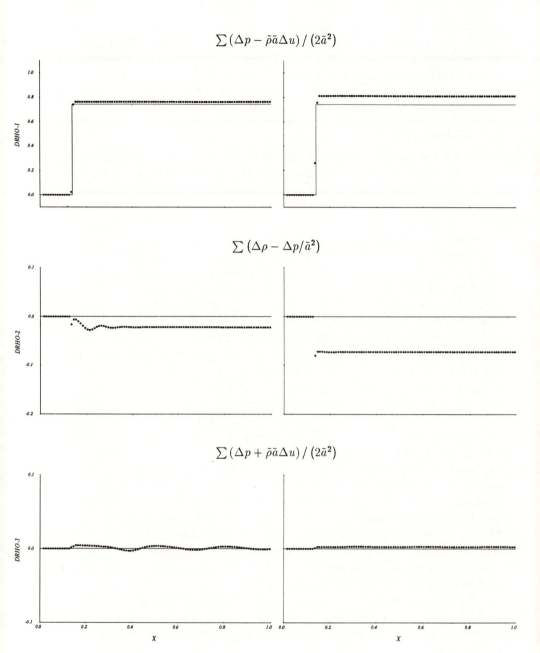

Fig. 3: Full Euler equations, Roe's and Osher's schemes: $u - a$ characteristics (top), u characteristics (middle), and $u + a$ characteristics (bottom)

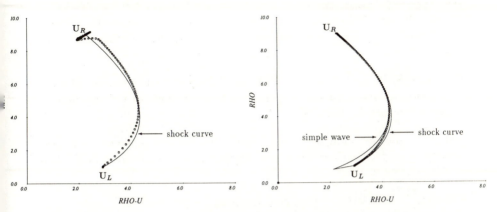

Fig. 4: Isothermal shock structure: Roe's scheme (left), Osher's scheme (right)

path near the post-shock state. Also, the ordering of the wavepaths in Osher's scheme is very important; if the order is reversed from the one he prescribes, the post-shock error becomes more pronounced.

The most important consequence of this error occurs when higher order accuracy is obtained through the use of TVD flux limiters. Since the TVD property (which strictly holds only in the scalar case) depends upon the underlying first order accurate scheme being truly monotone, and because the error in this case is so smooth, the noise generated at the shock is preserved for longer distances downstream. Thus the problem is accentuated by the use of flux limiting. It is suspected that this is a contributing cause of the slow convergence to steady state of TVD schemes that has been reported by some researchers.

5. Conclusions

To summarize, there are two main conclusions to the present work. First, for slowly moving shocks, some flux difference splitting schemes yield significant error behind the shock. The error is pronounced for schemes that exactly solve Riemann's problem at a Rankine-Hugoniot jump. This noise cannot be eliminated by appealing to TVD concepts, and may be a contributing factor to the slow convergence to steady state observed for such schemes. Second, this error can be explained in terms of the discrete shock structure of the particular scheme, and suggests that the approximate Riemann solvers that do not recognize a Rankine-Hugoniot jump may be better suited for shock capturing than an exact Riemann flux.

References

Colella, P. and Woodward, P. (1984). The piecewise parabolic method (PPM) for gas-dynamical simulation. *Journal of Computational Physics* 54 : 174-201

Godunov, S. K. (1959). A finite-difference method for the numerical computation of discontinuous solutions of the equations of fluid dynamics. *Math. Sbornik* 47 : 271-306

Osher, S. and Solomon, F. (1982). Upwind difference schemes for hyperbolic systems of conservation laws. *Mathematics of Computation* 38 : 339-374

Roberts, T. W. (1988). FFA Technical Note, in preparation

Roe, P. L. (1981). Approximate Riemann solvers, parameter vectors, and difference schemes. *Journal of Computational Physics* 43 : 357-372

Woodward, P. and Colella, P. (1984). The numerical simulation of two-dimensional fluid flow with strong shocks. *Journal of Computational Physics* 54 : 115-173

A Total Variation Diminishing scheme for computational aerodynamics

D. M. Causon

Manchester Polytechnic

1. Introduction

An improved Euler method is presented for computing steady high speed external aerodynamic flows. The method is pseudo time-dependent and uses operator-splitting in conjunction with a finite volume formulation. The resolution of captured shock waves is improved by the use of a total variation diminishing (TVD) version of the well-known MacCormack scheme and artificial compression techniques. Existing production code implementations of MacCormack's method can be updated quickly and easily by the addition of a simple subroutine.

Over the last fifteen years, substantial advances have been made in the numerical analysis of hyperbolic partial differential equations, particularly those arising in computational aerodynamics. A popular approach is to solve the Euler equations by time-marching, using an appropriate conservative finite-difference or finite-volume scheme to capture shock waves and contact discontinuities. In the early 1970's this approach provided the aerodynamicist with useful results for practical configurations which previously had been unobtainable theoretically. A notable advance around that time was the method of operator-splitting which enables a three-dimensional problem to be solved as a sequence of one-dimensional problems. This led to a focus on one-dimensional schemes. The early schemes were analysed using linear stability theory and those that were intended to capture shock waves were designed to be dissipative. However, the ability of these schemes accurately to resolve shock waves was impaired by the appearance of undershoots and overshoots around the shock profile. Artificial viscosity was added to the scheme, often quite liberally, in order to dampen them. Unfortunately, this procedure causes a marked loss of resolution. Although pressure and Mach number distributions may appear reasonably accurate, a spurious entropy layer tends to develop, emanating particularly from regions of high spatial gradient, like stagnation points and shock waves. In practice this makes the evaluation of wave drag and total pressure loss difficult and inaccurate; and these are key parameters. This can be related back to the relative crudity with which the dissipative term was added to the scheme. Invariably it had a global (and problem-dependent) coefficient, so too much damping was being applied. What was needed was to apply the term more selectively.

In the 1980's, Roe (1981), Osher (1984), van Leer (1984), Sweby (1984), Davis (1984), Chakravarthy (1985), Harten and Osher (1987) and Yee (1987) began to put the procedures on a more systematic footing. There was renewed interest in Godunov methods and Riemann solvers generally which resulted in substantial research activity. Out of this work emerged the total variation diminishing (TVD) schemes, so-named by Harten, who gave a set of theorems leading to conditions which, when satisfied by any three-point, conservative, second-order accurate finite difference scheme, are necessary and sufficient for the scheme to be TVD. As a sequel to this work, Davis (1984) showed that it is possible to put a classical Lax-Wendroff scheme in TVD form. This is accomplished by appending to the scheme a non-linear term which applies precisely the correct amount of artificial viscosity needed at each mesh point to limit overshoots and undershoots. Since there are many production computer codes in use which employ the MacCormack variant of the Lax-Wendroff scheme this is of especial interest, it being possible quickly and easily to bring the scheme to state of the art. This paper describes the modifications necessary to construct a TVD MacCormack scheme. A supporting theoretical derivation is given in Causon (1988).

2.   Formulation

Since the use of a body-fitted mesh is anticipated, the equations of motion are written in integral form as a prelude to discretisation by the finite volume method.

$$\frac{\partial}{\partial t} \int \int \int_{vol} \mathbf{U} dvol + \int \int_s H \cdot \hat{n} ds = 0 \qquad (2.1)$$

where $\mathbf{U} = [\rho, \rho w_1, \rho w_2, \rho w_3, e]^T$
$$H = \begin{bmatrix} \rho \mathbf{q} \\ \rho w_1 \mathbf{q} + p \mathbf{a}_1 \\ \rho w_2 \mathbf{q} + p \mathbf{a}_2 \\ \rho w_3 \mathbf{q} + p \mathbf{a}_3 \\ (e+p)\mathbf{q} \end{bmatrix},$$

and the flow velocity $\mathbf{q} = w_l \mathbf{a}_l$, where $\mathbf{a}_l$ are the Cartesian unit base vectors. We solve equations 2.1 using a factored sequence of one-dimensional difference operators, where each component operator relates to its respective co-ordinate direction. Further details may be found in (Causon and Ford 1985).

3. TVD MacCormack finite volume scheme (TVDM)

The MacCormack finite volume operator $L_1(\Delta t)$ is:

$$\mathbf{U}_{ijk}^{\overline{n+1}} = \mathbf{U}_{ijk}^n - \Delta t \left(H_{ijk}^n \mathbf{S}_{i+\frac{1}{2}} + H_{i-1jk}^n \mathbf{S}_{i-\frac{1}{2}} \right) \tag{3.1}$$

$$\mathbf{U}_{ijk}^{n+1} = \frac{1}{2} \left[\mathbf{U}_{ijk}^n + \mathbf{U}_{ijk}^{\overline{n+1}} - \Delta t \left(H_{i+1jk}^{\overline{n+1}} \mathbf{S}_{i+\frac{1}{2}} + H_{ijk}^{\overline{n+1}} \mathbf{S}_{i-\frac{1}{2}} \right) \right] \tag{3.2}$$

where $\mathbf{U}_{ijk} = vol_{ijk}[\rho, \rho w_1, \rho w_2, \rho w_3, e]_{ijk}^T$ and $\mathbf{S}_{i\pm\frac{1}{2}}$ are the are vectors on opposite faces of the cell, normal to the surface x^1 =constant. Scheme 3.1-3.2 can be updated easily to total variation diminishing (TVD) form by appending to the right hand side of the "corrector" step 3.2 the term:

$$+ \left[G_i^+ + G_{i+1}^- \right] \Delta \mathbf{U}_{i+\frac{1}{2}}^n - \left[G_{i-1}^+ + G_i^- \right] \Delta \mathbf{U}_{i-\frac{1}{2}}^n \tag{3.3}$$

where for clarity we have suppressed subscripts j, k and:

$$\Delta \mathbf{U}_{i+\frac{1}{2}}^n = \mathbf{U}_{i+1}^n - \mathbf{U}_i^n, \qquad \Delta \mathbf{U}_{i-\frac{1}{2}}^n = \mathbf{U}_i^n - \mathbf{U}_{i-1}^n \tag{3.4}$$

$$G_i^\pm = G \left[r_i^\pm \right] = 0.5 C(\nu) \left[1 - \phi \left[r_i^\pm \right] \right], \tag{3.5}$$

$$C(\nu) = \begin{cases} \nu(1 - \nu), & \nu \le 0.5 \\ 0.25, & \nu > 0.5 \end{cases}, \tag{3.6}$$

$$\nu = \nu_i = \max_l \mid \lambda_l \mid \frac{\Delta t}{\Delta x^1}, \tag{3.7}$$

$$r_i^+ = \frac{\left[\Delta \mathbf{U}_{i-\frac{1}{2}}^n, \Delta \mathbf{U}_{i+\frac{1}{2}}^n \right]}{\left[\Delta \mathbf{U}_{i+\frac{1}{2}}^n, \Delta \mathbf{U}_{i+\frac{1}{2}}^n \right]}, \qquad r_i^- = \frac{\left[\Delta \mathbf{U}_{i-\frac{1}{2}}^n, \Delta \mathbf{U}_{i+\frac{1}{2}}^n \right]}{\left[\Delta \mathbf{U}_{i-\frac{1}{2}}^n, \Delta \mathbf{U}_{i-\frac{1}{2}}^n \right]} \tag{3.8}$$

where $(.,.)$ denotes the usual inner product on R^5 and :

$$\phi(r) = \begin{cases} \min(2r, 1), & r > 0 \\ 0, & r \le 0 \end{cases} \tag{3.9}$$

The procedure is as follows:

1. Calculate $r_i^\pm$ using (3.8).

2. Calculate ν_i from (3.7) and $C(\nu)$ from (3.6). Here,

$$max_l \mid \lambda_l \mid = u^1 + c,$$

where u^1 and c are local flow speed and sound speed respectively.

3. Calculate flux limiter functions $\phi [r_i^\pm]$ from (3.9).

4. Calculate $G_i^\pm$, G_{i+1}^-, G_{i-1}^+ using (3.5)

4.   TVD MacCormack scheme with artificial compression (TVD-MAC)

In order selectively to add artificial compression in regions where the numerical solution changes abruptly, we construct a split operator C_Δ which "compresses" a TVDM solution at any given time level. Here, C_Δ is defined as:

$$C_\Delta V_i = V_i - \frac{\lambda}{2} \left[\theta_{i+\frac{1}{2}} G_{i+\frac{1}{2}} - \theta_{i-\frac{1}{2}} G_{i-\frac{1}{2}} \right] \tag{4.1}$$

where:

$$V_i = L_1(\Delta t) U_i^n \tag{4.2}$$

is the solution obtained by applying the TVDM scheme to the solution at the previous time level, and again for clarity we have suppressed subscripts j, k. Also,

$$\theta_{i+\frac{1}{2}} = \max \left[\hat{\theta}_i, \hat{\theta}_{i+1} \right], \tag{4.3}$$

$$\hat{\theta}_i = \begin{cases} \mid \frac{|\Delta \rho_{i+\frac{1}{2}}| - |\Delta \rho_{i-\frac{1}{2}}|}{|\Delta \rho_{i+\frac{1}{2}}| + |\Delta \rho_{i-\frac{1}{2}}|} \mid, & \mid \Delta \rho_{i+\frac{1}{2}} \mid + \mid \Delta \rho_{i-\frac{1}{2}} \mid > \epsilon \\ 0, & \mid \Delta \rho_{i+\frac{1}{2}} \mid + \mid \Delta \rho_{i-\frac{1}{2}} \mid \leq \epsilon \end{cases} \tag{4.4}$$

$$\epsilon = 0.01 \max_i \mid \rho_{i+1} - \rho_i \mid, \qquad \rho_i = V_i^1 \tag{4.5}$$

$$G_{i+\frac{1}{2}} = G_{i+\frac{1}{2}}^m = g_i^m + g_{i+1}^m - \mid g_{i+1}^m - g_i^m \mid \text{sgn} \left[V_{i+1}^m - V_i^m \right] \tag{4.6}$$

$$g_i^m = \alpha_i(V) \left[V_{i+1}^m - V_{i-1}^m \right]$$

$$\alpha_i(V) = \max \left\{ 0, \min_{1 \leq m \leq M} \frac{\min \left[[V_{i+1}^m - V_i^m], [V_i^m - V_{i-1}^m] \text{sgn} \left[V_{i+1}^m - V_i^m \right] \right]}{\mid V_{i+1}^m - V_i^m \mid + \mid V_i^m - V_{i-1}^m \mid} \right\}$$

where $m = 1(1)M$, M being the number of components of the solution vector V.

Clearly, the artificial compression operator C_Δ can be applied more than once, if desired, in order further to "compress" the data. In practice, we have rarely found this to be necessary.

The M, TVDM and TVDMAC schemes have been used to solve the one-dimensional Riemann problem. At time $t = 0$ the left and right states are:

$$U_L = \begin{bmatrix} 0.445 \\ 0.311 \\ 8.928 \end{bmatrix}, \qquad U_R = \begin{bmatrix} 0.5 \\ 0.0 \\ 1.4275 \end{bmatrix}.$$

with increasing time, a mixing process takes place such that a rarefaction wave moves to the left and a contact discontinuity and shock wave move to the right. The results shown in Fig. 1 were obtained after 100 time steps with 140 cells and λ in (4.1) set to 1.0. Other computed solutions are given in (Harten 1983). It can be seen that scheme TVDM exhibits none of the oscillations of the M scheme and that the resolution of the contact discontinuity is improved dramatically with artificial compression. The results shown in Fig. 1 cannot be distinguished from those obtained from computationally more expensive TVD schemes based on Riemann solvers (see Harten 1983). Clearly, any production computer code employing the MacCormack method can be modified quickly and simply, as described above, to yield a high-resolution scheme. Further discussion of the use of artificial compression techniques can be found in (Harten 1983).

5. Applications

Fig. 2a illustrates an aircraft forebody geometry which is typical of the type of problem which we wish to solve. Here, we wish to predict the disturbance field around the forebody at proposed intake locations, as well as the usual forces and moments, at Mach numbers ranging from high subsonic to supersonic and at various angles of attack. A three-dimensional implementation of the scheme described earlier is used as the basis of our flow solver, together with pre- and post-processing routines for mesh generation, evaluation of forces and moments, graphical output and other ancilliary functions. It is a particular advantage of operator-splitting that one can substitute one (one-dimensional) scheme for another. So, although an applications code may be large and complex, at the lowest level performing the numerical calculations is a one-dimensional scheme. This enables us readily to update the method.

The forebody shown in Fig. 2a has an elliptical nose cone dipped at $5°$, sharp corners and flat sides to accommodate engine intakes. An analytical description of the body surface was formulated by fitting curves through the ordinates of particular cross sections. This formed the basis for the construction of a body-fitted H-O type grid in which the radial lines, emanating from the effective centre line of the body, were segmented such that the radial spacing at the body surface was approximately the same as the streamwise spacing. The latter spacing was uniform and corresponded to the stations at which the individual cross sections were defined. So constructed, the mesh had $57(x) \times 16(\theta) \times 10(r)$ cells and a converged solution was obtained after 800 time steps, commencing the calculation with an impulsive start at Mach 1.40 and $-5°$ angle of attack. Fig. 2b shows the isobar contours on the surface of the body. Unlike similar calculations using the MacCormack method, there were no parameters to be adjusted to control the amount of added dissipation for shock capturing. Fig. 3 shows a similar calculation for another forebody, typical of a civil aircraft. An analytical description of this geometry is given in the reference cited.

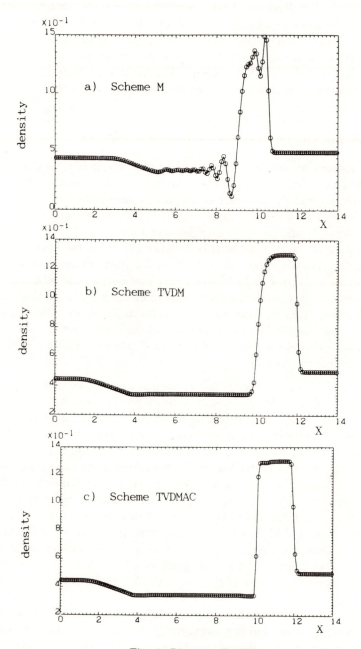

Fig. 1: Riemann Problem

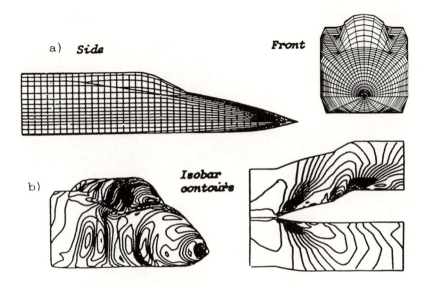

a) **Side**

Front

b)

Isobar contours

Fig. 2: Combat Aircraft Forebody at Mach 1.40 and $\alpha = -5^o$

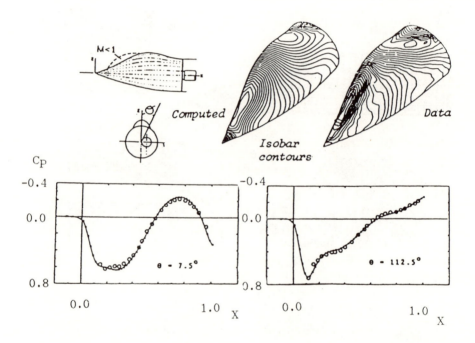

Fig. 3: Civil Aircraft Forebody from NASA TM 80062 at Mach 1.70 and $\alpha = -5°$

6. Conclusions

A high resolution version of the MacCormack method has been presented, which is more robust and has an improved shock capturing capability. The required modifications to the standard method are minor and will enable many existing production computer codes to take advantage of recent developments in one-dimensional schemes.

References

Causon, D. (1988). High resolution finite volume schemes and computational aerodynamics. In *Proceedings of the second international conference on hyperbolic problems.* Springer-Verlag.

Causon, D. and Ford, P. (1985). Numerical solution of the euler equations governing axisymmetric and three dimensional transonic flow. *The Aeronautical Journal*, 89 : 226–241.

Chakravarthy, S. (1985). A new class of high accuracy TVD schemes for hyperbolic conservation laws. AIAA Paper 85-0363

Davis, S. (1984). TVD finite-difference schemes and artificial viscosity. Paper CR 172373, NASA.

Harten, A. (1983). High resolution schemes for hyperbolic conservation laws. *Journal of Computational Physics*, 49 : 357–392.

Harten, A. and Osher, S. (1987). Uniformly high-order accurate non-oscillatory schemes. *SIAM Journal of Numerical Analysis*, 24 : 279–309.

van Leer, B. (1984). On the relation between the upwind differencing schemes of Godunov, Engquist-Osher and Roe. *SIAM Journal on Scientific and Statistical Computing*, 5 : 1–20.

Osher, S. (1984). Riemann solvers, the entropy condition and difference approximations. *SIAM Journal of Numerical Analysis*, 21 : 217–235.

Roe, P. (1981). Approximate Riemann solvers, parameter vectors and difference schemes. *Journal of Computational Physics*, 43 : 357–372.

Sweby, P. (1984). High resolution schemes using flux limiters for hyperbolic conservation laws. *SIAM Journal of Numerical Analysis*, 21 : 995–1011.

Yee, H. (1987). Upwind and symmetric shock capturing schemes. Paper TM 89464, NASA.

Properties of two computational methods for shallow water flow problems

Th. L. van Stijn
RWS, Rijswijk; The Netherlands.

P. Wilders
Delft University of Technology; The Netherlands.

G. A. Fokkema
RWS/Delft University of Technology; The Netherlands.

G. S. Stelling
Nederlandse Philips Bedrijven B.V.,
Eindhoven; The Netherlands.

1. Introduction

For the computation of velocities and water levels of tidal flows in shallow seas, estuaries and rivers often the vertically integrated, two dimensional shallow water equations are used. In this paper we discuss some accuracy and efficiency aspects of two computational methods for the resolution of the shallow water equations. These methods are an ADI technique and a fully implicit splitting method. For the sake of simplicity, the linear 'frozen coefficient', two dimensional shallow water equations will be treated. These read

$$\frac{\partial w}{\partial t} + \mathbf{A}\frac{\partial w}{\partial x} + \mathbf{B}\frac{\partial w}{\partial y} = 0 \qquad (1.1)$$

where $w = [\zeta, u, v]^{\mathsf{T}}, \mathbf{A} = \begin{bmatrix} 0 & H & 0 \\ g & U & 0 \\ 0 & 0 & U \end{bmatrix}, \mathbf{B} = \begin{bmatrix} 0 & 0 & H \\ 0 & V & 0 \\ g & 0 & V \end{bmatrix}.$

In (1.1) u and v are horizontal velocities of the water flow, whereas U and V determine the advective transport. H is the total local depth, ζ the water elevation with respect to a reference level and g is the acceleration of gravity. The first ζ-equation in (1.1) is the continuity equation. The remaining two (u and v)-equations render conservation of momentum. The physical phenomena that are resolved with equation (1.1) are wave propagation and linear advection. However, properties that are important in a realistic situation are non-linear advection, parametrization of bottom and wind stresses and a non-constant H also including dynamic drying and flooding algorithms. Furthermore, diffusion of momentum caused by turbulence is an important mechanism and the Coriolis force. There are several ways to construct a numerical approximation to the shallow water equations (1.1). These can be separated into classes according to their basic numerical methodology. The first class is that of the explicit methods. Examples are the methods of Praagman (1979) and of Wubs (1987). Generally, explicit methods are simple to implement. The finite difference method can be easily vectorizable, whereas the finite element method can be applied for unstructured grids. Both methods, however, are subject to a stability condition, which in case of (1.1) is reflected in a bound on the Courant number C_f, defined by

$$C_f = \Delta t \sqrt{\frac{2gH}{(\Delta x)^2 + (\Delta y)^2}} \tag{1.2}$$

Here Δx and Δy are the widths of the computational mesh and Δt the time step. Since the total depth H is involved in C_f, the deepest spot determines a global limitation on Δt, which can become quite restrictive. In shallow regions stability is restricted by bottom friction terms. The recent work of Wubs demonstrates that the optimal smoothing technique has a favourable effect on these stability limitations. A second class is given by the fully implicit methods. These methods oppose stability restrictions and are in this sense robust. Solving the large matrix system, that is fully coupled in ζ, u and v, is in many cases prohibitively expensive. Therefore, many of the industrial codes for the calculation of tidal flows make use of the Alternating Direction Implicit (ADI) splitting method.

2. The ADI technique and its consequences

When the ADI technique is applied in a straightforward manner to (1.1) the following time split system of equations emerges:

$$\begin{cases} S1 \; : \; (I + \tfrac{1}{2}\Delta t L_x^{(1)})w^{k+\frac{1}{2}} = (I - \tfrac{1}{2}\Delta t L_y^{(1)})w^k, \\[2mm] S2 \; : \; (I + \tfrac{1}{2}\Delta t L_y^{(2)})w^{k+1} = (I - \tfrac{1}{2}\Delta t L_x^{(2)})w^{k+\frac{1}{2}}. \end{cases} \tag{2.1}$$

Here w^k is the solution at time t_k , I is the unit matrix and

$$L_x^{(1)} = L_x^{(2)} = \mathbf{A}\frac{\partial}{\partial x}, \qquad\qquad L_y^{(1)} = L_y^{(2)} = \mathbf{B}\frac{\partial}{\partial y}.$$

A splitting method that is widely used in the Netherlands for practical problems in civil engineering is the composite ADI method of Stelling (1983). Here much attention has been paid to the proper expression for the advection terms in (1.1). The linear composite scheme has the form (2.1) with

$$L_x^{(1)} = \begin{bmatrix} 0 & HD_{0x} & 0 \\ gD_{0x} & 0 & 0 \\ 0 & 0 & US_x^{(u)} + VD_{1y} \end{bmatrix}, L_y^{(1)} = \begin{bmatrix} 0 & 0 & HD_{0y} \\ 0 & UD_{1x} + VS_y^{(c)} & 0 \\ gD_{0y} & 0 & 0 \end{bmatrix}.$$

$$(2.2)$$

In (2.2) D_{0x}, D_{1x} and $S_x^{(c)}$ are second order central finite difference operators to x on a staggered grid. $S_x^{(u)}$ is a second order upwind operator, such that $\frac{1}{2}[S_x^{(u)} + S_x^{(c)}]$ has a third order of consistency. The operator $L_x^{(2)}$ has the same expression as $L_x^{(1)}$ save for a central $S_x^{(c)}$ instead of $S_x^{(u)}$ in the advection term. Similarly in $L_y^{(1)}$ the operator $S_y^{(c)}$ is replaced by $S_y^{(u)}$ for $L_y^{(2)}$. The definitions of the finite difference operators can be found in Stelling (1983).

The original method is robust since it includes many non-linear physical phenomena and it is implemented in the software system WAQUA with which many practical experiments have been made. The method is unconditionally stable and has a fourth order dissipation which abates spurious oscillations. The method is efficient, since in $S1$ the v-equation is uncoupled from the remaining equations. The u-equation is substituted into the ζ-equation and in total two tri-diagonal systems need to be solved. This can be done very efficiently. A similar reasoning holds for $S2$. Recently it has become clear that the ADI technique can produce serious inaccuracies that are connected to the numerical integration in time. These inaccuracies emerge in complex geometric or bathymetric situations which frequently occur in practical applications. An explanation was given by Stelling et al. (1986) and is demonstrated with a simple example. When the flow in the channel of Fig.1 is calculated with a time step large enough, the analytical domain of influence will encompass the entire channel. The numerical domain of influence for one time step, however, is limited to the hatched area, as can be observed by imposing a disturbance in P. The resulting time step restriction degrades the efficiency of the method, especially e.g. in cases with a complex gully structure and with tidal flats.

3. A fully implicit splitting approach

To cope with the inaccuracies of the ADI method a fully implicit splitting technique was devised (Wilders et al. 1988). The equations that resulted

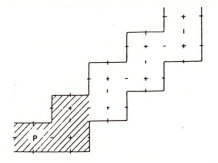

Fig. 1: A zig-zag channel

from the analysis can be put into the form (2.1) with

$$L_x^{(1)} = \begin{bmatrix} 0 & 0 & 0 \\ 0 & UD_{1x} + VS_y^{(u)} & 0 \\ 0 & 0 & US_x^{(u)} + VD_{1y} \end{bmatrix}, \quad L_y^{(1)} = \begin{bmatrix} 0 & HD_{0x} & HD_{0y} \\ gD_{0x} & 0 & 0 \\ gD_{0y} & 0 & 0 \end{bmatrix}.$$

(3.1)

Furthermore, $L_x^{(2)}$ has the same expression as $L_x^{(1)}$ but for a central $S_{\bullet}^{(c)}$ instead of an upwind operator $S_{\bullet}^{(u)}$ and $L_y^{(2)} \equiv L_y^{(1)}$. The above method is a second order perturbation to the trapezoidal rule, just like the ADI technique. It can be proved that (3.1) is unconditionally stable and has fourth order dissipation. The actual structure of the splitting (3.1) is an important aspect with respect to the efficiency. In stage $S1$ the ζ-equation is explicit, whereas the u and v-equations are uncoupled. In $S2$ the u and v-equations can be substituted explicitly into the ζ-equation, thus resulting in a total of three linear systems to be solved. The implicitness of (3.1) warrants the absence of the ADI inaccuracies. It is possible to extend the analysis to the non-linear equations while conserving the features of the fully implicit splitting technique (Wilders et al. 1988). The linear systems that result from (3.1) are non-symmetrical. It turns out that the Conjugate Gradients Squared algorithm of Sonneveld (1984) is superior to others in the resolution of the linear systems.

4. Result of numerical experiment

A test problem was introduced by Weare (1979) to demonstrate the ADI inaccuracies (see Fig.2). It consists of two channels, of which the first (I) serves as a reference and the second (II) is at an angle of 45° with respect to the coordinate axes. In the experiment the depth has a constant value of $10m$, whereas the sides have a length of $4000m$ and $400m$. In I we have

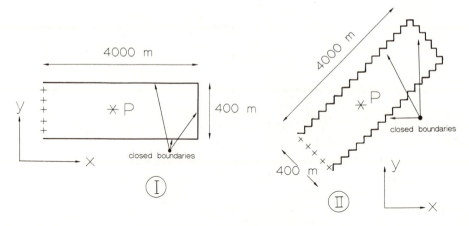

Fig. 2: Channel configurations I and II

C_f	ADI		fully implicit		ADI		fully implicit	
	a_ζ^I	a_u^I	a_ζ^I	a_u^I	a_ζ^{II}	a_u^{II}	a_ζ^{II}	a_u^{II}
12	0.3	4.5	0.6	11.1	1.3	32.9	0.8	16.3
24	0.4	8.3	0.6	9.3	5.6	56.4	0.8	14.0
48	0.4	8.6	0.4	8.2	20.4	99.7	0.3	5.8

Table 1: Values of the relative deviations for the horizontal channel I and the zig-zag channel II.

$\Delta x = \Delta y = 100m$, in II $\Delta x = \Delta y = 71m$. At the open boundary $\delta\Omega_1$ a wave is prescribed and the velocity component parallel to the boundary, viz.

$$\zeta = \cos(2\pi t/21600), \qquad u_\parallel = 0 \quad \text{at } \delta\Omega_1. \tag{4.1}$$

The flow quantities, that are functions of C_f are recorded in Table 1 with a deviation a_σ^i; $(i = I, II; \sigma = \zeta, u)$ with respect to a reference solution $\mathbf{R}$ generated for $C_f = 1.2$: a_σ^i is in per cent relative to max $\|\mathbf{R}\|$. The left hand part of Table 1, refering to I, shows that both methods can predict the flow very well in this case. In the right hand part of the table, relating to channel II, the strong C_f dependence of the ADI method is clear and the method in fact fails to predict the flow for large C_f. The fully implicit method renders acceptable results.

5.   Conclusions

Some properties of two computational methods for shallow water flow problems have been reviewed. These methods are a composite ADI technique and a fully implicit splitting method. Both methods are unconditionally stable and have a fourth order dissipation. The ADI technique is very efficient and robust. Many important practical situations have been calculated successfully. The time step of this technique, however, is limited due to a restricted numerical domain of influence. The fully implicit time splitting method is able to use much larger time steps. It performs well in test problems. The quality of the solver determines to a large extent the costs of the method. At present a Conjugate Gradients Squared algorithm is implemented.

References

Praagman, N. (1979). *Numerical Solution of the Shallow Water Equations by a Finite Element Method.* Thesis, Delft University.

Sonneveld, P. (1984). CGS, a fast Lanczos-type solver for non-symmetric linear systems, Report 84-16, Dept. of Math., Delft University; to be published in *SIAM Journal of Sc. and Statistical Computation.*

Stelling, G. S. (1983). *On the Construction of Computational Methods for Shallow Water Flow Problems.* Thesis, Delft University.

Stelling, G. S., Wiersma, A.K. and Willemse, J.B.T.M. (1986). Practical aspects of accurate tidal computations. *Journal of Hydraulic Engineering*, ASCE, 112:802–817.

Weare, T.J. (1979). Errors arising from irregular boundaries in ADI solutions of the shallow water equations. *International Journal for Numerical Methods in Engineering*, 14:921–931.

Wilders, P., Van Stijn, Th. L., Stelling, G. S. and Fokkema, G. A. (1988). A fully implicit splitting method for accurate tidal computations. To appear in *International Journal for Numerical Methods in Engineering.*

Wubs, F.W. (1987). *Numerical Solution of the Shallow Water Equations.* Thesis, University of Amsterdam.

Consistent boundary conditions for cell centered upwind finite volume Euler solvers

H. Deconinck and R. Struys

von Karman Institute for Fluid Dynamics
Rhode-Saint-Genèse, Belgium

1. Introduction

Two basically different approaches can be found in the litterature to introduce boundary conditions for the Euler equations. In a first approach, the unknowns along a boundary are determined from inside the domain by some space extrapolation technique under the constraint that the unknowns must satisfy the boundary conditions. Various degrees of sophistication have been used, e.g. the use of the normal momentum equation for pressure extrapolation at a solid wall (Rizzi 1978).

A second approach follows more closely the physics of the hyperbolic equations by discretizing compatibility equations for the outgoing characteristic variables, while replacing the ingoing characteristic information by the physical boundary conditions, see e.g. (Chakravarthy 1983).

In the present paper, we propose a variant of this second approach formulated in the context of flux vector or flux difference splitting methods. For a linear set of equations, the proposed treatment reduces exactly to characteristic boundary conditions. However, for the Euler equations, priority is given to a full consistency with the approximate Riemann solver used for the splitting of the flux at interior finite volume boundaries, while at the same time we require an exact fullfilment of the physical boundary conditions in weak form, e.g. zero mass flux through a solid wall.

The starting point is a consistency condition for a numerical flux function: the unknowns along the boundary are determined such that the numerical flux function based on the boundary and adjacent interior variables leads to a flux which satisfies the boundary condition.

For the explicit scheme this results in a nonlinear algebraic relation between the boundary unknowns U_B and the interior unknowns U_I. In the case of implicit schemes, a linearisation of the relation between the unknowns U_B and U_I around the solution at time level n gives an expression which closes the set of implicit relations.

2. The flux equation for the boundary point

In a finite volume technique, the flux balance over the boundaries of a finite volume is used to update variables within the volume (Hemker and Spekreijse 1986), (Anderson et al. 1986). The flux H through a given boundary is obtained from a numerical flux function. In the first order versions of such solvers the solution is assumed piecewise constant over these volumes, and the numerical flux depends only on the constant states U_I and U_J on both sides of the boundary, and on the boundary normal $\hat{n}$:

$$\tilde{H} = \tilde{H}(U_I, U_J, \hat{n}) \tag{2.1}$$

The precise form of this numerical flux function is characteristic for the flux-splitter which has been chosen, e.g. Steger-Warming, Van Leer, Roe, Osher-Engquist etc. For example, for flux vector splittings such as Van Leer or Steger-Warming, one has (discarding the dependence on $\hat{n}$ in the notation) :

$$\tilde{H} = \tilde{H}(U_I, U_J) = H^+(U_I) + H^-(U_J) \tag{2.2}$$

where the Jacobian of H^+ resp. H^- has only positive resp. negative eigenvalues, expressing propagation of information from the left resp. right towards the finite volume boundary. For consistency, the numerical flux function has to satisfy the relation

$$\tilde{H}(U, U) = H(U) \tag{2.3}$$

where $H(U)$ is the flux function of the Euler equations for a given normal direction $\hat{n}$:

$$H(U) = (\rho u_n, \rho u u_n + p n_x, \rho v u_n + p n_y, \rho u_n H)^T \tag{2.4}$$

When the finite volume boundary lies along the physical boundary of the flow domain, additional unknowns U_B are used along the boundary, for which no flux balance equation can be written. To find these additional unknowns, the consistency condition eq. (2.3) is extended to the boundary fluxes by requiring the following relation to be satisfied by the numerical flux through a physical boundary :

$$\tilde{H}(U, U) = H^*(U) \tag{2.5}$$

where $H^*(U)$ is the flux function of the Euler equations in which the constraints of the boundary conditions have been enforced, i.e. $H^*(U)$ satisfies the physical boundary conditions. For example, for a solid wall boundary one has :

$$H^*(U) = (0, p(U)n_x, p(U)n_y, 0)^T \qquad (2.6)$$

The unknowns U_B along the boundary are then determined such that

$$\tilde{H}(U_I, U_B) = H^*(U_B) \qquad (2.7)$$

In the above expressions, U_B is the vector of conservative variables at the wall. The left hand side is the numerical flux function depending on the particular Riemann splitter which has been chosen, while the right hand side is the flux function of the Euler equations satisfying the physical boundary conditions. A similar approach for the boundary conditions using the Osher scheme is given by Hemker and Spekreijse (1986). In this particular case, the treatment is equivalent to the use of characteristic boundary conditions because of the definition of the Osher splitting.

Equation (2.7) provides a nonlinear algebraic relationship between the interior unknown U_I and the boundary unknown U_B. It is the additional equation for the boundary unknowns which completes the system of flux balance equations.

In appendix A eq. (2.7) is worked out for the Van Leer splitting for a solid wall in an explicit scheme.

In an implicit scheme eq. (2.7) is taken at the new time level $n+1$. Then a linearisation around time level n gives :

$$\tilde{H}(U_I^n, U_B^n) + \frac{\partial \tilde{H}}{\partial U_I}\Delta U_I + \frac{\partial \tilde{H}}{\partial U_B}\Delta U_B = \frac{\partial H^*}{\partial U_B}\Delta U_B + H^*(U_B^n) \qquad (2.8)$$

This equation closes the set of implicit relations obtained from the Newton linearisation of the interior cell equations. The matrices $\partial \tilde{H}/\partial U_I$ and $\partial \tilde{H}/\partial U_B$ appear also in the scheme for the interior region. An example of $\partial H^*/\partial U$ for a solid wall is given in appendix B.

3.   Analysis of the 1D linear case

The consequences of this method are most easily analysed when applied to a 1D linear case. The system to be considered is :

$$\frac{\partial U}{\partial t} + A\frac{\partial U}{\partial x} = 0 \quad \text{or} \quad \frac{\partial U}{\partial t} + \frac{\partial H}{\partial x} = 0 \qquad (3.1)$$

The boundary conditions can be applied by imposing a number of characteristic ingoing variables V_*^{in}. The boundary condition equation (2.7) can be transformed to characteristic variables V using the matrices of the left (L) and right (L^{-1}) eigenvectors of A. The characteristic variables V are split into an ingoing and outgoing group. Taking the normal in an outward direction, one obtains :

$$V = \begin{pmatrix} V^{in} \\ V^{out} \end{pmatrix} = LU$$

$$A = L^{-1}DL, \; A^+ = L^{-1}D^+L, \; A^- = L^{-1}D^{-1}L, \; H^+ = A^+U, \; H^- = A^-U$$

with

$$D = \begin{pmatrix} D^{in} & 0 \\ 0 & D^{out} \end{pmatrix}; \; D^+ = \begin{pmatrix} 0 & 0 \\ 0 & D^{out} \end{pmatrix}; \; D^- = \begin{pmatrix} D^{in} & 0 \\ 0 & 0 \end{pmatrix}$$

and eq. (2.7) becomes

$$D^+V_I + D^-V_B = D \begin{pmatrix} V^{in}_* \\ V^{out}_B \end{pmatrix}$$

which leads to

$$V^{out}_B = V^{out}_I \text{ and } V^{in}_B = V^{in}_*$$

The method therefore boils down to characteristic boundary conditions.

A similar result is obtained if the boundary conditions are imposed by fixing some primitive variables. For example in the case of the Euler equations at a subsonic outlet we could specify the pressure.

Splitting the variables U in two groups, the imposed variables and the unknown part,

$$U = \begin{pmatrix} U^{imp} \\ U^{unk} \end{pmatrix}$$

the boundary conditions are specified by giving U^{imp} in B. The transformation matrix L couples U^{imp} and U^{unk} to both the ingoing and outgoing characteristics V^{in} and V^{out}.

$$\begin{pmatrix} V^{in} \\ V^{out} \end{pmatrix} = \begin{pmatrix} L^{in\ imp} & L^{in\ unk} \\ L^{out\ imp} & L^{out\ unk} \end{pmatrix} \begin{pmatrix} U^{imp} \\ U^{unk} \end{pmatrix}$$

Eq. (2.7) now leads to

$$D^{out}L^{out\ imp}U^{imp}_B + D^{out}L^{out\ unk}U^{unk}_B = D^{out}L^{out\ imp}U^{imp}_I + D^{out}L^{out\ unk}U^{unk}_I$$

which is solved for the unknown part of U_B. This means that in order to have a well posed boundary condition, $L^{out\ unk}$ must be invertible.

In the case that all variables at B are specified, as well as the case that none of the variables at B are given, the same result is obtained as with characteristic boundary conditions.

4. Application on test cases

The method described above has been implemented in both an explicit and an implicit scheme. The test case chosen was the flow through a channel with on the lower surface a circular bump (Rizzi and Viviand 1981). In Fig. 1, the result is shown for the upwind scheme where the in and outflow boundaries are treated with first order space extrapolation, while at the solid walls, the pressure was extrapolated from the interior, using the normal momentum equation (Rizzi 1978). In Fig. 2, the boundary condition described in this paper is used. As can be seen, the result for the Mach isolines is the same. The erronous entropy generation on the lower boundary with the pressure extrapolation boundary conditions is much larger than with the upwind boundary conditions. The convergence history for the present method for the explicit scheme and some implicit schemes with the upwind boundary conditions is given in Fig. 3.

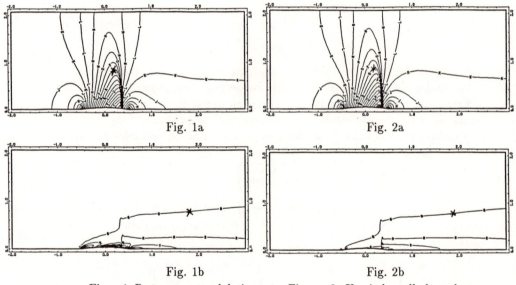

Fig. 1a   Fig. 2a

Fig. 1b   Fig. 2b

Fig. 1 Pressure extraplolation at   Fig. 2 Upwind wall boundary
solid wall (Rizzi 1978)   condition

Figs. 1a, 2a Isomachlines for the channel with the circular bump on the lower wall. $M_{inlet}=0.85$. *=sonic line. Increment=0.25.

Figs. 1b, 2b Isentropy lines. The entropy in units R is related to the freestream conditions. *=freestream. Increment=0.01.

RES

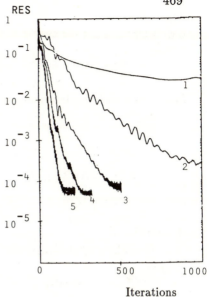

Fig. 3: Convergence history for the channel with the circular bump on the lower wall for explicit and implicit schemes (72 × 72 cells)

$$1 = \text{explicit (CFL=1)}$$
$$\left.\begin{array}{l} 2 = \text{point Jacobi} \\ 3 = \text{line Jacobi} \\ 4 = \text{point Gauss--Seidel} \\ 5 = \text{line Gauss--Seidel} \end{array}\right\} \text{CFL} = 10^5$$

Iterations

5. Conclusions

The proposed boundary treatment is consistent with the flux splitter chosen and reduces to characteristic boundary conditions for the case of a linear hyperbolic system of equations.

Infinite CFL-numbers were possible due to the consistency of the boundary treatment with the interior scheme. The numerical entropy production along the wall for the new boundary conditions is compared with the results for a non characteristic boundary treatment showing improved accuracy for the upwind boundary method.

Appendix A

The upwind solid wall boundary condition for an explicit scheme with van Leer flux vector splitting.

For the numerical flux function of the van Leer splitting, one has (Anderson et al. 1986) :

$$\tilde{H}(U_I, U_B, \hat{n}) = \begin{pmatrix} A_I + A_B \\ n_x(A_I T_I + A_B T_B)/\gamma - n_y(A_I V_I + A_B V_B) \\ n_y(A_I T_I + A_B T_B)/\gamma + n_x(A_I V_I + A_B V_B) \\ \frac{A_I}{2}[\frac{T_I^2}{\gamma^2-1} + V_I^2] + \frac{A_B}{2}[\frac{T_B^2}{\gamma^2-1} + V_B^2] \end{pmatrix}$$

where , (with u^t and u^n the tangential and normal velocity components) :

$$A_I = \frac{1}{4}\rho_I c_I (M_I + 1)^2, \ T_I = (\gamma - 1)u_I^n + 2c_I, \ V_I = u_I^t, \ M_I = \frac{u_I^n}{c_I}$$

and

$$A_B = -\frac{1}{4}\rho_B c_B (M_B - 1)^2, \; T_B = (\gamma - 1)u_B^n - 2c_B, \; V_B = u_B^t, \; M_B = \frac{u_B^n}{c_B}$$

For a solid wall the boundary condition is imposed by setting the normal velocity to zero as given in eq. (2.6), substituting in eq. (2.7), one obtains:

$$A_I = -A_B, \; 2A_I T_I = \gamma p_B, \; V_I = V_B, \; T_I = -T_B$$

which can be solved to give : ·

$$u_B^n = 0, \; u_B^t = u_I^t, \; c_B = c_I[1 + M_I(\gamma - 1)/2], \; \rho_B = \rho_I(M_I + 1)^2/[1 + M_I(\gamma - 1)/2]$$

Appendix B

The jacobian of the flux with the solid wall boundary conditions imposed

At a solid wall the normal velocity is zero. Taking this constraint into account, the flux $H^*(U_B)$ is given by eq. (2.6).

The derivatives have to be taken with respect to the conservative variables $U_B = (\rho, \rho u, \rho v. \rho e)_B^T$ and the jacobian then becomes :

$$\frac{\partial H^*}{\partial U_B} = (\gamma - 1) \begin{pmatrix} 0 & 0 & 0 & 0 \\ \frac{u^2 + v^2}{2} n_x & -u n_x & -v n_x & n_x \\ \frac{u^2 + v^2}{2} n_y & -u n_y & -v n_y & n_y \\ 0 & 0 & 0 & 0 \end{pmatrix}$$

References

Anderson, W. K. et al. (1986). A comparison of finite volume flux vector splitting for the Euler equations. *AIAA Journal*, 24(9):1453–1460.

Chakravarthy, S. (1983). Euler equations - Implicit schemes and boundary conditions. *AIAA Journal*, 21(5):699.

Hemker and Spekreijse (1986). *Applied Numerical Mathematics*, 2:475–493.

Rizzi, A. (1978). Numerical implementation of solid body boundary conditions for the Euler equations. *Z.A.M.M.*, 58(7):T301–T304.

Rizzi, A. and Viviand, H., editors (1981). *Notes on numerical fluid mechanics*, volume 3, Numerical methods for the computation of inviscid transonic flows with shock waves. Vieweg.

An optimistic reappraisal of computational techniques in the supercomputer era

G. Moretti

Polytechnic University
Farmingdale, NY 11735, USA

1. Introduction

In the early sixties, I was challanged by the request of evaluating two- and three-dimensional shock layers about reentry vehicles at supersonic and hypersonic speeds. To reduce computational time to a minimum without loss of accuracy, I opted for a technique composed of two parts: an interior point calculation, in which the Euler equations in quasi-linear form were discretized on a non-orthogonal grid and integrated by a two-level scheme, and the calculation of the bow shock, fitted as a boundary, using the Rankine–Hugoniot conditions and carrying information from the inside along a characteristic (Moretti 1966). Later on, I replaced the interior point with the MacCormack scheme (1969), that was simpler and faster. The underlying philosophy was just the opposite to a brute-force approach. The grid was normalized between body and shock, the method of characteristics inspired the calculation of boundary conditions, and the shock was fitted and treated separately from the other nodes.

The accuracy of the proceedure was proved by amazingly good results, obtained on extremely coarse grids (Fig. 1). Blunt body calculations could be made on the computers of the time, at a reasonable cost (6 minutes on an IBM 7094, using a 7×14 mesh; 15 seconds on the same machine, using a 2×4 mesh; such times being reduced by a factor of 6 on a CDC 6600).

Unfortunately, the blunt body problem is too simple, geometrically as well as aerodynamically. Severe topological difficulties arose when we tried to apply the technique of shock-fitting, considering the shocks as boundaries, to more complicated problems (Grossman 1973, Marconi 1973).

Meanwhile, supercomputers came into use. Memory and speed were boosted by orders of magnitude. Apparently, the reasons that had stimulated a search for efficiency and thrift and the consequent generation of shock-fitting techniques are gone. Shock-capturing obtained by different, well-known techniques (Woodward 1984, Klopfer 1988, Loehner 1987) has already produced a large number of very good results.

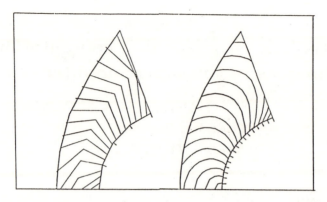

Fig. 1: Constant Mach number lines. On left: 2×4 mesh. On right: 10×16 mesh

Should we, then, proclaim the demise of shock-fitting? Almost paradoxically, I found that shock-fitting, if properly handled, works at its best on supercomputers and it may claim to have advantages over shock- capturing. In this paper, I present arguments in support of the above statement and some results of practical applications. For brevity's sake, no details of the technique will be given.

2.   Qualities of an ideal code

To begin with, what is the goal we aim to reach in the sector of numerical gas dynamics for inviscid flows? We would surely like to have a code for general purposes, that is, capable of handling complicated, unsteady flow fields in the presence of bodies of arbitrary shape, in two and three dimensions. We would like to have a code acting, as much as possible, by brute force. We would like to have a vectorizable code, as fast as possible (so that minutes, not hours, of supercomputer time suffice to analyse unsteady flows about complex geometries, flows about bodies in relative motion, and flutters).

3.   Computational grids

Even before choosing a computational technique, the problem of choosing a computational grid must be faced. Ideally, one would like to have an orthogonal grid, with all rigid surfaces as grid lines. This is highly Utopian. In two dimensions, orthogonal grids obtained by conformal mapping (whether analytic or numerical) may be difficult to generate. Even worse, the size of the cells is far from uniform, so that resolution is lost around concave corners and the time step size is severely limited by the size of cells around convex corners. In three dimensions, orthogonality is unreachable except in elementary cases.

Adaptive grids, coupled with shock-capturing codes, are in fashion. I cannot express opinions on the difficulty of generating a grid that is self-adaptive both to complicated rigid boundaries and to regions of strong gradients. I would only like to say that adapting a grid to shocks is a step toward shock-fitting, without abandoning shock-capturing techniques. Now, shock-capturing codes are generally cumbersome and slow, because the ability to capture shocks must be exercised even where there are no shocks at all (in which case they may also introduce some minor inaccuracy). Adaptive grids add another slowing factor since, at every step, the regions of high gradients must be defined and a new grid may have to be generated. In my opinion a shock-fitting technique on a fixed grid is, at least, much faster than a shock-capturing technique on an adaptive grid.

Setting adaptive grids aside, the generation of a fixed grid may still be a delicate problem. With the introduction of a simple integration scheme (not depending on fine-tuning parameters) plus shock-fitting, grid generation remains the only part of a program requiring a very peculiar skill, particularly for three-dimensional problems. I have explored the possibility of using Cartesian grids throughout. In so doing, the basic integation scheme reaches an utmost simplicity. Boundary conditions along rigid walls which do not lie along Cartesian grid lines can be treated by properly adapting the boundary routine and, as in the case of shocks, by executing the procedure using single arrays of information at chosen points on the boundaries. Current results on selected two-dimensional problems are very encouraging. See, for example, Fig. 2 where lines of constant Mach number for a subsonic flow in a channel with a bump are plotted. The calculation has been made, using the λ-scheme, on a Cartesian grid of 150×30 intervals. (We may note a slight deterioration in the results downstream of the bump, an effect that is not unusual for such a strong perturbation). Cartesian grids have also been used to evaluate the transonic flow past a NACA 0012 airfoil at $M = 0.805$ and no incidence, using a Cartesian grid of 128×64 intervals around the profile and coarser Cartesian grids outside. In this case, the imbedded shock has been fitted. Isobars and the fitted shock are shown in Fig. 3 on two of the overlapping grids. Details and results will be found in a paper by the present author and A. Dadone, to appear shortly in the Proceedings of the 2nd International Conference on Hyperbolic problems, Aachen, March 1988.

4. Practical advantages of shock-fitting

The following considerations do not depend on the choice of grid.

The basic idea of shock-fitting consists of separating the calculation at grid nodes from the shock calculation. The architecture of a supercomputer, such as the Cray X-MP, is properly exploited. Indeed, both proce-

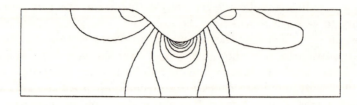

Fig. 2: Subsonic flow in a channel computed with a Cartesian grid

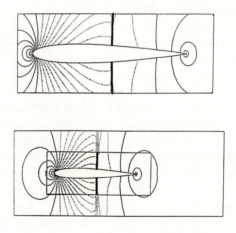

Fig. 3: Isobars around a NACA 0012 airfoil, computed with a Cartesian grid

dures can be vectorized separately; on a grid containing N nodes in each direction, the former updates N^2 or N^3 nodes (for the two-dimensional and three- dimensional problems, respectively), but the latter updates only a number of points of the order of N or N^2.

Not being encumbered by shock-capturing needs, the integration at grid nodes can be performed using a high-efficiency code, based on the concept of characteristics, with entropy as one of the unknowns. The code is at the same time simple and accurate, particularly at sonic transitions and along boundaries.

5. A shock-fitting black-box

Shock-fitting consists of two parts: detection of new shocks, and updating of values, including shock velocity and position. The latter, in turn, requires application of the Rankine–Hugoniot conditions and some special provision to avoid using differences taken across a shock when computing grid nodes. All these procedures are coded in a shock-fitting black-box, which is always the same, regardless of the problem or the computational grid. The most relevant feature of the black-box is a cross-referencing device between grid points and shock points, which allows the neighbourhood of one shock point to be explored without having to organise shock points into shock lines or shock surfaces, in this way eliminating all topological difficulties. Working on a supercomputer, the computational grids are generally so fine that two nodes bracketing a shock can be assumed to represent the low-pressure and high-pressure points on the shock; the code is simpler and faster than when the same technique had to be applied to a coarse mesh and additional information evaluated on the shock itself.

The only part of the "black-box" requiring a scanning of the entire flow field is the test for detection of new shocks, which does not involve many computations. The remaining subroutines scan the arrays of shock point information only.

Details on the shock-fitting technique can be found in papers by the present author to appear shortly in the AIAA Journal and the Computer and Fluids magazine. In the same papers, applications are shown to transonic flows in ducts and intakes (with formation and disappearance of oblique shocks and Mach reflections), transonic airfoils, complex Mach reflections, muzzle flow, plume formation and flow in curved channels.

6. Orders of accuracy

In many problems, no sizeable difference is found between results obtained using a first-order or a second-order scheme. The advantage of first-order schemes is obvious, since the computational time can be cut in half. The possibility has to be kept in mind, in view of further improvements in supercomputers in the near future. Particularly in three-dimensional applications, it may be convenient to increase the number of grid nodes and use a first-order scheme.

7. Conclusions

The λ-scheme-plus–shock-fitting technique works on a supercomputer, such as the CRAY X-MP, taking full advantage of vectorization. The shock fitting black-box, still in an evolutionary phase, is full of redundancies. Nevertheless, the computational time that it requires is generally about 1/4 of the time required by a two-level integration step for nodal points, and never gets higher than 1/2 of such time in the worst of cases. The latter, per se, is very small. To give an idea of the computational speed, 1000 steps on a 151×31 mesh require 22.1 seconds in a shockless problem and 30.9 seconds in the presence of a moderately complicated shock pattern.

I agree with Woodward and Colella (1984) that there is a certain "conservation of difficulty" principle but, I would add, within the framework of a certain style of computing. As they say, the MacCormack scheme is about six times faster than their PPM, but it requires eight times as many computations to reach the same degree of accuracy. Nevertheless, if we change the approach, we can defy the principle. The two- level λ-scheme is more than twice as fast as the MacCormack scheme (which uses the equations in conservation form) and, with shock-fitting, it takes, in the worst of the tested cases, about 75% of the time required by the MacCormack scheme. To reach the same degree of accuracy as the PPM scheme, it requires a coarser grid (75% of the number of nodes in each direction, that is, a reduction in CPU to .42). Considering the factors, .75 (λ-to-MacCormack) and 6 (MacCormack-to-PPM), the total CPU time drops to 5% of PPM time. The same results obtained by the PPM in 30 minutes can be obtained in 1.5 minutes if the λ-plus-shock-fitting technique is used.

In view of future extensions to three-dimensional problems, it seems to me that these figures should not be overlooked.

References

Klopfer, G. and Yee, H. (1988). Viscous hypersonic shock on shock interaction on blunt cowl lips. Paper 0233, AIAA.

Loehner, R. (1987). Efficient FEM-based algorithm for unsteady compressible flows. *Advances in Computational Methods for Partial Differential Equations*, 6:147–56.

MacCormack, R. (1969). The effect of viscosity in impact cratering. Paper 354, AIAA.

Marconi, F. and Salas, M. (1973). Computation of three-dimensional flows about aircraft configurations. *Computers and Fluids*, (1):185–95.

Moretti, G. and Abbett, M. (1966). A time-dependent computational method for blunt body flows. *AIAA Journal*, (4):2136–41.

Woodward, P. and Colella, P. (1984). The numerical simulation of the two-dimensional fluid flow with strong shocks. *Journal of Computational Physics*, (54):115–73.

Evaluation of a parallel conjugate gradient algorithm

R. W. Leland[1] and J. S. Rollett

ICFD, Oxford University Computing Laboratory, Oxford

1. Introduction

When fluid dynamics problems are discretized, the usual result is a large, sparse linear system of equations

$$Au = b \tag{1.1}$$

whose subsequent solution dominates the computational load. Conjugate gradient (CG) type methods when used with appropriate preconditioning schemes have proved effective in reducing both the time complexity and memory requirement of this solution phase on serial computers. It is natural, then, to investigate reworking the basic CG algorithm into parallel form so as to take advantage of recent advances in computer architecture. A parallel CG algorithm has been proposed (Van Rosendale 1983), but apparently not tried. This paper describes our experience testing Van Rosendale's algorithm (VR) and a modification we found necessary.

2. Standard CG algorithm

A suitable version of the standard CG algorithm is (Reid 1971):

Choose an arbitrary u_0.
Set $r_0 = b - Au_0$.
Set $p_0 = r_0$.
Repeat until the residual, r_n, converges satisfactorily:

$$\alpha_n = r_{n-1} \cdot r_{n-1} / p_{n-1} \cdot Ap_{n-1}$$

$$u_n = u_{n-1} + \alpha_n p_{n-1}$$

$$r_n = r_{n-1} - \alpha_n Ap_{n-1}$$

---

[1]Supported by the Rhodes Trust.

$$\beta_n = r_n \cdot r_n / r_{n-1} \cdot r_{n-1}$$

$$p_n = r_n + \beta_n p_{n-1}.$$

Note that if there are N degrees of freedom, and d non-zeroes per matrix row, the weighted vector sums require $2N$ operations, the scalar products $2N - 1$ operations and the matrix by vector multiplications $N(2d - 1)$ operations, so the matrix by vector products will clearly dominate the serial time complexity of each iteration.

3. Parallel CG algorithm

On a parallel machine with enough processors to allocate one per matrix row, matrix by vector products require $2d - 1$ time. Scalar products, however, require $\log_2(N) + 1$ parallel time steps because following multiplication in parallel, N numbers must be combined into one by the binary addition operator. We assume small d and large N, hence the scalar products now dominate the computational complexity. If $O(N)$ iterations are required for convergence without preconditioning, then it would seem that parallel CG was at best an $O(N \log(N))$ process.

Van Rosendale's insight was that the scalar product calculations for several iterations can be overlapped. At first the algorithm's data dependencies seem to rule this out. For example, we cannot generate $r_n \cdot r_n$ needed for calculation of β_n until we know α_n, which in turn requires $r_{n-1} \cdot r_{n-1}$.

Consider, however, the formula for the residual

$$r_n = r_{n-1} - \alpha_n A p_{n-1}. \tag{3.1}$$

If we know r_{n-1}, p_{n-1} from the previous iteration, but do not yet know α_n, we can still perform most of the work in calculating $r_n \cdot r_n$ since we can find all of the scalar products on the right hand side of

$$r_n \cdot r_n = (r_{n-1} \cdot r_{n-1}) - 2\alpha_n(r_{n-1} \cdot A p_{n-1}) + \alpha_n^2(A p_{n-1} \cdot A p_{n-1}). \tag{3.2}$$

If we subsequently learned the value of α_n, we could evaluate $r_n \cdot r_n$ with only a few extra operations, negligible work compared with that of evaluating the scalar products at step $n - 1$ for large N.

Equation (3.1) can be used several more times in conjunction with the search vector recursion $p_n = r_n + \beta_n p_{n-1}$ to express $r_n \cdot r_n$ in terms of scalar products at, say, step $n - k$. An independent processor could begin calculating these at that step, and if k were large enough it would finish by step n, leaving only the summation of $O(k)$ terms to be performed, an $O(\log(k))$ parallel process. Meanwhile other useful work could progess, including calculation of like scalar products for use at steps greater than

n. This idea can be extended to cover all of the scalar products required in the CG algorithm.

Appropriate choice of the *look ahead* parameter, k, will fully overlap the scalar product calculations and minimize the delay associated with finding α_n, β_n. If k is too large the summation is too costly, but if it is too small we wait for the scalar product result. The correct balance is $k = \log_2(N)$, which gives each iteration parallel complexity $O(\log(\log(N)))$ and makes CG an $O(N \log(\log(N)))$ parallel process if no preconditioning is used.

Our version of VR goes like this. To calculate α_n, we let $m = n - 1$ in

$$
\begin{aligned}
r_m \cdot r_m = {} & \sum_{i=0}^{2k-2} a_i(r_{m-k} \cdot A^i r_{m-k}) + \sum_{i=0}^{2k-1} b_i(r_{m-k} \cdot A^i p_{m-k}) \\
& + \sum_{i=0}^{2k} c_i(p_{m-k} \cdot A^i p_{m-k})
\end{aligned}
\tag{3.3}
$$

$$
\begin{aligned}
p_m \cdot A p_m = {} & \sum_{i=0}^{2k-2} d_i(r_{m-k} \cdot A^i r_{m-k}) + \sum_{i=0}^{2k-1} e_i(r_{m-k} \cdot A^i p_{m-k}) \\
& + \sum_{i=0}^{2k} f_i(p_{m-k} \cdot A^i p_{m-k}).
\end{aligned}
\tag{3.4}
$$

The $a_i \ldots f_i$ are polynomials in the parameters

$$
\{\alpha_m, \ \alpha_{m-1}, \ \ldots, \ \alpha_{m-k+1}, \ \beta_m, \ \beta_{m-1}, \ \ldots, \ \beta_{m-k+1}\}
\tag{3.5}
$$

and can be found either explicitly or implicitly by repeated substitution into the right hand sides of (3.3) and (3.4) using the recurrences

$$
r_l \cdot A^i r_l = (r_{l-1} \cdot A^i r_{l-1}) - 2\alpha_l(r_{l-1} \cdot A^{i+1} p_{l-1}) + \alpha_l^2(p_{l-1} \cdot A^{i+2} p_{l-1})
\tag{3.6}
$$

$$
\begin{aligned}
r_l \cdot A^i p_l = {} & (r_{l-1} \cdot A^i r_{l-1}) - 2\alpha_l(r_{l-1} \cdot A^{i+1} p_{l-1}) + 2\beta_l(r_{l-1} \cdot A^i p_{l-1}) \\
& + \alpha_l^2(p_{l-1} \cdot A^{i+2} p_{l-1}) - \alpha_l\beta_l(p_{l-1} \cdot A^{i+1} p_{l-1})
\end{aligned}
\tag{3.7}
$$

$$
\begin{aligned}
p_l \cdot A^i p_l = {} & (r_{l-1} \cdot A^i r_{l-1}) - 2\alpha_l(r_{l-1} \cdot A^{i+1} p_{l-1}) + 2\beta_l(r_{l-1} \cdot A^i p_{l-1}) \\
& + \alpha_l^2(p_{l-1} \cdot A^{i+2} p_{l-1}) - 2\alpha_l\beta_l(p_{l-1} \cdot A^{i+1} p_{l-1}) \\
& + \beta_l^2(p_{l-1} \cdot A^i p_{l-1}).
\end{aligned}
\tag{3.8}
$$

When $l = m - k$ we have reached the base of the recursion, and need to evaluate

$$
\left\{r_{m-k} \cdot A^i r_{m-k}\right\}_{i=0}^{2k-2}, \quad \left\{r_{m-k} \cdot A^i p_{m-k}\right\}_{i=0}^{2k-1}, \quad \left\{p_{m-k} \cdot A^i p_{m-k}\right\}_{i=0}^{2k}
\tag{3.9}
$$

by some other means. We do this using the recurrence relations

$$A^i r_l = A^i r_{l-1} - \alpha_l A^{i+1} p_{l-1} \tag{3.10}$$

$$A^i p_l = A^i r_l - \beta_l A^i p_{l-1} \tag{3.11}$$

so that in principle we would only need to generate high powers of A using repeated matrix by vector multiplication once (when $l = 0$), effort which could be counted as overhead. In the next section we explain that in practice we needed to do this more often.

Note that we require only the vector sets

$$\left\{ A^i r_l \right\}_{i=0}^{k}, \quad \left\{ A^i p_l \right\}_{i=0}^{k+1} \tag{3.12}$$

since, for example $r_{m-k} \cdot A^{2k} r_{m-k} = A^k r_{m-k} \cdot A^k r_{m-k}$. Notice also that we will need to do one matrix by vector multiplication per iteration (as in CG) to get $A^{k+1} p_l$.

At the end of step $n-k$ we have available α_{n-k}, β_{n-k}, u_{n-k}, r_{n-k}, p_{n-k}, so we can launch an independent process to update (3.12) using (3.10) and (3.11) and hence determine (3.9). The process can then begin evaluating $r_{n-1} \cdot r_{n-1}$ and $p_{n-1} \cdot A p_{n-1}$ by filling in terms in (3.3) and (3.4). It will quickly be suspended because it does not know

$$\{ \alpha_{n-k+1}, \ \alpha_{n-k+2}, \ \ldots, \ \alpha_{n-1}, \ \beta_{n-k+1}, \ \beta_{n-k+2}, \ \ldots, \ \beta_{n-1} \} \tag{3.13}$$

but if the process is communicating with similar processes launched at steps $n-k+1$, $n-k+2$, $\ldots$, $n-1$, then these parameters will gradually be provided. By the start of step n, they will all be known and there will be only a relatively short delay while the terms are summed. A similar process is launched to find $r_n \cdot r_n$ and thereby calculate β_n, enabling us to update to u_n, r_n and p_n.

4. Results

We implemented VR in *Pascal* on a sequential computer using IEEE 64 bit real numbers and tested it on a system derived from a uniform central finite difference discretization of the problem

$$\frac{\partial^2 u}{\partial x^2} + 3 \frac{\partial^2 u}{\partial y^2} = 0, \qquad \Omega = [0, \pi] \times [0, \pi] \tag{4.1}$$

$$
\begin{aligned}
u &= \sin(y), & x &= 0; & u &= \cos(x/2), & x &= \pi \\
u &= \sin(x/2), & y &= 0; & u &= 0, & y &= \pi
\end{aligned}
\tag{4.2}
$$

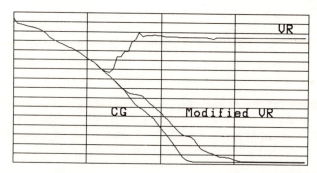

Fig. 1: Method comparison, $N = 64$. The horizontal axis shows iteration 0 to 80, the vertical $\log_{10}(\|\text{error}\|_2/\|\text{solution}\|_2)$ ranging from 2 to -15

We used a chequerboard mode of unit amplitude in superposition with $x(x-1)y(y-1)$ normalized to peak at unity as initial data, and assumed Gaussian elimination gave the correct solution to the discrete problem.

Fig. 1 shows that while CG achieves nearly full machine accuracy, VR is at best some ten orders of magnitude less accurate. In Modified VR we formed the vector sets (3.12) directly rather than recursively on each iteration, and good accuracy returned, although marginally delayed.

5.   Conclusions

We can probably blame the disappointing accuracy of VR on the recursive updates (3.10), (3.11), which isolate the algorithm from its recent history and accentuate loss of orthogonality in the residual and search vectors. The situation will probably get much worse for N of sufficient size to warrant the method's use because this isolation will increase and because such problems are generally of poorer condition as well. Hence VR as originally presented is probably not viable.

Modified VR reaches full accuracy after a modest number of additional iterations. Unfortunately it returns us to an $O(N \log(N))$ parallel complexity unless we have $O(\log(N))$ processors per degree of freedom, which is unrealistic. However if about $\log_2(\log_2(N))/d \log_2(N)$ direct updates will suffice, then $O(N \log(\log(N)))$ parallel complexity can be retained with $O(N)$ processors in total, which is potentially realizable. This should certainly be investigated.

Preconditioning is another important area for investigation since it may reduce the required frequency of direct updates. While polynomial preconditioning is not cost effective on a serial machine, it will probably be appropriate in this context since it parallelizes much more directly than more standard factorization schemes.

If the method can be succesfully adapted using no more than about one processor per degree of freedom, it will be useful primarily in solving systems generated by very local e.g. finite difference schemes since it requires $N > 2^{2d}$. We must await a significant increase in the number of processors which can operate in parallel in order to apply the method to, say, a 3D finite element problem where typically $d = 250$.

References

Reid, J. K. (1971) On the method of conjugate gradients for the solution of large sparse systems of linear equations. In *Large Sparse Sets of Linear Equations*, J. K. Reid, editor, Academic Press, London.

Van Rosendale, J. (1983) Minimizing inner product data dependencies in conjugate gradient iteration. In *1983 International Conference on Parallel Processing*, H. J. Siegel *et al.*, editors, IEEE.

Mixed finite elements for highly viscoelastic flows

J.-M. Marchal

University of Louvain-la-Neuve   Place du Levant,2
B–1348 Louvain-La-Neuve,Belgium

1. Introduction

All polymer solutions and melts are characterized by a non-Newtonian behaviour which is often entirely different from what has been learnt over many years of Newtonian fluid mechanics. The elasticity of the fluid for example, i.e. the memory of its past deformation, plays an important role. Many constitutive equations have been proposed over the past 30 years to describe viscoelastic materials. Our present interest is in fluids of the differential type, i.e. fluids whose stress tensor is expressed as a function of the deformation history by means of a differential equation. The associated mathematical problems are strongly non-linear, and the use of numerical methods is indispensible for solving non-Newtonian flows. The numerical simulation of polymer processes requires efficient and accurate algorithms for solving flow problems based on viscoelastic constitutive equations. The development of numerical methods for calculating viscoelastic flow is now about ten years old, and several reviews are available (see e.g. Crochet et al. 1984, Keunings 1988). The elasticity of the flow is characterized by a non-dimensional number, e.g. the Weissenberg number We $= \lambda V/L$, where λ is a typical relaxation time of the fluid, and V, L are characteristic velocity and length of the flow. As long as We < 1, viscoelastic solutions differ little from their Newtonian counterpart as far as kinematics is concerned. Early developments have shown however that it was extremely difficult to obtain solutions for moderate values of We; the so-called "high Weissenberg number problem" has given rise to intensive research. The problem has been attacked with an arsenal of the most available numerical methods: finite elements and finite differences, boundary elements and spectral/finite element methods (most references may be found in Keunings (1988)).

Various conjectures have been proposed for justifying the loss of convergence of these numerical methods for large values of We. Over the last couple of years, however, a consensus has been reached over the fact that the major reason lies in the loss of accuracy of the numerical methods. Most viscoelastic flows are characterized either by thin stress boundary layers (the flow around a sphere) or by strong stress singularities (the flow through a 4:1 contraction for example).

We have designed a new stress-velocity-pressure finite element based on stress subrefinement and a Streamline-Upwind integration of the hyperbolic constitutive equation; this technique allows us to compute viscoelastic flows at (very) high Weissenberg numbers and to predict important kinematic deviations form the Newtonian flow. The method, described in detail in Marchal and Crochet (1987) is summarized below, and two simulations are presented.

2.   The flow of Oldroyd-B fluids

Solving a steady-state differential Oldroyd-B problem on a domain Ω of boundary Γ is equivalent to finding $(\mathbf{T}_1, \mathbf{v}, p)$ such that

$$\mathbf{T}_1 + \text{We}\overset{\triangledown}{\mathbf{T}}_1 - \delta(\nabla\mathbf{v} + \nabla\mathbf{v}^T) = 0, \qquad \text{on } \Omega$$

$$\nabla.(\mathbf{T}_1 + (1 - \delta)(\nabla\mathbf{v} + \nabla\mathbf{v}^T) - p\mathbf{I}) = \text{Re } \mathbf{v}.\nabla\mathbf{v}, \qquad \text{on } \Omega$$

$$\nabla.\mathbf{v} = 0, \qquad \text{on } \Omega \quad (2.1)$$

$$\mathbf{v} = \mathbf{g} \text{ on } \Gamma, \text{ constrained by } \int_\Gamma \mathbf{g}.\mathbf{n}d\gamma = 0, \; p = 0, \text{ in } \mathbf{x} = \mathbf{x}^0,$$

$$\mathbf{T}_1 = \mathbf{h}, \qquad \text{on } \Gamma-$$

where $\mathbf{T}_1, \mathbf{v}$ and p are respectively the viscoelastic stress tensor, the velocity vector and the pressure, and $\Gamma-$ is the inflow boundary. We is the Weissenberg number, Re is the Reynolds number and δ is a viscosity ratio. In most polymer applications, $\text{Re} \ll 1$, so that neglecting the inertia term $(\text{Re} = 0)$ is a common assumption, although recent work by Joseph et al. (1987) has revealed that even a small non-vanishing Reynolds number can entirely modify the nature of the flow in the presence of elasticity. The system (2.1) is of mixed elliptic-hyperbolic type, and the limit $\delta = 1$ corresponds to the so-called upper-convected Maxwell model, which is a singular perturbation of (2.1). $\overset{\triangledown}{\mathbf{T}}_1$ is the upper convected or contravariant derivative of $\mathbf{T}_1$, defined as:

$$\overset{\triangledown}{\mathbf{T}}_1 = \mathbf{v}.\nabla\mathbf{T}_1 - \nabla\mathbf{v}^T.\mathbf{T}_1 - \mathbf{T}_1.\nabla\mathbf{v}. \qquad (2.2)$$

3.   The finite element technique

The finite element discretization of (2.1) follows a classical approach: a weak form of (2.1) is written and finite element approximation subspaces $\mathbf{T}^h, \mathbf{V}^h$ and $\mathbf{P}^h$ are introduced for the stresses, velocities and pressure. The discrete Oldroyd-B problem then reduces to finding $(\mathbf{T}_1^h, \mathbf{v}^h, p^h) \in \mathbf{T}^h\mathbf{x}\mathbf{V}^h\mathbf{x}\mathbf{P}^h$ such that

$$\left\langle \mathbf{T}_1^h + \text{We}\overset{\triangledown}{\mathbf{T}}_1^h - \delta(\nabla\mathbf{v}^h + \nabla\mathbf{v}^{h^T}); \mathbf{S}^h \right\rangle = 0, \qquad \forall \mathbf{S}^h \in \mathbf{T}^h (3.1)$$

$$\left\langle \mathbf{T}_1^h + (1-\delta)(\nabla\mathbf{v}^h + \nabla\mathbf{v}^{h^{\mathrm{T}}}) - p^h\mathbf{I}; \nabla\mathbf{u}^h + \nabla\mathbf{u}^{h^{\mathrm{T}}}\right\rangle = f(\mathbf{u}^h), \ \forall\mathbf{u}^h \in \mathbf{V}^h (3.2)$$

$$\left\langle \nabla.\mathbf{v}; q^h\right\rangle = 0, \qquad\qquad\qquad \forall q^h \in \mathbf{P}^h (3.3)$$

where $\langle \cdot; \cdot \rangle$ is th L^2 inner product over Ω (the case of the natural boundary conditions has not been included above for simplicity). Conformity when We > 0 requires that $\mathbf{T}^h$ and $\mathbf{V}^h$ be C^0 at least, whilst $\mathbf{P}^h$ only has to be C^{-1}.

The discrete problem (3.1–3.3) is non-linear, and its complete analysis remains to be done; however, one can understand its nature by studying a simpler case, the stress-velocity-pressure Stokes problem, obtained when We $= 0$ and $\delta = 1$.

3.1   *The stress-velocity-pressure Stokes problem*

The varitional principle corresponding to the Stokes problem is written as follows: find $(\mathbf{T}_1^h, \mathbf{v}^h, p^h) \in \mathbf{T}^h \mathrm{x}\mathbf{V}^h \mathrm{x}\mathbf{P}^h$ such that

$$
\begin{array}{llll}
a(\mathbf{T}_1^h, \mathbf{S}^h) + b(\mathbf{v}^h, \mathbf{S}^h) & & = \quad 0, & \forall \mathbf{S}^h \in \mathbf{T}^h \\
b(\mathbf{u}^h, \mathbf{T}_1^h) & +c(p^h, \mathbf{u}^h) = f(\mathbf{u}^h), & & \forall \mathbf{u}^h \in \mathbf{V}^h \qquad (3.4) \\
+c(q^h, \mathbf{v}^h) & & = \quad 0, & \forall q^h \in \mathbf{P}^h
\end{array}
$$

where

$$a(\mathbf{T}_1^h, \mathbf{S}^h) = \langle \mathbf{T}_1^h; \mathbf{S}^h\rangle, \ \ b(\mathbf{v}^h, \mathbf{S}^h) = -\langle \nabla\mathbf{v}^h + \nabla\mathbf{v}^{h^{\mathrm{T}}}; \mathbf{S}^h\rangle,$$

$$c(p^h, \mathbf{u}^h) = 2\langle p^h; \nabla.\mathbf{u}^h\rangle.$$

The problem (3.4) appears to be constrained, the primal unknowns being the stresses; the divergence free equation is a constraint on the velocity field and the momentum equation is a constraint on the constitutive equation. Finding an element satisfying the Brezzi-Babuska conditions with C^0 *stresses* appears to be non-trivial; we have tested a family of elements based on a sub-refinement of the primal unknowns. The velocity field being C^0-biquadratic and the pressure C^0-bilinear, the stress field is C^0-bilinear on a 2x2, 3x3 or 4x4 sub-mesh (see Fig. 1). These three elements have been tested on the well-known stick-slip problem. Whilst the 2×2-stress element gave very poor results, the 3×3-stress element was much more accurate, and the 4×4-stress solution was almost identical to its velocity-pressure counterpart. *The 4×4 sub-refinement been chosen regardless of the computational cost.*

3.2   *Viscoelastic problems*

The viscoelastic stick-slip problem has then been computed. Not surprisingly the Galerkin formulation (3.1) gave rather poor results, even with

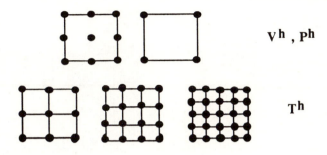

Fig. 1: A new family of viscoelastic elements

the 4×4 stress sub-mesh. Therefore, we implemented in a first step a consistent Streamline Upwind/Petrov Galerkin (SUPG, Brooks and Hughes 1982) formulation of (3.1). The weighting function $\mathbf{S}^h$ is replace by $\hat{\mathbf{S}}^h$ defined as

$$\hat{\mathbf{S}}^h = \mathbf{S}^h + \overline{\mathbf{k}}\mathbf{v}^h.\nabla\mathbf{S}^h;$$

where $\overline{\mathbf{k}}$ is of the order of the element size h.

Although the SUPG technique gave accurate results for calculating stresses on the basis of a known velocity field, it completely *failed* in the case of coupled stress-velocity-pressure calculations. Therefore, we have tested a non-consistent (O(h)) method where only the term $\mathrm{We}\,\mathbf{v}^h.\nabla\mathbf{T}_1^h$ is weighted by $\hat{\mathbf{S}}^h$ in (3.1), the other terms being weighted by $\mathbf{S}^h$. The method, which can be interpreted in terms of streamline diffusion, has been labelled Streamline Upwind (SU).

The "SU 4×4" element has been found to perform outstandingly well on the viscoelastic stick-slip problem, as well as on alternate problems. The use of a (very) refined stress sub-mesh keeps the artificial diffusion at an acceptable level; however, the discrepancy between the coupled "SUPG 4×4" and the "SU 4×4" results is not yet totally understood.

4.   The 4:1 axisymmetric contraction problem

The flow of an Oldroyd-B fluid through an axisymmetric abrupt 4:1 contraction has been computed as a function of the Weissenberg number, choosing $\delta = 8/9$. The fluid sticks at the wall of the contraction. Defining the Weissenberg number as

$$\mathrm{We} = \frac{\lambda V}{6L},$$

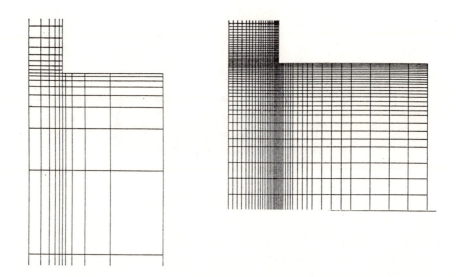

Fig. 2: Macro-mesh and stress sub-mesh for the 4:1 contraction problem. Zoom in the corner region

where V and L are respectively the average velocity and the radius in the downstream small tube, the simulation has been pursued up to We = 60 (whilst former methods where typically limited at We = 3 ... 5). Fig. 2 shows an enlargement of one of the macro-meshes that we have used for these computation, as well as the stress sub-mesh. Fig. 3 presents the development of the corner vortex due to elasticity; the presence of such vortices has been confirmed by calculation on the three different meshes. For a comparison with experimental results, see Boger (1987) and Debbaut et al. (1988).

5.   The flow of an Oldroyd-B fluid around a sphere

A sphere of radius R moves at a constant speed relatively to a cylinder of radius $2R$. We are interested in predicting the resultant force on the sphere as a function of the Weissenberg number, defined as

$$\text{We} = \frac{\lambda V}{R},$$

where V is the relative velocity between the sphere and the cylinder. Defining the drag factor χ_e as the relative deviation from the Newtonian drag, and selecting $\delta = 8/9$, we have computed the solution as a function of We up to We = 3.5 (the calculation has then been stopped by lack of interest).

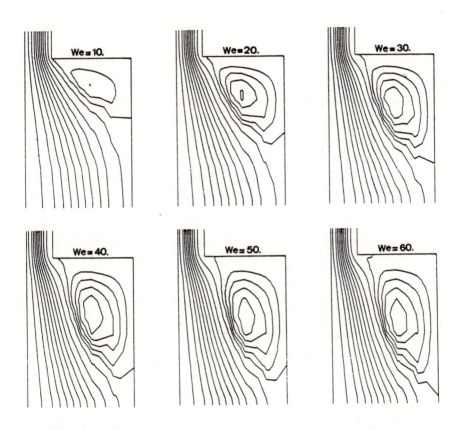

Fig. 3: Development of the corner vortex due to elasticity

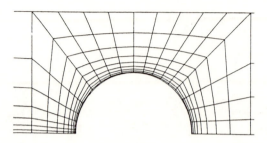

Fig. 4: Zoom on the macro-mesh used for the simulation of flow around a sphere

One of the three macro-meshes used for this calculation is presented in Fig. 4. The computed relationship between χ_e and We is presented in Fig. 5. Interestingly, the curve presents a minimum for We = 1.1; this behaviour has been conjectured by Barnes and Walters (1980) on qualitative grounds.

6.   Conclusion

The improvement of the numerical techniques and the development of a new mixed macro-element allows us to compute flows for values of the Weissenberg number well beyond 1, and to predict important kinematic deviations from the Newtonian flow. Lack of convergence has not limited our numerical investigations for flows of an Oldroyd-B fluid through an abrupt contraction and around a sphere. The failure of the coupled consistent "SUPG 4×4" element still remains an open question. Moreover, further analysis will be required in order to obtain accurate viscoelastic simulations at a lower computational cost.

References

Barnes, H. and Walters, K. (1980). Anomolous extensional-flow effects in the use of commercial viscometers. In *Proc. VIII Int. Cong. on Rheology*, Astarita, G., Marucci, G., and Nicolais, L., editors, pages 45–62, N.Y. Plenum Press.

Boger, D. (1987). Viscoelastic flow through contractions. *Ann. Rev. Fluid Mech.*, (19):157–182.

Brooks, A. and Hughes, T. (1982). Streamline-upwind/Petrov-Galerkin formulations for convection dominated flows with particular emphasis on the compressible Navier-Stokes equations. *Comp. Meth. Appl. Mech. Eng*, (32):199–259.

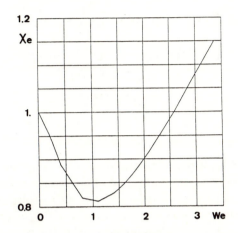

Fig. 5: Relative deviation from the Newtonian drag

Crochet, M., Davies, A., and Walters, K. (1984). *Numerical simulation of non-Newtonian Flow*. Elsevier.

Debbaut, B., Marchal, J., and Crochet, M. (1988). Numerical simulation of highly viscoelastic flows through an abrupt contraction. To appear in *Journal of non-Newtonian Fluid Mechanics*.

Joseph, D., Matta, J., and Chen, K. (1987). Delayed die swell. *Journal of non-Newtonian Fluid Mechanics*, (24):31–65.

Keunings, R. (1988). Simulation of viscoelastic fluid flow. In *Fundamentals of Computer Modelling for Polymer Processing*, Tucker, C., editor. Carl Hanser Verlag.

Marchal, J. and Crochet, M. (1987). A new nixed finite element for calculating viscoelastic flow. *Journal of non-Newtonian Fluid Mechanics*, (26):77–114.

Multiphase flow—a self-consistent approach

D. F. Fletcher and A. Thyagaraja

Culham Laboratory

Abingdon, Oxon

1. Introduction

The calculation of multiphase flow is of considerable importance in the nuclear and process industries. Typical applications include the study of two-phase flow in boilers and condensers. In such applications the flow is often in complex geometries and a wide variety of flow regimes may occur. Most of the physics of the problem is contained in the constitutive relations for heat transfer, inter-phase drag and wall drag. The computation of such flows is now a relatively routine matter, although the concept of qualitative consistency is often missing from the schemes used.

We use the term qualitative consistency to mean: constructing a finite difference model so that it has the same physical characteristics as the real-world problem (which has already been translated into a system of partial differential equations). Examples of qualitative consistency are making sure that the volume fractions, used to measure occupancy, are always positive and sum to unity and ensuring that the entropy increases monotonically across a shock.

In this paper we will *not* be concerned with the question of ill-posedness of the multiphase flow equations (Stewart and Wendroff 1984) but rather with practical methods to obtain a consistent solution to the problem of interest. We have been using multiphase flow models to study vapour explosions which can occur when a hot liquid is poured into a cooler volatile liquid (Cronenberg 1980). Our study of this phenomenon has led to the development of two related codes. The first, a transient, two-dimensional, incompressible flow code, models the initial mixing phase. The second, a transient, one-dimensional compressible flow code, models detonations. The principles employed in the construction of these codes were:

- They must be robust, stable and ensure qualitative consistency;

- They must be able to work with virtually arbitrary constitutive relations, such as inter-phase drag and heat transfer, since these quantities are not well-known;

- There is no great need for high order accuracy in the finite difference scheme since many of the constitutive relations are only known approximately;

- They must be relatively cheap to use so that the effect of uncertain parameters can be scoped.

In the next two sections we will describe the principles we employed to construct self-consistent schemes to study mixing and detonations.

2. Incompressible multiphase flow

In this section we describe the principles employed to construct a finite difference code to model the behaviour of a hot liquid (melt) poured into a cold volatile liquid (water). The code is transient, 2D (cylindrically symmetric), incompressible and models three components, namely melt (M), water (W) and steam (S). It is fully described in Thyagaraja and Fletcher (1988) and Fletcher and Thyagaraja (1987a). To illustrate the types of equations to be solved we will present the equations for conservation of mass and momentum for the water phase. Conservation of water mass gives:

$$\frac{\partial}{\partial t}(\alpha_W) + \frac{1}{r}\frac{\partial}{\partial r}(r\alpha_W U_W) + \frac{\partial}{\partial z}(\alpha_W V_W) = -\dot{m}_W \qquad (2.1)$$

and conservation of momentum in the radial and axial directions gives:

$$\rho_W\frac{\partial}{\partial t}(\alpha_W U_W) + \rho_W\frac{1}{r}\frac{\partial}{\partial r}(r\alpha_W U_W^2) + \rho_W\frac{\partial}{\partial z}(\alpha_W U_W V_W) =$$

$$-\alpha_W\frac{\partial \bar{p}}{\partial r} - \alpha_W\frac{\partial \chi}{\partial r} + K_{SW}^r(U_S - U_W) + K_{MW}^r(U_M - U_W) + F_{\dot{m}W}^r \qquad (2.2)$$

and

$$\rho_W\frac{\partial}{\partial t}(\alpha_W V_W) + \rho_W\frac{1}{r}\frac{\partial}{\partial r}(r\alpha_W U_W V_W) + \rho_W\frac{\partial}{\partial z}(\alpha_W V_W^2) =$$

$$-\alpha_W\frac{\partial \bar{p}}{\partial z} + g\alpha_W\alpha_M(\rho_M - \rho_W) + g\alpha_W\alpha_S(\rho_S - \rho_W)$$

$$+K_{SW}^z(V_S - V_W) + K_{MW}^z(V_M - V_W) + F_{\dot{m}W}^z. \qquad (2.3)$$

The symbols are defined in the nomenclature. The terms on the RHS of the momentum equations represent the effect of the pressure gradient force, buoyancy forces due to gravity, inter-phase drag and the evaporation reaction force. The buoyancy force is rather more complicated than may be expected at first sight due to the introduction of a reduced pressure (see section 2.2).

In addition to the above equations we also have the constraint

$$\alpha_W + \alpha_M + \alpha_S = 1 \qquad (2.4)$$

and a number of additional equations of the same form as equation (2.1) to model the change in melt length-scale and melt temperature.

In the next sections we will describe the solution procedure and the principles employed to construct the finite difference scheme. The equations were finite differenced on the usual staggered grid. As far as possible we use explicit methods because of their simplicity and speed.

2.1   Evolution of the volume fractions

The three mass conservation equations can be used to time advance the volume fractions. However, there is an arbitrary choice as to which two of the species should be time-advanced using the differential equations and which should be calculated using equation (2.4). In the problems of interest to us $\dot{m}_M = 0$ and $\dot{m}_W = -\dot{m}_S = \alpha_W \alpha_M / \tau$, where τ is a rate parameter depending on the local melt temperature, particle size etc. Thus we chose to time advance the melt equation first (since it was the easiest) and then to time advance the water equation, treating the source term implicitly.

By using upwind differencing to model convective terms we ensure that the volume fractions are always non-negative and that in the absence of source terms the mass of each species is conserved exactly (Thyagaraja, Fletcher and Cook 1987). This requires that the Courant number, defined as

$$C = \max(\frac{V_{i,j}\Delta t}{\Delta z}, \frac{U_{i,j}\Delta t}{\Delta r}) \qquad (2.5)$$

is less than 1/4. By using the velocities at the old time no iteration is needed. The source term in the water equation is treated implicitly, which ensures that $\alpha_W \geq 0$. Note that the only stability requirement is that the Courant number is small enough but accuracy requires that

$$\frac{\Delta t}{\tau} \ll 1 \qquad (2.6)$$

i.e. the time-step must be short enough to resolve the fastest rate of change adequately. α_S is then obtained using equation (2.4).

2.2   Solution of the momentum equations

We solve the momentum equations by using the pressure at the old time and then correcting the pressure field using a pressure correction scheme (see Patankar and Spalding (1972) for an explanation of the principles). Since we intend to use a pressure correction scheme we first introduce a reduced pressure, defined by

$$\bar{p} = p - \chi = p - \int_z^H (\rho_W \alpha_W + \rho_S \alpha_S + \rho_M \alpha_M)g\,dz \qquad (2.7)$$

where H is a reference height (taken to be the height of the mixing vessel). The new pressure $\bar{p}$ contains only the hydrodynamic part of the pressure and is the correct part of the pressure for use in a pressure correction scheme. Without the introduction of a reduced pressure the buoyancy term on the RHS of equation (2.3) would have been $-\rho_W \alpha_W g$, so that introducing the reduced pressure complicates the momentum equations slightly.

Because the different species have very different densities ($\rho_M/\rho_S \approx 10^4$) each species has a very different velocity and there are large drag forces present. We ensure that these are modelled in a stable manner by treating these terms implicitly and solving for all three velocity components simultaneously at a given node. This ensures that no pseudo-forces are introduced and has proved to be a very robust solution procedure. Care has to be taken to ensure that if a species is absent from any cell its velocity is well defined. This was achieved by adjusting the coefficients in the drag terms to ensure that if a species is absent the finite difference scheme sets the velocity of that species to that of the dominant species.

The other terms in the momentum equations require no special practices and details are given in Thyagaraja and Fletcher (1988).

2.3 Pressure correction scheme

It now only remains to describe the pressure correction scheme. We constructed an elliptic constraint equation by summing all the mass conservation equations. This equation was then used to correct the pressure in the usual manner using Newton's method. Because our flows have a dominant direction (steam rises carrying droplets of melt and water with it) it was possible to avoid solving the full elliptic problem by using a block-correction scheme to get the pressure right in vertical blocks and then to distribute this correction within blocks. Thus we only had to invert tridiagonal matrices. This method was found to work well and to converge in typically two iterations.

2.4 Boundary conditions and applications

At solid boundaries the normal component of the velocity was set to zero. At the outlet we either assumed developed flow (i.e. we set the radial velocity to zero, the axial velocity gradients of the melt and water species to zero and then calculated a uniform steam outlet velocity from the mass conservation constraint equation) or set a constant outlet pressure. In many situations the top boundary was split into an inflow region, for which the melt velocity, temperature etc. were specified and an outlet region where the steam produced in the mixing process could escape.

This completes the description of the incompressible flow code. There is insufficient space in this paper to present example calculations but these can be found in Thyagaraja and Fletcher (1988) and Fletcher and Thyagaraja (1987a) and (1987b). These simulations include the modelling of experiments and examining the effect of parameters, such as ambient pressure and mass scale, on mixing.

3.   Compressible multiphase flow

In this section we describe a one-dimensional, two-component, compressible multiphase flow code developed to study detonations. For each species we must solve mass, momentum and energy conservation equations of the form given below:

$$\frac{\partial}{\partial t}(\alpha_1\rho_1) + \frac{\partial}{\partial z}(\alpha_1\rho_1 V_1) = \dot{m}_1 \tag{3.1}$$

$$\frac{\partial}{\partial t}(\alpha_1\rho_1 V_1) + \frac{\partial}{\partial z}(\alpha_1\rho_1 V_1^2) = -\alpha_1\frac{\partial p}{\partial z} + K_{12}(V_2 - V_1) + F_{\dot{m}} \tag{3.2}$$

$$\frac{\partial}{\partial t}(\alpha_1\rho_1(e_1 + \tfrac{1}{2}V_1^2)) + \frac{\partial}{\partial z}(\alpha_1\rho_1 V_1(e_1 + \tfrac{p}{\rho_1} + \tfrac{1}{2}V_1^2))$$

$$= -p\frac{\partial\alpha_1}{\partial t} + R_{12}(T_2 - T_1) + \Phi + Q. \tag{3.3}$$

The source terms Φ and Q represent drag and mass transfer work and enthalpy production due to "combustion", respectively. Details are given in Fletcher and Thyagaraja (1987c).

3.1   General principles

We solve the equations in conservation form and the energy equation is written in stagnation form so that the Rankine-Hugoniot equations are contained within the conservation equations. We difference the equations in such a manner that the Rankine-Hugoniot equations are also contained in the finite difference form of equations (3.1) → (3.3). This gives the scheme good shock-capturing properties.

We again use a staggered finite difference grid. Equation (3.1) is time-advanced in the same manner as described in section 2.1 to get $\alpha\rho \ (= \tilde{\rho})$ at the new time. Equation (3.2) is then time advanced using the old pressure field to obtain the new velocity field.

The mass flux used in the momentum equation, which is required at a density node, is set equal to the average of the fluxes at the velocity nodes (used in the solution of equation (3.1)). This ensures that the same mass flux is used in each equation and was found to be essential to obtain good predictions of shock jumps. Details are given in Fletcher and Thyagaraja (1987d). The energy equation was time advanced in a similar manner to the density equations. The main difference between our scheme and other schemes is in the calculation of the pressure field.

3.2 Determination of the pressure

In compressible flow the pressure is obtained from an Equation of State (EOS). There is no need to introduce an elliptic constraint to determine the pressure (and the volume fractions as these are still to be determined) as is done by, for example, Spalding (1980) and Harlow and Amsden (1975). Consider the procedure for the case of ideal gases, for which

$$p = \rho_1 R_1 T_1. \tag{3.4}$$

Multiplying equation (3.4) by α_1 and adding this to a similar equation for species 2 and using the fact that $\alpha_1 + \alpha_2 = 1$ gives

$$p = \tilde{\rho}_1 R_1 T_1 + \tilde{\rho}_2 R_2 T_2. \tag{3.5}$$

Thus the pressure is obtained directly from the values of $\tilde{\rho}$ (calculated from the mass conservation equation) and T (obtained from e by using a suitable caloric equation). This procedure is particularly simple if the EOS is linear in the density but if this is not the case the EOS can be linearized and iteration can be used. The remaining quantities (the thermodynamic density and the volume fractions) can now be calculated.

This completes the specification of the procedures employed to develop the compressible flow code. Example calculations for shock tubes and steady-state detonations are given in Fletcher and Thyagaraja (1987c) and (1987d). A study of detonations in two-component mixtures is presented in Thyagaraja and Fletcher (1987).

4. Conclusions

In this paper we have described the numerical procedures developed in order to study both incompressible and compressible multiphase flow. These procedures have produced numerical schemes which are robust and qualitatively consistent. We have used these models to study vapour explosions and they have allowed us to gain some new insight into the phenomena of mixing and detonations in multi-component systems.

Nomenclature

e	internal energy
g	acceleration due to gravity
K	constant in drag term
$\dot{m}$	mass source term
p	pressure
$\bar{p}$	reduced pressure
Q	energy source term
R	constant in temperature equilibration term; gas constant

r radial coordinate
T temperature
t time
U, V radial and axial velocity components
z axial coordinate

Greek Symbols

α volume fraction
ρ density
$\tilde{\rho}$ macroscopic density $(= \alpha\rho)$
τ characteristic timescale
Φ work term in energy equation

References

Cronenberg, A. W. (1980). Recent developments in the understanding of energetic molten fuel coolant interactions. *Nucl. Safety,* 21:319-337.

Fletcher, D. F. and Thyagaraja, A. (1987a). Numerical simulation of two-dimensional transient multiphase mixing. *Proc. 5th Int. Conf. Numerical Methods in Thermal Problems V.* 945-956. Pineridge Press.

Fletcher, D.F. and Thyagaraja, A. (1987b). A method of quantitatively describing a multi-component mixture. *PhysicoChem. Hydrodynam.,* 9:621-631.

Fletcher, D. F. and Thyagaraja, A. (1987c). Multiphase flow simulation of shocks and detonations, Part I. *Culham Laboratory Report: CLM-R279.*

Fletcher, D. F. and Thyagaraja, A. (1987d). Some calculations of shocks and detonations for gas mixtures. *Culham Laboratory Report: CLM-R276.* (To appear in Computers and Fluids.)

Harlow, F. H. and Amsden, A. A. (1975). Numerical Calculations of Multiphase Fluid Flow. *J. Comput. Phys.,* 17:19-52.

Patankar, S. V. and Spalding, D. B. (1972). A calculational procedure for heat, mass and momentum transfer in three-dimensional parabolic flows. *Int. J. Heat Mass transfer,* 15:1787-1806.

Spalding, D. B. (1980). Numerical computation of multi-phase fluid flow and heat transfer. Recent advances in numerical methods in fluids, Vol. 1, Eds. Taylor, C and Morgan, K. Pineridge Press.

Stewart, H. B. and Wendroff, B. (1984). Two-phase flow: models and methods. *J. Comput. Phys.,* 56:363-409.

Thyagaraja, A. and Fletcher, D. F. (1987). Multiphase flow simulation of shocks and detonations, Part II. *Culham Laboratory Report: CLM-R280.*

Thyagaraja, A. and Fletcher, D. F. (1988). Buoyancy-driven, transient, two-dimensional thermo-hydrodynamics of a melt-water-steam mixture. *Comput. Fluids,* 16:59-80.

Thyagaraja, A., Fletcher, D. F. and Cook, I. (1987). One dimensional calculations of two-phase mixing flows. *Int. J. Numer. Methods Eng.,* 24:459-469.

Chebyshev collocation methods for the solution of the incompressible Navier–Stokes equations in complex geometries

Timothy N. Phillips and Andreas Karageorghis

Department of Mathematics
University College of Wales, Aberystwyth

1. Introduction

Spectral methods are most easily applied to problems defined in rectangular or circular regions in which case Chebyshev or Fourier series expansions, respectively, are appropriate. However, the natural choice of spectral expansion functions for a problem defined in a general irregular region is computationally difficult to determine, unwieldy and inefficient to use and needs to be computed for each new irregular region. This apparent failure of spectral methods has been overcome by means of domain decomposition techniques in certain situations (Orszag 1980). Of course, in a general irregular region, the finite element method is much easier to implement than a spectral technique. However, for many problems in computational fluid dynamics the use of spectral methods in conjunction with domain decomposition techniques appears to be attractive since they combine the flexibility of finite element methods with the accuracy of spectral methods (Patera 1984; Karageorghis and Phillips 1987).

In this study we investigate the laminar flow through an abruptly contracting channel. The spectral element method is quite naturally applied to this problem by dividing the flow region into two semi-infinite domains as shown in Fig. 1. The semi-infinite elements are treated in one of two ways: domain truncation or mapping. The former technique has been extensively used and involves truncating each of the semi-infinite elements at a finite distance from the contraction. Alternatively algebraic-type mappings are used to transform each of the semi-infinite elements into finite ones through a transformation of one of the coordinate directions (Grosch and Orszag 1977). An advantage of mapping techniques is that it is unnecessary to impose artificial boundary conditions at a large but finite distance from the contraction. This is particularly important in non-Newtonian flow where the behaviour of the flow is dependent on the entry and exit lengths

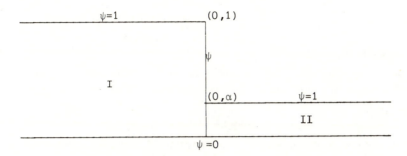

Fig. 1: Upper half of contraction geometry

(Crochet et al. 1984). This effect is more pronounced for highly elastic fluids for which the corresponding stress field requires a longer distance over which to relax.

2. The governing equations

The incompressible Navier-Stokes equations are

$$(\mathbf{v}.\nabla)\mathbf{v} = -\nabla p + (Re)^{-1}\nabla^2\mathbf{v}, \tag{2.1}$$
$$\nabla.\mathbf{v} = 0, \tag{2.2}$$

where $\mathbf{v} = (u, v)$ is the velocity vector, p is the pressure and Re is the Reynolds number. The introduction of a stream function, $\psi(x, y)$, defined by

$$u = -\frac{\partial \psi}{\partial y}, \quad v = \frac{\partial \psi}{\partial x}, \tag{2.3}$$

means that the continuity equation (2.2) is automatically satisfied and (2.1) becomes

$$\nabla^4\psi - Re\left[\frac{\partial \psi}{\partial y}\frac{\partial}{\partial x}(\nabla^2\psi) - \frac{\partial \psi}{\partial x}\frac{\partial}{\partial y}(\nabla^2\psi)\right] = 0. \tag{2.4}$$

We solve (2.4) in a 1:α contraction geometry. We have no-slip constraints on the upper channel wall. The flow is assumed to be symmetric about the line $y = 0$ so that only the upper half of the channel flow need be considered. Thus along $y = 0$ we impose the conditions

$$\psi = 0, \quad \frac{\partial^2\psi}{\partial y^2} = 0. \tag{2.5}$$

We impose Poiseuille flow upstream and downstream of the constriction which means that

$$\psi \ \rightarrow \ \frac{1}{2}y(3 - y^2) \text{ as } x \rightarrow -\infty, \tag{2.6}$$

$$\psi \ \rightarrow \ \frac{1}{2}\left(\frac{y}{\alpha}\right)\left[3 - \left(\frac{y}{\alpha}\right)^2\right] \text{ as } x \rightarrow \infty. \tag{2.7}$$

The governing equation (2.4) is nonlinear and is solved in an iterative manner using a Newton-type method to linearize it (Phillips 1984). Let us rewrite (2.4) as

$$L(\psi) = 0, \tag{2.8}$$

where L is a nonlinear operator. Suppose that ψ^* is some approximation to the solution of (2.8). We replace L by its linearization about ψ^* and then solve the linearized problem

$$L'(\psi^*).\phi = -L(\psi^*), \tag{2.9}$$

where $L'(\psi^*)$ is the Fréchet derivative of L at ψ^* defined by

$$\begin{aligned} L'(\psi).\phi \ = \ & \nabla^4\phi - Re\left[\frac{\partial\psi}{\partial y}\frac{\partial}{\partial x}(\nabla^2\phi) - \frac{\partial\psi}{\partial x}\frac{\partial}{\partial y}(\nabla^2\phi)\right. \\ & + \ \left.\frac{\partial}{\partial x}(\nabla^2\psi)\frac{\partial\phi}{\partial y} - \frac{\partial}{\partial y}(\nabla^2\psi)\frac{\partial\phi}{\partial x}\right]. \end{aligned} \tag{2.10}$$

The new approximation to the solution is thus $\psi^* + \phi$.

3.   Spectral approximation and discretization

If the mapping approach is followed each of the semi-infinite elements is mapped onto a finite rectangle. For example, region I is mapped onto the unit square under the transformation

$$z = \frac{x}{(x + L_I)}, \ L_I < 0. \tag{3.1}$$

In the transformed region we seek an approximation to the stream function of the form

$$\psi^I(z, y) = g(y) + \sum_{m=0}^{M}\sum_{n=4}^{N} a_{mn}T_m^*(z)P_n(y), \tag{3.2}$$

where $T_m^*(z)$ is the Chebyshev polynomial defined on $[0, 1]$ and

$$P_n(y) = T_n^*(y) + \alpha_n T_3^*(y) + \beta_n T_2^*(y) + \gamma_n T_1^*(y) + \delta_n T_0^*(y).$$

The constants α_n, β_n, γ_n and δ_n are chosen so that the boundary conditions along the top of the channel and the axis of symmetry are automatically satisfied. Similarly, region II is mapped onto the rectangle $[0,1] \times [0,\alpha]$ under the algebraic mapping

$$z = \frac{x}{(x + L_{II})}, \quad L_{II} > 0, \tag{3.3}$$

and $\psi^{II}(z, y)$ has a representation similar to (3.2).

At each step of the Newton process, the linearized equation (2.9) is solved using the spectral collocation method. The equation is collocated at certain points in the mapped regions related to the extrema of the Chebyshev polynomial of highest degree used in the representations. The solutions in the two regions are matched by imposing C^3 continuity conditions in a collocation sense across the interface. In fact, these are the natural boundary conditions for the associated variational principle.

Suppose we know a^*_{mn} and b^*_{mn}, the coefficients of ψ^* in the mapped regions I and II, respectively, then the collocation method applied to (2.9) results in a linear system of algebraic equations for $(\delta a)_{mn}$ and $(\delta b)_{mn}$, the coefficients of ϕ in the mapped regions I and II, respectively. The coefficients of the updated representations are therefore $a^*_{mn} + (\delta a)_{mn}$ and $b^*_{mn} + (\delta b)_{mn}$.

4. Direct method of solution

We take advantage of the structure of the spectral element matrix in constructing direct solution techniques. We consider an application of the capacitance matrix technique of Buzbee et al. (1971). One should note that the interface conditions in the spectral element method are much more complex than in the finite difference method. Typically they involve all the expansion coefficients in adjoining elements. The rows of the spectral matrix corresponding to these conditions are full.

We write the spectral element equations for $\delta \mathbf{a}$ and $\delta \mathbf{b}$ in the partitioned form:

$$\left(\begin{array}{c|c|c} E & F & 0 \\ \hline G & H & R \\ \hline 0 & P & Q \end{array} \right) \left(\begin{array}{c} \mathbf{x} \\ \mathbf{y} \\ \mathbf{z} \end{array} \right) = \left(\begin{array}{c} \mathbf{a} \\ \mathbf{b} \\ \mathbf{c} \end{array} \right) \tag{4.1}$$

The first and last blocks of rows correspond to collocation in regions I and II, respectively, while the middle block corresponds to the interface conditions. We write (4.1) in the natural form suggested by the partitioning

$$E\mathbf{x} + F\mathbf{y} \qquad\qquad = \mathbf{a}, \tag{4.2}$$
$$G\mathbf{x} + H\mathbf{y} + R\mathbf{z} = \mathbf{b}, \tag{4.3}$$
$$P\mathbf{y} + Q\mathbf{z} = \mathbf{c}. \tag{4.4}$$

We write $\mathbf{x}$ and $\mathbf{z}$ in terms of $\mathbf{y}$ by premultiplying (4.2) and (4.4) by the inverses of E and Q, respectively, i.e.

$$\mathbf{x} = E^{-1}\mathbf{a} - E^{-1}F\mathbf{y}, \ \mathbf{z} = Q^{-1}\mathbf{c} - Q^{-1}P\mathbf{y}.$$

Eliminating $\mathbf{x}$ and $\mathbf{z}$ from (4.3) we obtain the following system for $\mathbf{y}$:

$$(H - GE^{-1}F - RQ^{-1}P)\mathbf{y} = \mathbf{b} - GE^{-1}\mathbf{a} - RQ^{-1}\mathbf{c}, \qquad (4.5)$$

where the coefficient matrix on the left-hand side of (4.5) is known as the capacitance matrix. This system, which is much smaller than the full system, can be solved efficiently using the Crout factorization technique. Further details of such techniques can be found in (Phillips and Karageorghis 1987).

5. Numerical results

The use of the capacitance matrix technique produced significant savings in computational cost and time over the original method. The efficiency of the method is greater when the system of equations becomes larger. In figures 2–7 we show contours of the stream function for $Re = 0, 10, 25, 50, 75, 100$ for a 2:1 contraction. The Newton process was terminated when the maximum difference between successive coefficients was less than 10^{-8}. Six steps of the Newton process were sufficient for convergence from a zero solution for $Re > 0$. The degree of the Chebyshev polynomials used to obtain these solutions was 25 in each direction. The plots shown are for the mapping method with $L_I = -1.5$ and $L_{II} = 1.5$. The method is insensitive to values of these parameters in the proximity of the chosen ones. The performance of the truncation method was more susceptible to changes in the truncation lengths.

The results agree qualitatively with those of Dennis and Smith (1980) who use many more degrees of freedom. The plots show the size of the vortex decreasing from $Re = 0$ until around $Re = 50$ and thereafter slowly increasing.

Contours of the stream function

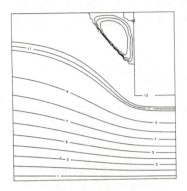

Fig. 2: $Re = 0$

Fig. 3: $Re = 10$

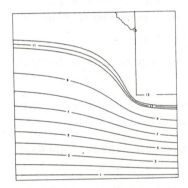

Fig. 4: $Re = 25$

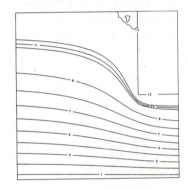

Fig. 5: $Re = 50$

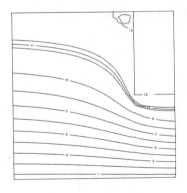

Fig. 6: $Re = 75$

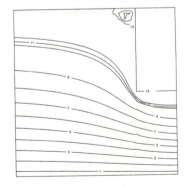

Fig. 7: $Re = 100$

References

Buzbee, B. L., Dorr, F. W., George J. A. and Golub G. H. (1971). The direct solution of the discrete Poisson equation on irregular regions, *SIAM Journal of Numerical Analysis* 8 : 722-736.

Crochet, M. J., Davies, A. R., and Walters, K. (1984). *Numerical simulation of non-Newtonian flow*, Elsevier, Amsterdam.

Dennis, S. C. R. and Smith, F. T. (1980). Steady flows through a channel with a symmetrical constriction in the form of a step, *Proc. R. Soc. Lond.* A372 : 393–414.

Grosch, C. E. and Orszag, S. A. (1977). Numerical solution of problems in unbounded regions: coordinate transforms, *J. Comput. Phys.* 37 : 70–92.

Karageorghis, A. and Phillips, T. N. (1987). Spectral collocation methods for Stokes flow in contraction geometries and unbounded domains. To appear in *J. Comput. Phys.*

Orszag, S. A. (1980). Spectral methods in complex geometries, *J. Comput. Phys.* 37 : 70–92.

Patera, A. T. (1984). A spectral element method for fluid dynamics: laminar flow in a channel expansion, *J. Comput. Phys.* 54 : 468–488.

Phillips, T. N. (1984). Natural convection in an enclosed cavity, *J. Comput. Phys.* 54 : 365–381.

Phillips, T. N. and Karageorghis, A. (1987). Efficient direct methods for solving the spectral collocation equations for Stokes flow in rectangularly decomposable domains. To appear in *SIAM J. Sci. Stat. Comput.*

Far field boundaries and their numerical treatment: an unconventional approach

S. Karni

College of Aeronautics
Cranfield Institute of Technology

1. Introduction

The motivation for the present approach to the problem of far field boundary treatment lies in a thorough numerical study (Karni 1987a) which exposed the inefficiency of a variety of b.c.'s in absorbing outgoing waves. Even in ideal cases when the b.c.'s should perfectly absorb outgoing waves, strong reflections may occur and one need not set extremely severe conditions to actually observe them. In section 2 we briefly survey the problem of far field boundaries and its traditional numerical treatment. In section 3 we introduce the new approach, namely slowing down the outgoing waves and in section 4 we present some new numerical results.

2. Traditional far field boundary treatment

When seeking to obtain the numerical solutions of problems in physically unbounded domains, one first needs to convert the problem to a finite region. This can sometimes be done by coordinate mapping techniques. Alternatively, one can truncate the computational domain at some finite distance and introduce far field boundaries. This, however, is at the expense of imposing b.c.'s to ensure a unique solution. The outer boundaries should ideally give rise to zero reflection of incident waves back into the domain of computation. To prevent spurious reflected waves from being generated at the boundary, the b.c.'s are made to be compatible with the outgoing part of the solution. Carefully designed b.c.'s not only improve the overall order of accuracy during the transient phase, but also increase convergence rate to steady state. If the solution is required only in a small part of the flow field (e.g. around the areofoil) applying such b.c.'s usually enables the bringing in of the outer boundaries, thereby reducing computational costs.

The b.c.'s can be matched to the outgoing part of either the analytic or the numerical solution. Both approaches have their limitations. The analytical approach relies on the theory of characteristics which is only available in 1D. In multi-dimensional flows one usually resorts to 1D arguments normal to the boundary, to asymptotic expansions etc. (Bayliss and Turkel 1981, Engquist and Majda 1977). More importantly, replacing the analytically derived b.c.'s by numerical approximations results in reflections. The efficiency of the b.c.'s in absorbing the waves then greatly depends on the combination of interior scheme and boundary procedures, and on local numerical parameters (CFL number, grid irregularities etc). In the direct numerical approach, the b.c.'s are designed to absorb waves of a particular (usually low) frequency, and are then hoped to perform well for neighbouring frequencies. Their efficiency is, as expected, very sensitive to wave number and grid irregularities. The advantage of this approach is that it enables the absorption of non-physical parasitic modes, which are entirely scheme dependent and cannot be dealt with by any analytical approach(Lindman 1975, Higdon 1986).

3.   The proposed far field treatment

3.1   The concept of slowing down the outgoing waves

Any choice of far field b.c.'s should be consistent with the uniform free stream flow field. Since disturbances propagate at a finite speed there is a period of time during which those b.c.'s are 'inactive'. They do not play a role until the first disturbances, generated at the aerofoil, say, have reached the boundary, and if the solution is only required in the vicinity of the aerofoil, until they have propagated back. This period of time can be prolonged by setting the outer boundaries far away from the areofoil, which typically results in large computational domains and highly stretched grids on which 2nd order accuracy is usually lost. Alternatively, one can modify the set of governing equations so that in the far field the outgoing waves are slowed down. A suitable choice of slowing down rate can ensure that within a given time T (possibly $T \to \infty$) the outgoing waves will not reach the boundary, hence will not 'reflect back. At the same time it is proposed not to alter the propagation of the incoming waves. This latter requirement is particularly important in problems of genuine time dependence (e.g. a vibrating areofoil). It should be mentioned that the major difficulty, mainly in multidimensional flows, is to distinguish between outgoing and incoming waves. Once this is established, other far field modifications are possible (e.g. selectively damping outgoing waves only) but these shall not be discussed here.

3.2   Slowing down the waves in 1D

Consider the 1D wave equation

$$\left(\frac{\partial}{\partial t} - c_1 \frac{\partial}{\partial x}\right)\left(\frac{\partial}{\partial t} + c_1 \frac{\partial}{\partial x}\right)\phi = 0 \qquad \text{on } (0, X) \tag{3.1}$$

Slowing down the family of, say, right going waves is equivalent to replacing (3.1) by the interface problem

$$
\begin{aligned}
\left(\frac{\partial}{\partial t} - c_1 \frac{\partial}{\partial x}\right)\left(\frac{\partial}{\partial t} + c_1 \frac{\partial}{\partial x}\right)\phi &= 0 \qquad \text{on } (0, x_0) \\
\left(\frac{\partial}{\partial t} - c_1 \frac{\partial}{\partial x}\right)\left(\frac{\partial}{\partial t} + c_2 \frac{\partial}{\partial x}\right)\phi &= 0 \qquad \text{on } (x_0, X) \\
[\phi]_{x=x_0} = [\frac{\partial \phi}{\partial x}]_{x=x_0} &= 0 \qquad c_2/c_1 < 1
\end{aligned}
\tag{3.2}
$$

Similarly a $N \times N$ hyperbolic system

$$\vec{U}_{t_-} + A\vec{U}_x = 0 \qquad \text{on } (0, X) \tag{3.3}$$

is replaced by the interface problem

$$
\begin{aligned}
\vec{U}_t + A\vec{U}_x &= 0 \qquad \text{on } (0, x_0) & (3.4) \\
\vec{U}_{t_-} + A^*\vec{U}_x &= 0 \qquad \text{on } (x_0, X) & (3.5) \\
[U]_{x=x_0} &= 0 & (3.6)
\end{aligned}
$$

An obvious requirement is that no reflection should occur from the interface $x = x_0$. Standard transmission analysis shows that (2.2) gives rise to a reflected wave of amplitude

$$R = \frac{c_2 - c_1}{c_2 + c_1} \tag{3.7}$$

(3.4) however ensures full transmission if A and A^* share the same (complete) set of e-vectors. The e-values are then chosen suitably to slow down the right going waves only.

3.3 Slowing down the waves in 2D

The nature of wave propagation is now rather more complicated and so is the concept of slowing down the waves. Information may now travel in an infinite number of directions. Consider the 2D wave equation

$$\phi_{tt} - \phi_{xx} - \phi_{yy} = 0 \tag{3.8}$$

As illustrated below, the envelope of wave fronts forms a circle, reflecting the isotropic nature of acoustic wave propagation. This is precisely the

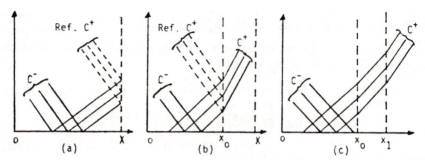

Fig. 1: c^+ waves partially reflected at the boundary (a), at the interface (b), and fully transmitted across the interface (c). c^- waves unaltered

property we want to discard, for we want outgoing waves only to travel at a reduced speed. In view of the 1D transmission analysis, we consider N *times* N first order linear hyperbolic systems:

$$\vec{U}_t + A\vec{U}_x + B\vec{U}_y = 0 \qquad (3.9)$$

which is to be replaced in the far field by

$$\vec{U}_{t-} + A^*\vec{U}_x + B^*\vec{U}_y = 0 \qquad (3.10)$$

with A^* and B^* suitably chosen to yield the distorted cone illustrated below. The matrices to analyze for a 1D disturbance in any general direction θ are

$$\begin{aligned} M(\theta) &= A\cos\theta + B\sin\theta \\ M^*(\theta) &= A^*\cos\theta + B^*\sin\theta \end{aligned} \qquad (3.11)$$

If $M(\theta)$ and $M^*(\theta)$ share the same set of (θ–dependent) e-vectors, full transmission of any 1D disturbance in a general direction is ensured. If A and B commute, they can be diagonalized simultaneously. In this case, consider matrices A^* and B^* which are θ–independent, can be constructed which will yield full transmission across the interface in a slowing down process. In the general case, however, it is not possible to preserve the e-vectors in all directions θ, unless A^* and B^* are made θ–dependent. Yet, it is possible to preserve the e-vectors in one preferred direction, and choose this direction to coincide with the radial direction. Suffcently far away from the aerofoil, disturbances travel radially and will therefore be slowed down without generating reflections.

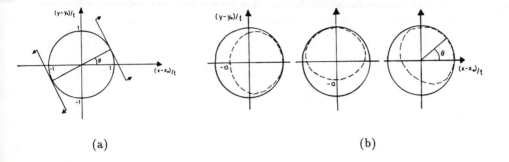

Fig.2: Envelope of wave fronts emerging from (x_0, y_0)
  (a): unmodified isotropic case
  (b): modified non-isotropic case

3.4 General features of the modification

The modification is performed on the general level of the PDE's which can then be discretized in a variety of ways. The speed of the outgoing waves is controlled by one (possibly 2) slowing down functions, when all set to 1 the original PDE's are recovered. In practice, this makes the proposed modification easy to apply in a uniform manner, over both an inner region (function set to 1) and an outer boundary layer. Further details regarding transmission analysis, stability and conservation aspects may be found in Karni (1987b).

4. Numerical tests

Test A

The governing set of equations is the 1D Euler equations. A high pressure is being fed through the left hand boundary into a uniform state of rest, giving rise to a right running shock wave followed by a contact discontinuity. In Fig. 3, the right going waves undergo (a) a sudden (b) a gradual change of propagation speed without generating any reflected waves from the interface.

Test B

The governing set of equation is the 2D linearised Euler equations. Initial high pressure is centered about the origin, which generates an outgoing pressure wave. This wave decays like $1/\sqrt{r}$ as it moves away from the origin. Fig. 4 shows typical reflections from the outer boundary when free stream pressure is imposed. The reflected wave is of small amplitude but focusses very strongly as it converges towards the origin. Fig. 5 shows various slowing down rates of the outgoing wave, which prevent it from reaching the boundary. Although the outgoing waves propagate at a reduced, physically incorrect, speed, they do not send any reflections towards the centre and leave a quiet region behind them which is physically correct.

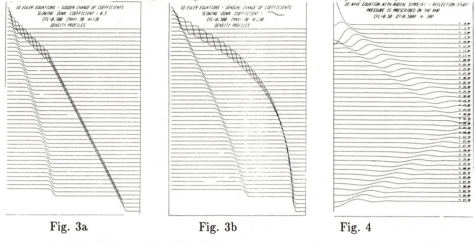

Fig. 3a	Fig. 3b	Fig. 4

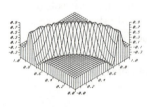 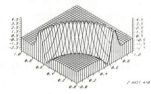 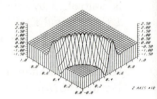

Fig. 5

References

Bayliss, A and Turkel, E. (1981). ICASE Report No. 81-27.

Enquist, B and Majda, A. (1977). *Mathematics of Computation*, 31 : 629-651.

Lindman, E. L. (1975). *Journal of Computational Physics*, 18 : 66-78.

Higdon, R. L. (1986). *Mathematics of Computation*, 47 : 437-459.

Karni, S. (1987a). Cranfield College of Aeronautics Report No. 8711.

Karni, S. (1987b). Cranfield College of Aeronautics Report No. 8721.

Moving element methods for time dependent problems

M. J. Baines

Department of Mathematics, University of Reading

1. Introduction

With the increase in complexity of the phenomena now being modelled by computational methods, detailed resolution of physical features on a uniform mesh has become a considerable strain on resources. The idea of using a numerical representation in which nodes are adaptive and are positioned only where needed is highly attractive and of increasing importance. As yet however there is no consensus as to the strategy to be used to achieve this goal. Two main ways have been proposed. (i) enrichment /deletion of nodes. (ii) moving grids. The first of these makes use of an appropriate monitor function and seeks to position the nodes so as to provide uniform accuracy in the solution. The second relies on an algorithm for moving the nodes in order to obtain the best accuracy possible for (usually) a given number of nodes. It is the latter approach which is discussed here.

In approximately solving time-dependent partial differential equations in this context we seek an algorithm for the evolution of the grid as well as that of the solution. Fixing the grid is one option: at the other extreme it may be possible to move the grid in a Lagrangian way so that the solution remains fixed in time (as in the case of the equation $u_t + au_x = 0$). In general however the best algorithm will lie somewhere between these two extremes. We illustrate this point below with a detailed examination of the Moving Finite Element (MFE) method (Miller 1981; Miller 1984; Wathen and Baines 1985; Carlson and Miller 1986) which uses the same procedure to generate both the movement of the grid and of the solution, namely residual minimisation (see below).

An alternative to residual minimisation in defining the nodal movement is equidistribution of the error and this has been used as a node movement strategy (Adjerid and Flaherty 1986). Another strategy is the use of mechanical analogies such as springs to control the velocities of the nodes (Nakahashi and Diewert 1985; Palmerio et al. 1986; Catherall 1988). All of these methods have been used with varying degrees of success but without suggesting that the optimal strategy has yet been found .

2.   The MFE method and characteristics

In this section we analyse the MFE method, which seeks a uniform treatment of grid and solution. This method uses the method of residual minimisation, which may be described in relation to the fixed finite element method as follows.

Consider the approximate solution of the partial differential equation

$$\frac{\partial u}{\partial t} = Lu \tag{2.1}$$

(where Lu is a spatial differential operator) in the Ritz form

$$U = \sum U_j \phi_j \tag{2.2}$$

where ϕ_j are basis functions and U_j are time-dependent coefficients. In general, because of the approximation, if (2.2) is subsitituted into (2.1) there results a non-zero residual

$$R \equiv \frac{\partial U}{\partial t} - LU \tag{2.3}$$

Minimising the L_2 norm $\|R\|$ over the coefficients $\dot{U}_j$ results in the set of equations

$$< \phi_i, \frac{\partial U}{\partial t} - LU >= 0, \qquad \forall i \tag{2.4}$$

which is the standard Galerkin weak form of (2.1).

Suppose now that we allow the ϕ_j to depend on t as well as U_j. Then minimisation of $\|R\|$ over the time rates of change of U_j *and* nodal positions leads to a larger set of weak equations for the evolution of both the solution and the basis functions. This is the MFE strategy. A framework for discussing this situation may be set up as follows, see (Mueller and Carey 1985).

Consider a transformation from x, t space to ξ, τ space in which $t = \tau$, $x = \hat{x}(\xi, \tau)$ and $u = \hat{u}(\xi, \tau)$. As long as the transformation is non-singular we have by the chain rule

$$\frac{\partial u}{\partial t} = \dot{u} - u_x \dot{x} \tag{2.5}$$

for each ξ, where $\dot{u} = \frac{\partial \hat{u}}{\partial \tau}$, $\dot{x} = \frac{\partial \hat{x}}{\partial \tau}$,   c.f. (Mueller and Carey 1985). Then (2.1) can be written

$$\dot{u} - u_x \dot{x} = Lu \tag{2.6}$$

Using the Ritz expansions (2.2) and $X = \sum X_j \phi_j$ (cf. (2.2)) in the minimisation of $\|R\|$ over the coefficients $\dot{U}_j$ and $\dot{X}_j$ leads to the double set of Galerkin equations

$$\left.\begin{array}{ll} < \phi_i, \dot{U} - U_x \dot{x} - LU > & = 0 \\ < -U_x \phi_i, \dot{U} - U_x \dot{x} - LU > & = 0 \end{array}\right\} \quad \forall i \qquad (2.7)$$

To illustrate the connection with characteristics we now consider the case $Lu = f(u_x)$. Extensions to more general functions are readily made (Baines 1987). In this case we have a single data parameter u_x, each value of which gives a straight line

$$\dot{u} - u_x \dot{x} - f(u_x) = 0 \qquad (2.8)$$

in $\dot{x}, \dot{u}$ space (Fig. 1). To determine $\dot{x}$ and $\dot{u}$ fully we take a general point $\dot{x}_0, \dot{u}_0$ in $\dot{x}, \dot{u}$ space and minimise the norm $\|R\|$ (as in the above procedure), which happens to be the "distance" of the point $\dot{x}_0, \dot{u}_0$ from the line (2.8). Because of the double minimisation (over $\dot{x}$ and $\dot{u}$), we obtain two equations for $\dot{x}_0, \dot{u}_0$. The solution $(\dot{x}_0, \dot{u}_0)$ is also that point of the line (2.8) which touches the envelope of (2.8) as u_x varies. Inspection of equation (2.8) shows that at this point

$$\dot{x} = -\frac{\partial f}{\partial u_x}, \qquad \dot{u} = f(u_x) - u_x \frac{\partial f}{\partial u_x} \qquad (2.9)$$

two of the characteristic equations for the nonlinear equation (2.8). The MFE method is a discretisation of this situation.

Since the main property of the method of characteristics is the equivalence between the partial differential equation and an ordinary differential equation along a particular trajectory, the compactness of finite elements is entirely appropriate here and leads to a local method (Baines 1986). Another property, that if Lu does not contain x explicitly then $\dot{u}_x = 0$, shows that the nodes move to preserve the slope. This will be referred to later.

3. The MFE method and Diffusion Operators

Let us now illustrate what happens when the above arguments are applied to second order operators.

For the linear heat equation we have

$$\left.\begin{array}{ll} & u_t = u_{xx} \\ \text{or} \quad & \dot{u} - u_x \dot{x} = u_{xx} \end{array}\right\} \qquad (3.1)$$

and minimisation of the residual $\|R\|$ (with x now used as the data parameter) leads to the nodal speed $\dot{x}$ being given by

$$-u_{xx} \dot{x} = u_{xxx} \text{ or } \dot{x} = -\frac{u_{xxx}}{u_{xx}} \quad (u_{xx} \neq 0) \qquad (3.2)$$

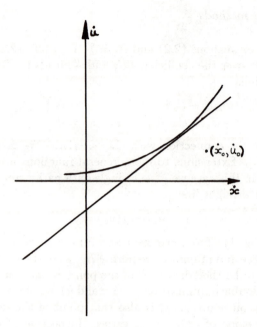

Fig. 1

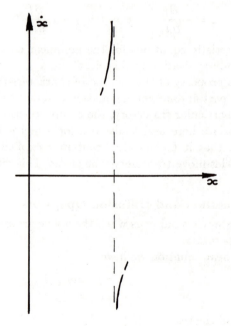

Fig. 2

Clearly the velocity $\dot{x}$ is singular at points for which $u_{xx} = 0$ (zero curavature). This was anticipated in the MFE method (Miller 1981) and is coped with using penalty functions or a special treatment (Wathen and Baines 1985). However it can also be shown that such points are cluster points, in violation of the MFE philosophy that nodes are moved towards regions where the action is, presumably high curvature regions. To see this result plot (3.2) against x (Fig. 2). Since

$$\frac{d\dot{x}}{dx} = \left[\frac{u_{xxx}}{u_{xx}}\right]^2 - \frac{u_{xxxx}}{u_{xx}} > 0 \text{ as } u_{xx} \longrightarrow 0 \tag{3.3}$$

it is clear that $\dot{x} > 0$ to the left of each asymptote and $\dot{x} < 0$ to the right of each asymptote, thus showing that points for which $u_{xx} = 0$ are cluster points. [By the same argument, if $\dot{x} = -u_{xx}/u_x$ nodes rush towards $u_x = 0$, see (Edwards 1985), while if $\dot{x} = +u_{xxx}/u_{xx}$ nodes rush away from $u_{xx} = 0$ points (see below)].

An explanation of why the $u_{xx} = 0$ points should be cluster points lies in the fact mentioned earlier that nodes move so as to preserve the slope u_x. In a pure diffusion problem the slope is always decreasing with time, its rate of decrease being greatest at points for which $u_{xx} = 0$, thus indicating a requirement for maximum stretching at such points.

The MFE method imitates this behaviour. However, if piecewise linear elements are used, special procedures are required for the treatment of the second derivative u_{xx} (and even more u_{xxx}). The result is a consistent approximation to u_{xx} but (partly because of the use of a central difference) gives an inconsistent approximation to u_{xxx} which in fact introduces a negative factor. Thus in practice nodes rush *away* from $u_{xx} = 0$ points, in accordance with expectation. In 2-D these arguments are less clear but there is evidence that without regularisation the method has significant deficiencies (Sweby 1986).

Successful application of the MFE method in diffusion problems has relied on the use of regularising penalty functions (Miller 1981; Miller 1984). It appears that in several instances the nodal velocitites are being determined principally by these penalty functions.

4. Conclusion

We have seen that if a particular approximation procedure (residual minimisation) is applied simultaneously to both the node movement and the solution movement, a sensible scheme is obtained for nonlinear first order equations but less so for second order equations. It appears therefore that some care is needed in choosing the algorithm for the grid in relation to the algorithm for the solution if optimal use of moving grids is to be made.

References

Adjerid, S. and Flaherty, J. E. (1986). A moving finite element method with error estimation and refinement for one-dimensional time dependent partial differential equations. *SIAM Journal of Numerical Analysis*, 23: 778–796

Baines, M. J. (1986). Locally adaptive moving finite elements. In *Numerical Methods for fluid dynamics II*, Morton, K. W. and Baines, M. J., editors. Oxford University Press.

Baines, M. J. (1987). Moving finite envelopes. Report 12/87, Department of Mathematics, University of Reading.

Carlson, N. and Miller, K. (1986). The gradient weighted moving finite element method in 2-d. Report PAM-347, Center for Pure and Applied Mathematics, University of California, Berkeley.

Catherall, D. (1988). A solution-adaptive-grid procedure for transonic flows around aerofoils. Report 88020, RAE, Farnborough.

Edwards, M. G. (1985). Mobile finite elements. Report 20/85, Department of Mathematics, University of Reading.

Miller, K. (1981). Moving finite elements part I (with R. N. Miller), part II. *SIAM Journal of Numerical Analysis*, 18: 1019–1057.

Miller, K. (1984). Recent results on finite element methods with moving nodes. In *Proceedings of ARFEC Conference, Lisbon*. Wiley.

Mueller, A. C. and Carey, C. F. (1985). Continuously deforming finite elements for transport problems. *International Journal of Numerical Methods*, 21: 2099.

Nakahashi, K. and Diewert, G. (1985). A self-adaptive grid method with application to aerofoil flow. AIAA Paper 85-1525.

Palmerio, B., Dervieux, A., and Periaux, J. (1986). Self-adaptive mesh refinements and finite element methods for solving the Euler equations. In *Numerical Methods for Fluid Dynamics II*, Morton, K. W. and Baines, M. J., editors. Oxford University Press.

Sweby, P. K. (1986). Private communication. University of Reading.

Wathen, A. J. and Baines, M. J. (1985). On the structure of the moving finite element equations. *IMA Journal of Numerical Analysis*, 5: 161.

Non-existence, non-uniqueness and slow convergence in discrete conservation laws

P. L. Roe

B. van Leer

Cranfield Institute of Technology

University of Michigan, U.S.A.

1. Introduction

It is often expedient to solve steady flow problems, governed by some equation $M(\mathbf{u}) = \mathbf{0}$, by seeking the solution at large times of a (pseudo-) unsteady problem

$$\mathbf{u}_t + P(\mathbf{u})M(\mathbf{u}) = \mathbf{0}. \tag{1.1}$$

Here P is some preconditioning operator (see for example, Turkel (1987)). If P is the identity, we solve a real time–dependent problem. A discrete version of (1.1) is (without preconditioning)

$$\mathbf{v}_t = \mathbf{R}(\mathbf{v}) \tag{1.2}$$

where $\mathbf{v}$ is the vector of discrete unknowns, and $\mathbf{R}$ is the vector of residuals $R_i = \partial v_i/\partial t$ corresponding to some chosen spatial discretisation.

Convergence to the steady state is conventionally analysed by linearising (1.2) about the steady state, i.e.

$$\mathbf{R}(\mathbf{v}) = J(\mathbf{v}_\infty)(\mathbf{v} - \mathbf{v}_\infty) \tag{1.3}$$

where J is the Jacobian matrix $\partial \mathbf{R}/\partial \mathbf{v}$. If the marching process is

$$\mathbf{v}^{n+1} = \mathbf{v}^n + K\mathbf{R}(\mathbf{v}^n) \tag{1.4}$$

(where K converts residuals to actual changes, and is most simply $I\Delta t$) then

$$\mathbf{v}^{n+1} - \mathbf{v}^\infty = \mathbf{v}^n - \mathbf{v}^\infty + KJ(\mathbf{v}_\infty)(\mathbf{v}^n - \mathbf{v}^\infty) \tag{1.5}$$

and convergence proceeds at a rate governed by the largest eigenvalue of $I + KJ$.

This analysis is only relevant if a steady solution $\mathbf{v} = \mathbf{v}^\infty$ satisfying $\mathbf{R}(\mathbf{v}^\infty) = \mathbf{0}$ exists and if the convergence process is actually dominated by the asymptotic phase in which (1.5) applies. This paper is motivated by the wish to explain why some spatial discretisations are observed to converge more slowly than others. In particular, CFD practitioners have the impression that accuracy and fast convergence are to some extent in conflict. That conflict does not arise when we study sufficiently simple problems, as will be demonstrated in Section 2. Section 3 describes test problems that may be more representative of real cases. Section 4 investigates their discrete structure. Section 5 offers some remedies.

The numerical method used throughout is a simple upwind scheme that yields, in our examples, second–order accurate steady solutions with optimally resolved shocks.

2.   Model Problems with Fast Convergence

Both examples in this section are almost trivial. They are discussed to contrast their behaviour with the examples in section 3.

2.1   Problem No. 1

Consider the scalar conservation law

$$u_t + f(u)_x = 0. \tag{2.1}$$

The analytic solution is that in smooth regions u is constant along straight characteristic lines $dx/dt = f'(u) = a(u)$. Shock curves $x = x_s(t)$ satisfy the jump relationship

$$\frac{[f]}{[u]} = \frac{dx_s}{dt} \tag{2.2}$$

(where [ ] denotes the jump across the shock) and also the entropy condition

$$a(x_s^+) \leq \frac{dx_s}{dt} \leq a(x_s^-). \tag{2.3}$$

See Jeffrey (1976).

To construct a solution in $(x,t) \in [-1,1] \times [0,\infty]$ we specify a left–hand boundary condition $u[-1,t] = u_L$ where u_L is a constant such that $a(u_L) > 0$, and arbitrary initial data $u(x,0) = u_0(x)$. The normal outcome is that after a transient phase lasting for finite time the boundary data will overwhelm the initial data (Fig. 1a). Abnormal outcomes result from unusual choices of the flux function $f(u)$ or the initial data, but these do not concern us.

A   numerical   scheme   on   a   regular   mesh $(i\Delta x, n\Delta t)$ with $u_i^n = u(i\Delta x, n\Delta t)$ and $f_i^n = f(u_i^n)$ is to compute, for all i at a given n,

$$\phi_{i+\frac{1}{2}}^n = \frac{f_{i+1}^n - f_i^n}{\Delta x}. \tag{2.4}$$

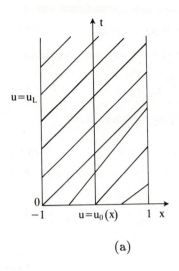

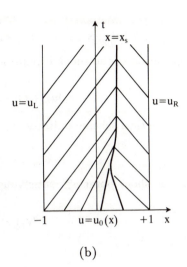

(a)   (b)

Fig. 1: (a) *Normal* evolution of solution with one sided boundary condition.
(b) *Normal* evolution with two sided boundary condition

Then $\Delta t \phi_{i+\frac{1}{2}}^n$ is subtracted from u_i^n if the average wave speed $a_{i+\frac{1}{2}}^n$ is negative, otherwise from u_{i+1}^n. Here

$$a_{i+\frac{1}{2}}^n = \frac{f_{i+1}^n - f_i^n}{u_{i+1}^n - u_i^n}. \tag{2.5}$$

Since u_i may receive a signal from its left, right, both or neither neighbours, R_i takes one of the four values $(\phi_{i-\frac{1}{2}}, \phi_{i+\frac{1}{2}}, (\phi_{i-\frac{1}{2}} + \phi_{i+\frac{1}{2}}), 0)$.

Normally the numerical scheme converges very fast to the steady solution $u_i^\infty \equiv u_L$, especially if preconditioned with a local time step, $\Delta t_i \propto 1/a_i$. Clearly this problem is too simple to be a paradigm of the convergence question.

2.2 Problem No. 2

Here we again solve (2.1), but subject to conditions on both boundaries, i.e.

$$u(-1, t) \equiv u_L; \quad u(1, t) \equiv u_R. \tag{2.6}$$

The steady state is again uniform unless u_L, u_R form a shock pair, i.e. $f(u_L) = f(u_R), a_L > 0 > a_R$. Then the normal outcome is a steady state with a shockwave at $x = x_s$ (see Fig. 1b), namely

$$u(x, \infty) \equiv u_L, \ x < x_s; \ u(x, \infty) \equiv u_R, \ x > x_s. \tag{2.7}$$

The position of the shock is determined from the conservation condition

$$(1 + x_s)u_L + (1 - x_s)u_R = \int_{-1}^{1} u_0(x)dx. \tag{2.8}$$

The numerical scheme imitates this by producing a steady state

$$u_i^{\infty} \equiv u_L, \ i < s; \ u_i^{\infty} \equiv u_R, \ i > s \tag{2.9}$$

with one shock point u_s^{∞}, satisfying

$$a_{s-\frac{1}{2}} > 0 > a_{s+\frac{1}{2}} \tag{2.10}$$

and such that

$$\Sigma u_i^{\infty} = \Sigma u_i^0. \tag{2.11}$$

For convex flux functions, such a solution exists uniquely. Equation (2.11) implies that

$$u_s^{\infty} = \Sigma u_i^0 - N(u_L + u_R) + s(u_R - u_L) \tag{2.12}$$

if the mesh runs from $i = -N$ to $i = N$. Then if the shock position s is changed between consecutive integer values, u_s changes by $(u_R - u_L)$. Therefore, there is a unique choice of s that places u_s between u_L and u_R, as required (by convexity) to satisfy (2.10). For such a solution, $\phi_{i+\frac{1}{2}}^{\infty} \equiv 0$, except if $i = s, s - 1$. At the shock

$$\phi_{s-\frac{1}{2}}^{\infty} + \phi_{s+\frac{1}{2}}^{\infty} = 0. \tag{2.13}$$

Note however, that (2.13) is satisfied for any choice of u_s^{∞} between u_L and u_R, so that in a sense it yields no information. It is (2.11) that determines the value of u_s. However, (2.11) is an identity satisfied at every time step; as the pre– and post–shock flows converge they squeeze the shock point toward its final value. It is easy to show that the Jacobian has repeated eigenvalues, equal either to a_L or $-a_R$, and convergence is again very fast. This model, like the first, is too simple to encounter difficulties.

3. Model Problems with Slow Convergence

The analytic problem is now

$$u_t + f_x = q(u, x) \tag{3.1}$$

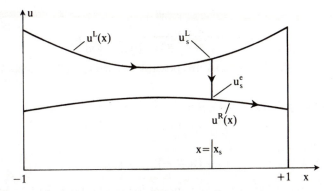

Fig. 2: Analytical solution to equation 3.1

subject to boundary conditions $u(-1,t) \equiv u_L, u(1,t) \equiv u_R$, where u_L, u_R do not necessarily form a shock pair, but for which $a_L > 0 > a_R$. The steady solution has left and right branches $u^L(x), u^R(x)$ satisfying $a(u)u_x = q(u,x)$ and one of the boundary conditions. A steady shock may be found at $x = x_s$ if $f^L(x_s) = f^R(x_s)$; see Fig. 2. A solution of this equation may exist only for some range of u_R corresponding to each u_L. Such a shock solution must also satisfy the entropy condition (2.3) and a condition of stability against small displacements

$$\frac{d}{dx_s} \frac{[f]}{[u]} \leq 0 \qquad (3.2)$$

which is easily shown to imply

$$\frac{q(u_s^R, x_s) - q(u_s^L, x_s)}{u_s^R - u_s^L} \leq 0. \qquad (3.3)$$

We have studied two specific examples.

3.1 Problem No. 3
Our first awkward example is the inviscid Burgers' equation with a specific source term

$$u_t + (\frac{1}{2}u^2)_x = -ku(u^* - u)x \qquad (3.4)$$

where u^*, k are constants. Steady solutions to this equation are

$$u - u^* = C \exp(\frac{1}{2}kx^2). \qquad (3.5)$$

It may be shown that $u_L > u_R$ is needed to satisfy the entropy condition, and the shock position may be deduced as

$$x_s^2 = 1 - \frac{2}{k} \ln(\frac{u^*}{u^* - \frac{1}{2}[u_L + u_R]}). \tag{3.6}$$

Note that two shock positions are possible, of which (3.3) selects the positive one.

3.2 Problem No. 4
This is

$$u_t + (u + \frac{1}{u})_x = -k\frac{x}{u}. \tag{3.7}$$

This can be regarded as a scalar model for nozzle flow. In fact at steady state u can be identified with the Mach number in an isothermal nozzle with area distribution $A = A_0 \exp(kx^2/2)$. There is a sonic point if $u = 1$, and the steady shock relationship is $u_-u_+ = 1$. Steady shock solutions exist under similar conditions to those in Problem No. 3.

4. The Structure of Discrete Solutions
We modify (2.4) by including the source term,

$$\tilde{\phi}_{i+\frac{1}{2}}^n = \frac{f_{i+1}^n - f_i^n}{\Delta x} - q_{i+\frac{1}{2}}^n, \tag{4.1}$$

where $q_{i+\frac{1}{2}}^n$ is some average of the source term over $(i, i+1)$. The precise form of this average may be very important. The converged solution $\tilde{\phi}_{i+\frac{1}{2}}^\infty = 0$ will be second order accurate in smooth regions because (4.1) is centred on the point $(i + \frac{1}{2})$; see Roe (1986). At shock points we have again

$$\tilde{\phi}_{s-\frac{1}{2}} + \tilde{\phi}_{s+\frac{1}{2}} = 0. \tag{4.2}$$

Unlike (2.13), equation (4.2) does contain information concerning u_s, through the source term. This shock equation replaces the conservation equation (2.11), which is no longer valid with sources present. The chief difference stemming from this replacement is that we can no longer prove existence and uniqueness for the discrete solution; in fact, these properties no longer hold.

To demonstrate this, we can attempt to construct the steady solution directly. Consider the following discretisation of (3.4)

$$\tilde{\phi}_{i+\frac{1}{2}} = \frac{1}{2}(u_{i+1}^2 - u_i^2) + \frac{k}{8}(u_i + u_{i+1})(2u^* - u_i - u_{i+1})(x_i + x_{i+1}). \tag{4.3}$$

Cancelling the factor $u_i + u_{i+1}$ leaves a linear relationship between u_i and u_{i+1}. This relationship can be marched numerically from u_{-N} towards the right, or from u_N towards the left, creating both branches (u_i^L and u_i^R) of the discrete solution. Then in turn we select each position $i = i_c$ as a candidate for the insertion of the shock point. That is, we attempt to find s and u_s such that

$$\tilde{\phi}_{s-\frac{1}{2}}(u_{s-1}^L, u_s) + \tilde{\phi}_{s+\frac{1}{2}}(u_s, u_{s+1}^R) = 0 \qquad (4.4)$$

under the entropy condition

$$a_{s+\frac{1}{2}} < 0 < a_{s-\frac{1}{2}} \qquad (4.5)$$

and the numerical stability condition

$$\frac{d}{du_s}(\tilde{\phi}_{s-\frac{1}{2}} + \tilde{\phi}_{s+\frac{1}{2}}) \geq 0. \qquad (4.6)$$

The condition (4.6) guarantees that the numerical solution is stable against rounding error when marched forward in time.

Note that (4.4) is no longer linear, but quadratic in u_s. The following possibilities may occur, at each trial location $i = i_c$.

(i) Equation (4.4) has no real root. Therefore no steady shock exists at $i = i_c$. However, a method which seeks for a *minimum* overall residual could converge to this solution.

(ii) The quadratic has real roots, u_{s1} and u_{s2}. Since a quadratic has gradients of different signs at each root, just one of the roots satisfies (4.6). This root may not satisfy (4.5).

(iii) There is just one position $i = i_c$ where a value u_s satisfying all of (4.4), (4.5), (4.6) can be found.

(iv) There are several such positions.

Numerical experiments show that all of these possibilities do occur. (i) and (ii) indicate non–existence and are more probable (experimentally and heuristically) on coarse grids. (iv) indicates non–uniqueness, and is more probable on fine grids.

Table 1 shows the possible shock locations for Problem 3, with $u^* = 1.0, k = 0.5, u_L = 9.0, u_R = 9.26$ on various grids. It seems that if the numerical solutions do converge to the true solution, it can only be in this rather peculiar sense: that on a sufficiently fine grid at least one solution exists, and collectively the set of solutions on a finer grid are closer to the

Range of Mesh Points	Numerical Shock Locations $(i_c$ satisfying (i) – (iv))
$[-5, 5]$	1
$[-25, 25]$	-24, -23, 13, 14
$[-50, 50]$	-42, -41, -40, 31, 32
$[-100, 100]$	-77, -76, 66, 67, 68

Table 1: Numerical shock locations in a particular experiment. The exact shock location is at $x = 0.667$.

analytic solution. Observe that sometimes a numerically stable solution is found close to an analytically unstable solution, i.e. a falsely stable solution.

It is convenient to think of this behaviour in the language of dynamical systems; see Sanz–Serna and Vadillo (1986). The set of values u_i constitute a $(2N + 1)$–dimensional phase space **u**. Under the differential operator

$$\dot{u}_i = R_i \qquad (4.7)$$

each starting point in phase space (initial data) describes a path that may leave the system, or else arrive at an attractor (steady state or limit cycle). Each attractor has its basin of attraction (set of initial states that lead to it). Numerical experience leads us to suppose that falsely stable solutions have very small basins of attraction.

Limit cycle behaviour is observed in cases (i) and (ii). The shock oscillates slowly between two adjacent cells, leaving all cells in their converged state. For a nonlinear system of equations this oscillation would be expected to radiate noise, as described by Roberts in this volume (p.442), but for a scalar problem only two points are involved, so the phase space is easily graphed. Fig. 3 is an example of such a limit cycle. We have used Problem 4 with $k = 0.2$, $u_L = 2.0$, $u_R = 0.4$, and $N = 5$. A time step of $0.15\Delta x$ is representative of explicit time–marching, but gives a result close to the semi–discrete form (4.7). The distinctive shape of the cycle, including the 45° corners, has a fairly simple explanation that the reader is invited to discover. Although eleven points is very few to use in a one–dimensional calculation, it might well be representative of the mesh ahead of a blunt body with a detached shock.

We have found that the problem of false stability can be overcome by reinterpreting the shock point. At any time during the calculation we identify a shock point $i = s$ as that for which (4.5) holds. Then, for the purpose of evaluating the source term, we assume that cell s contains

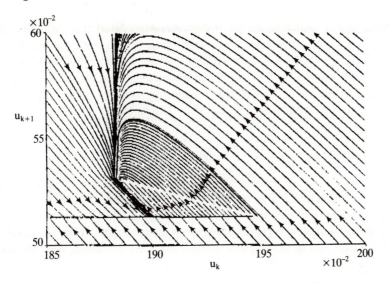

Fig. 3: Paths followed by the solutions in phase space. A limit cycle is entered from *most* initial data. No stable solution exists.

the states u_{s-1}, u_{s+1} in proportions $\lambda : 1 - \lambda$ where λ is obtained from conservation as

$$\lambda = \frac{u_{s+1} - u_s}{u_{s+1} - u_{s-1}}. \tag{4.8}$$

Then we write

$$q_{s-\frac{1}{2}} + q_{s+\frac{1}{2}} = (\frac{1}{2} + \lambda)q_{s-1} + (\frac{3}{2} - \lambda)q_{s+1} \tag{4.9}$$

and use (4.9) to evaluate the source terms in (4.4). This leads to

$$\frac{d}{du_s}(\tilde{\phi}_{s-\frac{1}{2}} + \tilde{\phi}_{s+\frac{1}{2}}) = (q_{s-1} - q_{s+1})\frac{d\lambda}{du_s} = \frac{(q_{s+1} - q_{s-1})}{(u_{s+1} - u_{s-1})}. \tag{4.10}$$

The numerical stability condition (4.6) is therefore satisfied under a discrete version of (3.3). This is an interesting use of the *recovery* arguments, used elsewhere in this volume (p.137) by Morton.

5. Accelerated Convergence

Even when a unique solution exists, convergence towards it may be excruciatingly slow. This is widely seen to be a practical problem with high resolution upwind schemes, despite the favourable verdict reached by Engquist and Gustafsson (1987) regarding shock–free one–dimensional problems.

Our model problems appear to have encapsulated this difficulty. The example we use is Problem 4, set as previously but with $u_R = 0.45$, and with $\Delta t/\Delta x = 0.15$. A simple explicit scheme takes around 3000 steps to reduce the residual by four orders of magnitude, and although local time steps do little to help, it is worth considering why not. One simple way to implement local time stepping would be through

$$\Delta t_i = \frac{k\Delta x}{\max[|a|_{i-\frac{1}{2}}, |a|_{i-\frac{1}{2}}]} \tag{5.1}$$

where k is a safety factor slightly less than unity. This would give Δt_s the same order of magnitude as any other Δt, but the last stages of convergence are marked by the fact that R_s is extremely small. (The other residuals may be even smaller; typically they are well converged while the shock cell is still finding its level, or even its position.) At the shock cell, we would like to use

$$\Delta t_s = \frac{k\Delta x}{|s|} \tag{5.2}$$

where s is the (very small) shock speed. Such a procedure would work for homogeneous problems, but we have found it very liable to instability when source terms are present. Instead, we adopt the following procedure for making large changes to u_s. We again identify the shock cell by (4.5), and then solve the equation

$$\tilde{\phi}_{s-\frac{1}{2}}(u_{s-1}, u_s) + \tilde{\phi}_{s+\frac{1}{2}}(u_s, u_{s+1}) = 0 \tag{5.3}$$

with u_{s-1}, u_{s+1} held fixed at the values thay have just been assigned in the current time step. The value u_s obtained from (5.3) is checked for stability, and, if it passes, overwrites the value of u_s just found.

It has proved prudent to modify this procedure in two ways. If the modified value of u_s no longer satisfies (4.5), the shock 'wants to leave' cell s. This may result in a large excursion that causes instability. Therefore, u_s is constrained to remain within the bounds implied by (4.5). Also, u_s is not permitted to change in the opposite sense to that implied by the time marching residual R_s. With both changes (local time steps and shock acceleration) the time to convergence is reduced by two orders of magnitude. Moreover, the convergence is no longer dominated by the shock behaviour and all residuals decay at similar rates.

Conclusions

We have presented two simple model problems that appear to reproduce some of the difficulties observed with practical high resolution schemes. For some of these (slow convergence and false stability) we have indicated cures that may generalise to more realistic situations. Others (non– existence and non–uniqueness) we have merely explored, but feel that our simple setting is a promising arena in which to continue to struggle.

Acknowledgements

The first author is grateful to Professor Avram Sidi for a stimulating conversation that focussed his interest on problems of convergence. Programming assistance has been provided by Creigh McNeil.

The work described in this paper was commenced while the authors were in residence at ICASE, NASA, Langley.

References

Engquist, B., Gustafsson, B. *Steady–state computations for wave propagation problems. Math. Comp.*, 49:39–64.

Jeffrey, A. *Quasilinear hyperbolic systems and waves. Research notes in Mathematics.* Vol. 5. Pitman Press.

Roe, P. L. *Upwind differencing schemes for hyperbolic conservation laws with source terms. Lecture Notes in Mathematics.* Vol. 1270. C. Carasso et al, editors. Springer Press.

Turkel, E. *Preconditioned methods for solving the incompressible and low–speed compressible equations. J. Comp. Phys.* 72:277–298.

Sanz–Serna, J. M., Vadillo, F. *Nonlinear instability, the dynamic approach. Research Notes in Mathematics.*, Vol. 140. D. F. Griffiths, E. A. Watson editors. Longman.